Statistics of Extremes and Records in Random Sequences

Statistics of Extremes and Records in Random Sequences

Satya N. Majumdar
Grégory Schehr

OXFORD
UNIVERSITY PRESS

OXFORD
UNIVERSITY PRESS

Great Clarendon Street, Oxford, OX2 6DP,
United Kingdom

Oxford University Press is a department of the University of Oxford.
It furthers the University's objective of excellence in research, scholarship,
and education by publishing worldwide. Oxford is a registered trade mark of
Oxford University Press in the UK and in certain other countries

Published in the United States of America by Oxford University Press
198 Madison Avenue, New York, NY 10016, United States of America

British Library Cataloguing in Publication Data
Data available

Library of Congress Control Number: 2024933499

ISBN 9780198797333

DOI: 10.1093/9780191838781.001.0001

Printed and bound by
CPI Group (UK) Ltd, Croydon, CR0 4YY

Contents

Preface

Extreme value statistics (EVS) is a truly interdisciplinary subject, spanning from statistics and mathematics on one side to physics of disordered systems on the other. It has tremendous importance and practical applications in a wide variety of fields, such as climate science, finance, spin glasses, random matrices and so on. One of the basic questions in EVS is to understand how the maximum or minimum of a time series of size N fluctuates from one sample to another. The EVS is well understood when the entries of the time series under study are independent and identically distributed (IID), and this is the subject of the classical theory of EVS covered in many books and monographs, mostly in the statistics and mathematics literature. However, more recently, EVS started to play a very important role in statistical physics, in particular in disordered systems (spin glasses and polymers in a disordered medium), random matrices, random walks, fluctuating interfaces and so forth. It turns out that in many physical systems the entries of the underlying time series are actually strongly correlated and the classical EVS is no longer applicable. This has led to a plethora of activities in both the statistical physics and mathematics communities over the last few decades. While there have been a few recent review articles on the EVS of correlated variables, what has been missing is a pedagogical book with examples illustrating the basic tools and techniques that can be useful to a student or a non-expert starting to work in this interesting and rapidly developing field. The purpose of this book is to provide an introductory monograph on this subject with a style adapted for a graduate student who only has a basic knowledge of probability theory and statistical mechanics. Even though the style is more adapted to physicists, we hope that it will be easy to follow for statisticians and mathematicians as well.

In this book we have tried to present the basic idea and the tools using two simple models of time series: (i) an IID sequence where there is no correlation between the entries, and (ii) a random walk sequence, where the entries are strongly correlated. The EVS and related observables can be computed exactly for both these examples, as shown in detail later in the book. The techniques used will hopefully be useful for readers tackling questions in time series appearing in other problems.

The material presented here resulted from our research on this and related subjects over the last few decades, done in collaboration with several students, postdocs and colleagues, whom we acknowledge in the next section.

Acknowledgments

We are fortunate enough to have a long list of collaborators, with whom we have worked over the last few decades on problems related to EVS and first-passage problems. This is a very good occasion to thank all of them: S. Agarwal, E. Agoritsas, G. Akemann, R. Allen, R. Allez, C. Appert-Rolland, J. Baik, A. Bar, M. Barma, T. Banerjee, A. Barrat, U. Basu, M. Battilana, D. ben-Avraham, I. Bena, E. Ben-Naim, O. Bénichou, B. Berkowitz, B. Besga, M. Biroli, R. Blythe, O. Bohigas, G. Borot, J.-P. Bouchaud, D. Boyer, A. J. Bray, J. Bun, I. Burenev, S. Bustingorry, P. Calabrese, P. Calka, I. P. Castillo, B. Chakraborty, A. D. Chepelianskii, S. Ciliberto, J. Cividini, M. Chupeau, A. Comtet, M. Constantin, S. J. Cornell, F. Cornu, L. F. Cugliandolo, F. D. Cunden, K. Damle, D. Das, C. Dasgupta, S. Das Sarma, D. S. Dean, C. De Bacco, B. De Bruyne, F. den Hollander, J. Desbois, C. Di Bello, A. Dhar, D. Dhar, E. Dumonteil, Y. Edery, M. R. Evans, B. Eynard, F. Faisant, A. Flack, P. J. Forrester, J. Franke, Y. V. Fyodorov, A. Gambassi, R. Garcia-Garcia, T. Gautié, T. Giamarchi, L. Giuggioli, G. Gouraud, C. Godrèche, A. Grabsch, G. Gradenigo, D. S. Grebenkov, J. Grela, M. Guéneau, S. Gupta, T. Hanney, A. K. Hartmann, H. J. Hilhorst, D. A. Huse, S. Iubini, M. J. Kearney, J. Kethepalli, A. B. Kolton, A. B. Kostinski, P. L. Krapivsky, S. Krishnamurthy, J. Krug, M. Kulkarni, A. Kundu, L. Kusmierz, B. Lacroix-A-Chez-Toine, A. Lakshminarayan, Y. Lanoiselée, H. Larralde, P. Leboeuf, V. Lecomte, P. Le Doussal, K. Liechty, R. Livi, M. Magoni, K. Malakar, K. Mallick, R. Mallikarjun, R. Marino, G. Marcado-Vasquez, O. C. Martin, R. J. Martin, I. Marzuoli, A. Mays, B. Meerson, P. Mergny, M. Meylahn, F. Mori, P. Mounaix, D. Mukamel, B. Mukherjee, A. Nagar, C. Nadal, S. Nechaev, N. O'Connell, H. Orland, G. Oshanin, S. Ouvry, I. Pagonabarraga, A. Pal, L. Palmieri, A. Pargellis, R. Paul, A. Perret, M. Poplavskyi, A. Ponsaing, M. Potters, R. Rajesh, J. Rambeau, K. Ramola, J. Randon-Furling, S. Redner, A. Reymbaud, H. Rieger, A. Rosso, S. Sabhapandit, T. Sadhu, R. Santachiara, L. Santen, A. Scardicchio, H. Schawe, V. M. Schimmenti, K. Sengupta, P. Singh, N. R. Smith, P. Sollich, J. C. Sunil, J. Szavits-Nossan, M. V. Tamm, M. Tarzia, C. Texier, S. Tomsovic, H. Touchette, L. Touzo, V. Tripathy, E. Trizac, G. Tucci, F. van Wijland, D. Venturelli, M. Vergassola, K. Vijay Kumar, D. Villamaina, P. Vivo, P. von Bomhard, G. Wergen, J. Whitehouse, K. J. Wiese, E. Williams, M. Yor, B. Yurke, R. K. P. Zia, R. M. Ziff, A. Zodage, A. Zoia.

We would also like to acknowledge useful discussions with A. Amir, R. Artuso, F. Aurzada, E. Aurell, G. Barraquand, E. Barkai, M. Bauer, G. Ben Arous, D. Bernard, E. Bertin, P. Biane, G. Biroli, E. Bogomolny, A. Borodin, P. Bougerol, J. Bouttier, E. Brunet, S. Burov, T. Burkhardt, Z. Burda, R. Butez, J. L. Cardy, M. E. Cates, D. Chafaï, D. Challet, H. Chaté, Y. Chen, R. Chetrite, T. Claeys, I. Corwin, S. R. Das, N. Davidson, A. Dembo, B. Derrida, P. Diaconis, P. Di Francesco, C. Donati-Martin, I. Dornic, L. Dumaz, B. Duplantier, A. Edelman, N. Feldheim, D. S. Fisher, S. Franz, S. Ghosh, O. Giraud, A. Godec, A. Guionnet, T. Halpin-Healy, A. Hardy, M. Henkel, D. Holcman, V. V. Ivanov, K. Jain, S. Janson, K. Johansson, Y. Kafri, M. Kardar, M. Katori, E. Katzav, D. Kessler,

A. Krajenbrinck, C. Krattenthaler, M. Krishnapur, J. Kurchan, T. Leblé, D. Levine, M. Lewin, J.-M. Luck, C. Maes, M. Maïda, G. Mandal, M. Marino, M. Marsili, G. Menon, R. Metzler, M. Mézard, D. Mitra, P. K. Mohanty, N. R. Moloney, C. Monthus, S. Munier, D. Najim, J. D. Noh, M. Nowak, R. Pandit, H. Park, J. Pitman, J. Quastel, Z. Rácz, S. Ramaswamy, K. Raschel, O. Raz, D. Remenik, S. Reuveni, H. Saleur, T. Sasamoto, S. Sastry, G. Semerjian, T. Seligman, P. Sen, S. Serfaty, Y. Shapir, Z. Shi, T. Simon, H. Spohn, J. Stavans, Sumedha, J. Tailleur, K. A. Takeuchi, J. Talbot, W. Tang, T. Thiery, M. Tierz, S. Torquato, C. A. Tracy, P. Viot, B. Virag, R. Voituriez, S. Wadia, D. Wang, O. Zeitouni and J.-B. Zuber.

We also thank our publishers, Oxford University Press, and in particular Sonke Adlung, for giving us the opportunity to write this book and for his immense patience over the last five years (with many missed deadlines!). We also thank Jodie Keefe for helping us prepare the final version of the book. It is also a great pleasure to thank Claudine Le Vaou for her administrative help at LPTMS during the writing of the book.

Finally, we are grateful to our respective families for their encouragement. Satya thanks Victoria for always being there in all weathers, and his sisters Anita and Kakali for their long-distance love and encouragement. Grégory would like to thank Aude, Capucine and Titien for being a permanent source of motivation as well as his parents and sisters for their constant encouragements.

Notation

N	Sample size
$X_{\max}$	Global maximum
$X_{\min}$	Global minimum
$Q_{\max}(x, N) = \text{Prob}(X_{\max} \leq x)$	Cumulative distribution of $X_{\max}$
$Q_{\min}(x, N) = \text{Prob}(X_{\min} \geq x)$	Cumulative distribution of the minimum
$P_{\max}(x, N)$	Probability distribution function of the maximum
$M_{k,N}$	kth maximum
$Q_{k,N}(x) = \text{Prob}(M_{k,N} \leq x)$	Cumulative distribution of $M_{k,N}$
$g_k = M_{k,N} - M_{k+1,N}$	Gap between consecutive maxima
$Q(x_0, N)$	Survival probability
$q(N) = Q(0, N)$	Survival probability starting at the origin
$F(x_0, N)$	First-passage time probability
M	Number of records
R_k	Record values
ℓ_i	Record ages
n and t	Running variables for random walk and Brownian motion
N and T	Length of the interval for random walk and Brownian motion

1
Introduction

Extreme events, such as earthquakes, tsunamis, tornadoes and so on, are rare events. They do not happen every day, but if/when they happen their effects can be devastating. Hence it is obviously important to estimate the magnitude of such extreme events, to predict the times at which they may occur, the frequency of their occurrence, etc. This is broadly the subject of extreme value statistics (EVS). This topic has been of interest for many years and there are several books in the mathematics and statistics literature that have emphasized the various applications of EVS in climate series, hydrology, sports, finance, etc. (*Katz et al.*, 2002; Gumbel, 2004; Novak, 2011). More recently, EVS has also found important applications in statistical physics, notably in the context of disordered and complex systems (Derrida, 1981; Bouchaud and Mézard, 1997; Amir, 2020*b*; Majumdar *et al.*, 2020). Since EVS has a naturally interdisciplinary flavor, it would be nice to have a monograph that explains the basic questions and techniques of EVS from both the statistics and the physics points of view. Moreover, it should be easily accessible for people working in the general complex systems area whoh may have just a basic knowledge of probability theory. The purpose of this book is to present such a unifying and easily accessible pedagogical picture of EVS and various associated questions and problems.

Let us discuss a few concrete examples first. In statistics, one typically has access to the data in the form of a time series. This time series could be, for example, the price of a stock over a certain period, the daily temperature/rainfall at a given weather station as a function of days, the height of a river as a function of time or the rate of daily contamination during a pandemic such as COVID-19 as a function of days. In the context of stock prices, one obviously would like to know when the stock price becomes a minimum (or maximum), and the associated minimal and maximal values of the stock. An agent is expected to sell their stock when the price is maximal and buy a stock when the price is minimal.

Another example comes from civil engineering. Imagine that you are an engineer and you are asked to construct a dam or a bridge over a river; you have at your disposal the time series providing the height of the river for the last 500 years (an example of the Nile river height near Cairo from AD 600–1200 is given in Fig. 1.1). Here, the time-averaged height of the river has little importance for deciding the

Statistics of Extremes and Records in Random Sequences. Satya N. Majumdar and Grégory Schehr, Oxford University Press.
© Satya N. Majumdar and Grégory Schehr (2024). DOI: 10.1093/9780191838781.003.0001

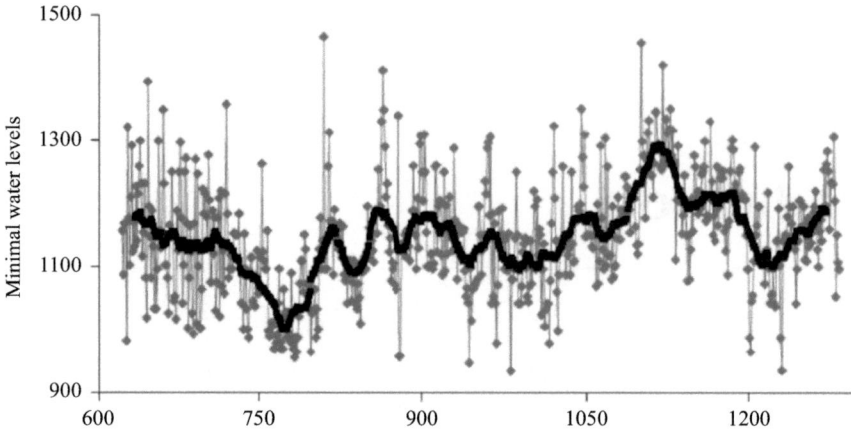

Fig. 1.1 Reproduced from Di Baldassarre *et al.*, (2011). Future hydrology and climate in the river nile basin: a review. *Hydrological Sciences Journal*, 56(2), 199–211.

height of the bridge. Instead, and more importantly, the spikes in river height (corresponding to years with big floods) play a more crucial role. This is because the bridge should be higher than the typical height of these spikes. These spikes indeed correspond to extreme values (or "outliers"), and the engineer had better have a very good idea of the statistics of the values of such spikes. Similarly, extreme values in temperature time series (corresponding to very hot or very cold days) or in rainfall time series (corresponding to extreme rainfall causing floods or no rainfall causing droughts) do play a crucial role in understanding global warming in climate studies.

Let us now be a little bit more precise about what we mean by the "statistics" of extreme values in such a time series. Suppose that $\{x_1, x_2, \ldots, x_N\}$ denote the entries of a time series with N entries. Extreme values may be either the maximum or the minimum of these entries, denoted by

$$X_{\max} = \max\{x_1, x_2, \ldots x_N\}, \tag{1.1}$$

$$X_{\min} = \min\{x_1, x_2, \ldots x_N\}. \tag{1.2}$$

In a given time series, these are just two numbers. However, if one considers a different realization (sample) of the time series, the values of $X_{\max}$ and $X_{\min}$ will be different. Thus, both $X_{\max}$ and $X_{\min}$ fluctuate from one sample to another sample of the time series and hence they are random variables. One of the goals of EVS is to estimate the statistics of $X_{\max}$ and $X_{\min}$. For example, what is the mean, the variance or even the full distribution of $X_{\max}$ and $X_{\min}$?

Let us now give a couple of examples from physics where such extreme values play important roles. Our first example is a simple random walk in d dimensions. The random walk is often used as a simple model for the motion of an animal when foraging for food. The maximum displacement of the walker from its origin gives

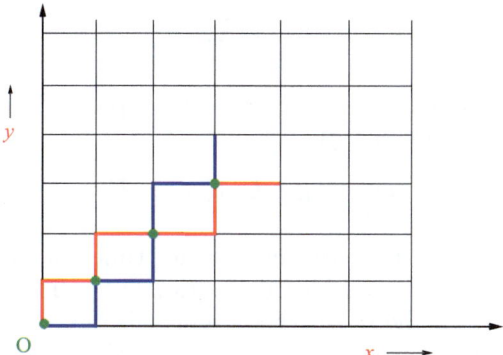

Fig. 1.2 Two different seven-step paths of a directed polymer (with north and east steps) on a square lattice that starts at the origin with a free endpoint. The two paths share four common sites: $(0, 0)$, $(1, 1)$, $(2, 2)$ and $(3, 2)$.

a measure of the territory covered by the animal in a given period (often known as "home range" in ecology), which is clearly an example of EVS. For example, in one dimension the positions of the walker at successive discrete times constitute a time series $\{x_1, x_2, \ldots, x_N\}$ up to N steps. Here, the entry x_k denotes the position of the walker at step k. In this case, $X_{\max}$ and $X_{\min}$ denote respectively the maximal displacements in the positive and negative directions, assuming that the walker starts at the origin. In this example, the span $S_N = X_{\max} - X_{\min}$ provides a measure of the territory covered by the walker in N steps (Hughes, 1995).

Another area where EVS plays an important role is in the physics of disordered systems, such as spin glasses, polymers in a random medium, and various combinatorial optimization problems. Let us consider the example of a directed polymer in two dimensions—a model that has played a central role in the statistical physics of disordered systems (Forster *et al.*, 1977; Huse and Henley, 1985; Kardar, 1985, 1987; Kardar *et al.*, 1986; Kardar and Zhang, 1987; Derrida and Spohn, 1988; Imbrie and Spencer, 1988; Cook and Derrida, 1989; Mézard, 1990; Fisher and Huse, 1991; Halpin-Healy and Zhang, 1995; Krug and Halpin-Healy, 1998); for a short review see Majumdar (2007b). Consider a two-dimensional lattice where at each lattice site i there is a quenched energy ε_i (see Fig. 1.2). These ε_i's are independent random variables from site to site, each drawn from a given distribution $\rho(\varepsilon)$. The ε_i do not fluctuate with time (i.e., they are quenched) and model the random medium. Now consider an n-step directed polymer that starts at the origin and moves with equal probability either to the north or the east neighbor. There are clearly $N = 2^n$ possible distinct polymer paths. The energy of a given polymer path is defined as the sum of all site energies ε_i through which the polymer path passes,

$$E_{\text{path}} = \sum_{i \in \text{path}} \varepsilon_i. \tag{1.3}$$

At zero temperature, the polymer will choose the "optimal" path that has the least energy (i.e., the ground state energy),

$$E_0 = \min_{\text{paths}} \{E_{\text{path } 1}, E_{\text{path } 2}, \ldots, E_{\text{path } N}\}. \tag{1.4}$$

Interpreting $x_i = E_{\text{path } i}$ as the ith entry of a "time series," E_0 is then clearly the minimum of a set of N variables as in Eq. (1.2). This ground-state energy E_0 depends, of course, on the disorder ε_is and will fluctuate from one sample of the disorder to another sample. Computing the statistics of E_0 is then clearly an extreme value problem (Majumdar and Krapivsky, 2000, 2003). Another example of a disordered system where the EVS plays an important role is in determining how the threshold force needed to depin an elastic line in a disordered medium fluctuates from one sample of disorder to another (Bolech and Rosso, 2004).

Yet another example that has played a central role in many areas of physics and mathematics is the study of the lowest or the largest eigenvalue of a random matrix (Tracy and Widom, 1994, 1996). More precisely, we consider an $N \times N$ Hermitian matrix $\mathbf{H}$ with entries H_{ij}. This can be viewed as the Hamiltonian of a quantum system with a finite-dimensional Hilbert space of size N. Now we want to make this matrix random to mimic the effects of disorder in the quantum Hamiltonian. For example, a disordered Hamiltonian on a lattice of N sites can be written as $\hat{H} = \sum_{i,j} H_{i,j} |i\rangle\langle j|$, where $\{|i\rangle\}$ with $i = 1, \ldots, N$ constitutes the site basis on the lattice. If the sum is restricted to nearest neighbors (e.g., with periodic boundary conditions), the random matrix $\mathbf{H}$ will be tridiagonal and this would correspond to the famous Anderson model (Anderson, 1958). On the other hand, if the sum extends over all pairs (i, j), this would correspond to a mean-field version of a disordered quantum lattice model. We then have to specify how to choose the probability distribution of the matrix elements H_{ij}. Consider for example the mean-field version. The Hermitian restriction implies that we can only choose freely the real and imaginary parts of the elements H_{ij} with $i \leq j$—the elements H_{ij} for $i > j$ are then automatically fixed by the Hermitian constraint. A natural choice would correspond to choosing $H_{i,j}$ with $i \leq j$ independently from a Gaussian distribution such that the joint distribution of the entries can be written as

$$P(\mathbf{H}) \propto \prod_{i,j} e^{-\frac{1}{2}|H_{ij}|^2} \propto e^{-\frac{1}{2}\text{Tr}(\mathbf{H}^2)}. \tag{1.5}$$

This is the famous Gaussian unitary ensemble of random matrix theory (RMT) (Mehta, 2004; Forrester, 2010; Livan *et al.*, 2018). The name originates from the fact that the particular form of the probability distribution in Eq. (1.5) is invariant under a unitary transformation $\mathbf{H'} \to \mathbf{U^\dagger H U}$, where $\mathbf{U}$ is an arbitrary unitary matrix. This particular ensemble has played a central role in various applications of RMT from nuclear physics to quantum chaos. Now, for any realization of this

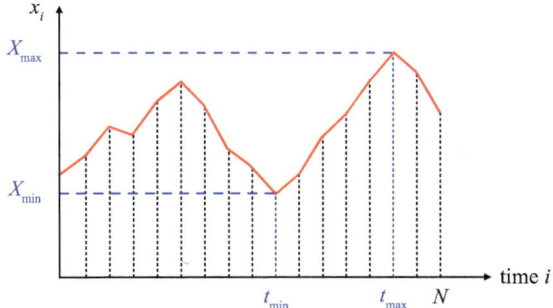

Fig. 1.3 Typical realization (schematic) of a time series x_i vs the discrete time i. $X_{\max}$ and $X_{\min}$ indicate the global maximum and global minimum up to time N. Similarly, $t_{\max}$ and $t_{\min}$ respectively indicate the times at which $X_{\max}$ and $X_{\min}$ occur.

random matrix, one can diagonalize it and obtain N real eigenvalues $\lambda_1, \lambda_2, \ldots, \lambda_N$. The hermiticity condition guarantees that the eigenvalues are real. The *ground-state energy* in this case is given by the lowest eigenvalue,

$$\lambda_{\min} = \min\{\lambda_1, \lambda_2, \ldots, \lambda_N\}. \tag{1.6}$$

We can thus interpret the λ_i as the entries x_i in Eq. (1.2). Clearly, again, $\lambda_{\min}$ will fluctuate from one realization of the H_{ij} to another, and determining the statistics of $\lambda_{\min}$ is clearly an extreme value problem.

So far, we have been discussing the statistics of the value of the maximum (or the minimum), i.e., the magnitude of an extreme event. An equally important question is when these extreme events occur. For example, in a given time series, the largest value of the time series may occur at "time" $t_{\max}$ (for a discrete-time series $\{x_i\}$, $t_{\max}$ is the label of the largest entry of the time series, see Fig. 1.3). Similarly, the smallest entry of the time series may occur at "time" $t_{\min}$. Estimating the statistics of $t_{\max}$ and $t_{\min}$ is another important extreme value question with many applications. For instance, if the time series represents the price of a stock, one would like to sell the stock when its price is maximum, i.e., at time $t_{\max}$, and buy the stock when its price is minimum, i.e., at time $t_{\min}$ (Shiryaev *et al.*, 2008; Majumdar and Bouchaud, 2008; Chicheportiche and Bouchaud, 2014; Mori *et al.*, 2019). In the context of climate data analysis, it is important to know when the hottest or the coldest temperature occurs. Let us give another example from the physics of disordered systems. Consider a single particle moving in a one-dimensional random potential $V(z)$, where z denotes the position of the particle on the real line. One can think of the value of the potential $V(z)$ as the entry of a time series where the position z plays the role of time (see Fig. 1.4). In this case, $z_{\max}$ and $z_{\min}$ denote the locations of the maximum and the minimum of the potential $V(z)$ respectively. At low temperature, the particle will be trapped around $z_{\min}$ in equilibrium and it is thus important to know the statistics of the equilibrium position of the particle. In addition, the difference between the extreme values $V(z_{\max}) - V(z_{\min})$ gives an

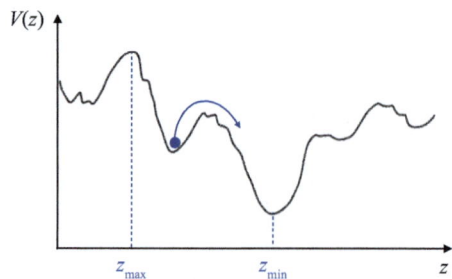

Fig. 1.4 Illustration of the activated dynamics of a particle in a one-dimensional random potential $V(z)$ vs z. Here, $z_{\max}$ and $z_{\min}$ respectively denote the location of the maximum and minimum of $V(z)$.

estimate of the highest barrier that the particle needs to cross in order to equilibrate. This provides important information on the transport properties of the particles in a disordered medium, and also characterizes its typically slow dynamics, as often observed in complex and glassy systems. Apart from the examples mentioned here, the statistics of $t_{\max}$ and $t_{\min}$ have found several interesting applications, some of which are discussed later in the book (see Chapter 5).

In the discussion above, $X_{\max}$ and $X_{\min}$ respectively denote the *global* maximum and minimum in a time series of size N. A natural generalization of this question, usually referred to as "order statistics," concerns the statistics of the second maximum, third maximum, etc. (David and Nagaraja, 2004; Arnold *et al.*, 2008). In the context of extreme events such as earthquakes, $X_{\max}$ corresponds to the magnitude of the biggest quake, while the second, third, etc. may correspond to the magnitude of before-/after-shocks of the principal earthquake. In other words, we order the entries of the time series in decreasing order, $\{x_1, x_2, \ldots, x_N\} \rightarrow \{M_{1,N} > M_{2,N} > \cdots > M_{N,N}\}$, and calculate the statistics of the ordered entries $M_{k,N}$ for different k. Note that $M_{1,N}$ corresponds to $X_{\max}$, i.e., the global maximum discussed above.

In the context of disordered systems, these ordered entries may represent the energy spectrum of the system. The last ordered entry in this example then corresponds to the ground-state energy, the second-last entry represents the first excited state and so on. In addition to the values of these ordered maxima, the gap between the values of two successive maxima (e.g., between the first and the second maxima) is another interesting observable (Schawe *et al.*, 2018). For instance, in the context of seismology, the empirical distribution of the gap between the magnitude of the largest and second largest earthquakes is known as the Båth's law (Båth, 1965; Vere-Jones, 1969). In the context of disordered systems, the energy gap between the first excited state and the ground state often controls the late-time relaxation dynamics of the system—hence it is important to know the statistics of this gap. Another question related to order statistics is how to estimate the number of ordered maxima in a given neighborhood of the value of the global maximum—a similar question is to estimate the number of excited states with energies close to the ground

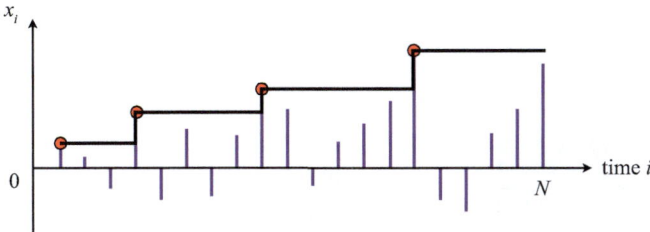

Fig. 1.5 Ilustration of the staircase structure of the records of a time series $\{x_i\}$; the records are indicated in by dots.

state. A simple observable to quantify this "crowding" near the value of the global maximum $X_{\max}$ is the so-called density of *near-extreme events* (Sabhapandit and Majumdar, 2007),

$$\rho(r, N) = \frac{1}{N - 1} \sum_{x_i \neq X_{\max}} \delta(r - (X_{\max} - x_i)), \qquad (1.7)$$

where the sum runs over all the entries excluding $X_{\max}$. This density, which is normalized to one, i.e., $\int_0^\infty \rho(r, N)\, dr = 1$, just counts the fraction of entries that have values between $X_{\max} - r$ and $X_{\max} - r + dr$. Note that $\rho(r, N)$ is a random variable, which fluctuates from one time series to another. A simple observable is the mean density of near-extreme events, obtained by averaging $\rho(r, N)$ over all possible realizations of the entries. We will denote this average by $\overline{\rho(r, N)}$. This quantity has several applications. For instance, it has been applied to characterize temperature time series (Sabhapandit and Majumdar, 2007). In the physics of disordered systems, it has been used to compute the response function of the zero-temperature dynamics of the spherical spin glass model (Fyodorov *et al.*, 2015).

Another interesting set of extreme observables corresponds to the so-called *records* (Bunge and Goldie, 2001; Arnold *et al.*, 2011). The study of the statistics of records in a time series is fundamental and important in a wide variety of systems, including climate studies, finance and economics, hydrology, sports, and others—for detailed references see Chapter 7. Consider any generic time series $\{x_i\}$ of N entries. A record happens at step k if the kth entry exceeds all previous entries, i.e., $x_k > \max\{x_1, x_2, \ldots, x_{k-1}\}$. The first natural question is how the actual record value evolves with time k. For example, when a record happens, with a record value R_k at step k, this value remains unchanged for a while until the current record gets broken by the next record, when the record value jumps to the new higher value. Thus, if we plot the record value as a function of time, it typically displays a staircase structure (see Fig. 1.5). Such a staircase structure has been observed in diverse complex systems such as the evolution of thermoremanent magnetization in spin glasses, the evolution of vortex density with increasing magnetic field in type-II disordered superconductors, avalanches of elastic lines in a disordered medium, the evolution of fitness in biological populations and in models of growing networks, among others (see Chapter 7 for details). For instance, when a domain wall in a

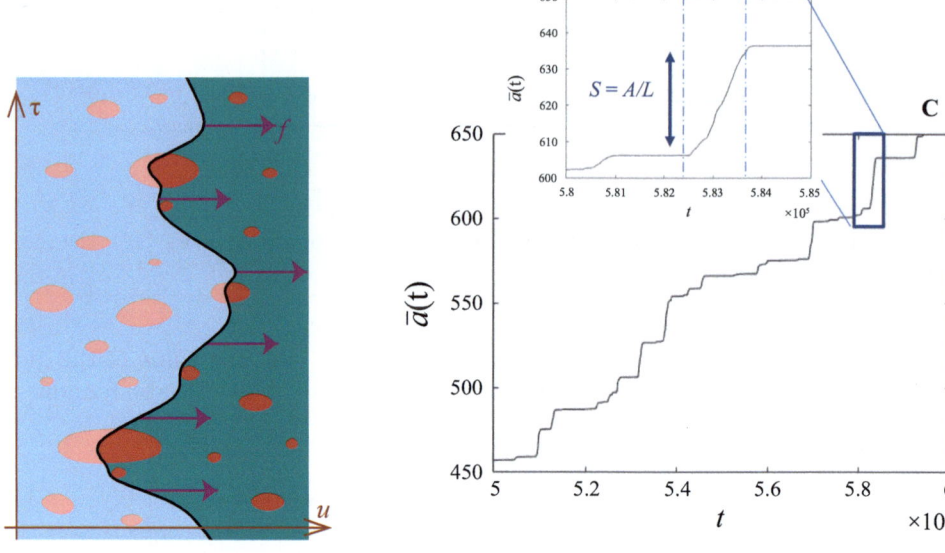

Fig. 1.6 Left panel reproduced from Fyodorov *et al.,* (2018). Exponential number of equilibria and depinning threshold for a directed polymer in a random potential. *Annals of Physics,* 397, 1–64. Right panel reproduced from Bonamy (2017). Dynamics of cracks in disordered materials. *Comptes Rendus Physique,* 18(5–6), 297–313.

disordered medium is driven by applying an increasing external magnetic field (see the left panel of Fig. 1.6), its center of mass remains immobile (pinned by disorder) for a while, and then, as the field increases further, an extended part of the wall gets depinned, giving rise to an avalanche and, consequently, the center of mass jumps over a certain distance. The position of the center of mass as a function of time (or increasing drive) displays a staircase structure as in the right panel of Fig. 1.6. The jump in the record value corresponds to the avalanche size. Record statistics has been extensively used to characterize such a staircase evolution. There are also other interesting extreme value questions related to record statistics. For instance, one can ask how many records occur in a given time series of size N. Similarly, the age of a record denotes its lifetime, i.e., the time during which a given record survives before being broken by the next record. Understanding the statistics of the number M of records and their ages $\{\ell_1, \ell_2, \ldots, \ell_M\}$ is an important issue in EVS that is discussed in Chapter 7.

Thus, summarizing, the general setting of the EVS problem can be described as follows. We have a set of random variables $\{x_i\}$ with $i = 1, 2, \ldots, N$ (see Fig. 1.3), where the index i may represent the real time in the context of a time series. In the context of disordered systems, on the other hand, it may label the configuration;

e.g., the ith polymer path in a random medium or the ith eigenvalue of a random matrix. We then need to have some estimate of how these N random variables are distributed, i.e., their joint distribution $P_{\text{joint}}(x_1, x_2, \ldots, x_N)$. This will turn out to be the main input that we need to compute the EVS. Estimating this joint distribution from the available data is a non-trivial problem in statistics and is part of the modelling of the system. However, assuming that this joint distribution $P_{\text{joint}}(x_1, x_2, \ldots, x_N)$ is given, computing the properties related to EVS is in general quite non-trivial, and this is where the techniques of statistical physics are useful, as we will see in the rest of the book.

As an example, suppose we want to compute the distribution of $X_{\max}$ as defined in Eq. (1.1), knowing the joint distribution $P_{\text{joint}}(x_1, x_2, \ldots, x_N)$. How does one progress? Instead of computing the PDF of $X_{\max}$, it turns out to be more convenient to consider the cumulative distribution

$$Q_{\max}(w, N) = \text{Prob}(X_{\max} \le w) = \int_{-\infty}^{w} P_{\max}(x, N)\, \mathrm{d}x, \tag{1.8}$$

where $P_{\max}(x, N)$ is the PDF of $X_{\max}$, i.e., $P_{\max}(x, N)\mathrm{d}x$ denotes the probability that $X_{\max} \in [x, x + \mathrm{d}x]$. If we know $Q_{\max}(w, N)$, we can compute the PDF $P_{\max}(w, N)$ by simply taking the derivative,

$$P_{\max}(w, N) = \frac{\partial Q_{\max}(w, N)}{\partial w}. \tag{1.9}$$

To compute the cumulative distribution $Q_{\max}(w, N)$, one makes the observation that the event "$X_{\max} \le w$" is equivalent to the event that each of the variables x_i is less than w, i.e.,

$$Q_{\max}(w, N) = \text{Prob}(X_{\max} \le w) = \text{Prob}[x_1 \le w, x_2 \le w, \ldots, x_N \le w]. \tag{1.10}$$

Knowing the joint distribution of the x_i, the last quantity can be written as

$$Q_{\max}(w, N) = \int_{-\infty}^{w} \mathrm{d}x_1 \int_{-\infty}^{w} \mathrm{d}x_2 \cdots \int_{-\infty}^{w} \mathrm{d}x_N \, P_{\text{joint}}(x_1, x_2, \ldots, x_N). \tag{1.11}$$

Computing this multiple integral is the main challenge in the calculation of the extreme value distribution. Indeed, this integral can be given a nice physical interpretation. Let us first define

$$E(x_1, x_2, \ldots, x_N) = -\ln P_{\text{joint}}(x_1, x_2, \ldots, x_N), \tag{1.12}$$

and rewrite Eq. (1.11) as

$$Q_{\max}(w, N) = \int_{-\infty}^{w} \mathrm{d}x_1 \int_{-\infty}^{w} \mathrm{d}x_2 \cdots \int_{-\infty}^{w} \mathrm{d}x_N \, \mathrm{e}^{-E(x_1, x_2, \ldots, x_N)}. \tag{1.13}$$

Thus, $Q_{\max}(w, N)$ can be interpreted as the partition function of a one-dimensional gas of N particles with positions $\{x_1, x_2, \ldots, x_N\}$ with the energy of a configuration

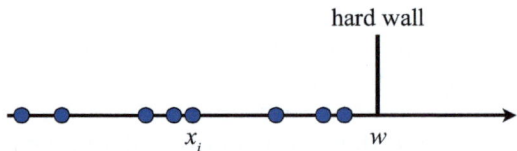

Fig. 1.7 The entries $\{x_i\}$ of a time series of size N can be interpreted as the positions of a gas of N particles on a line. The CDF of the maximum $X_{\max}$, i.e., $Q_{\max}(w, N)$, corresponds to the partition function of the gas restricted to be on the left of the hard wall at w.

given by $E(x_1, x_2, \ldots, x_N)$. Here, for convenience, we have set the inverse temperature $\beta = 1$. The only unusual thing in the partition function in Eq. (1.13) is that the gas is confined in the region $(-\infty, w]$, as if there is a hard wall at w (see Fig. 1.7). As we will see later, if the random variables x_i are correlated, this corresponds to an interacting gas of particles, whose partition function is in general hard to compute, even in one dimension. However, partition functions are central objects in equilibrium statistical physics and many techniques have been developed over the years to compute such partition functions. Examples include exact solutions in specific models, path-integral methods, the large-N saddle-point method, the renormalization group method and so on. Thus, statistical physics provides powerful tools to study EVS, as we demonstrate in this book.

There is yet another nice interpretation of $Q_{\max}(w, N)$ in the case when the random variables x_i represent the values of a stochastic process at discrete times. Then, $Q_{\max}(w, N)$ in Eq. (1.11) represents the probability that the process stays below the level w up to step N. In the statistical physics literature, this is usually called the *persistence* or *survival probability* (Redner, 2001; Bray *et al.*, 2013). The latter name comes from the fact that a trajectory of the process gets killed whenever it crosses the level w. Thus, one can think of the wall at w as an absorber of trajectories. The trajectories that contribute to $Q_{\max}(w, N)$ are those that survive the absorbing wall. In the mathematics literature, the survival probability $Q_{\max}(w, N)$ is sometimes referred to as the *barrier-crossing probability*, the barrier being the wall at w. A related quantity is the so-called *first-passage* probability $F(w, N)$ denoting the probability that a trajectory crosses the level w for the first time between steps $N - 1$ and N. It is simply related to the survival probability via

$$F(w, N) = Q_{\max}(w, N - 1) - Q_{\max}(w, N). \qquad (1.14)$$

This difference $Q_{\max}(w, N - 1) - Q_{\max}(w, N)$ is indeed the fraction of trajectories that survive the absorbing wall up to step $N - 1$, but not up to step N. First-passage properties of stochastic processes have numerous applications in physics, chemistry, biology and mathematics (Redner, 2001; Feller, 2008*a,b*; Bray *et al.*, 2013; Metzler *et al.*, 2014). Again, there are several tools and ideas that one can borrow from the study of stochastic processes in order to study the EVS.

In this book we study different extreme observables, as outlined above, for two principal models:

- IID (independent and identically distributed) random variables $x_1, x_2, \ldots, x_N$;
- RW (random walk) in discrete time and on the line. Here, x_i denotes the position of a random walker at step i. As explained later, the x_i happen to be strongly correlated in this case.

One of the main goals of this book is to investigate how the correlations between the entries x_i of the time series affect the properties of different extreme observables. We will use these two models as examples to illustrate the effects of correlations on the EVS.

The IID case. The simplest model corresponds to the case where the N random variables $x_1, x_2, \ldots, x_N$ are completely uncorrelated, each drawn independently from a distribution $p(x)$. In this case, the joint distribution factorizes,

$$P_{\text{joint}}(x_1, x_2, \ldots, x_N) = \prod_{i=1}^{N} p(x_i). \tag{1.15}$$

Many aspects of EVS in the IID case have been well studied in the statistics and mathematics literature and there are several books and monographs available. The factorization property in Eq. (1.15) simplifies the computation of extreme observables enormously. There is a nice interpretation of this factorization in terms of the "energy" of the gas of N particles defined in Eq. (1.12). The factorization property in Eq. (1.15) indicates that

$$E(x_1, x_2, \ldots, x_N) = -\sum_{i=1}^{N} \ln p(x_i). \tag{1.16}$$

This is just the energy of a gas of *non-interacting* particles. This is the reason why the partition function $Q_{\max}(w, N)$ in Eq. (1.11) for such a gas can be computed easily. In this book we will revisit this case of IID random variables and compute several extreme value observables from a more physical point of view, as mentioned above.

However, in most of the time series observed in nature, the underlying random variables are *correlated*. For example, in a typical temperature time series, if a day is hot, most likely the next day will also be hot. So there are clear correlations between the entries of the time series. In this correlated case, the factorization property in Eq. (1.15) no longer holds, and consequently computing the statistics of extreme observables becomes much harder. It is easy to see why from our discussion above: in this case the partition function $Q_{\max}(w, N)$ in Eq. (1.11) corresponds to an interacting gas of particles, for which it is in general hard to exactly compute the partition function. These interactions between the gas particles can be

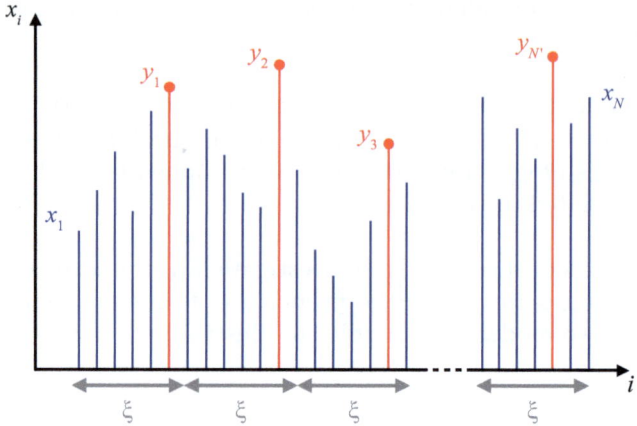

Fig. 1.8 Illustration of a set of weakly correlated random variables, characterized by a finite correlation length $\xi \ll N$, as in Eq. (1.17). The system is divided into $N' = N/\xi$ blocks of size ξ, which are roughly independent. Within each block, we denote by y_i, $i = 1, \ldots, N'$, the local maxima, which are essentially uncorrelated, since they belong to different blocks. Hence, for large N, the global maximum $M = \max[y_1, y_2, \ldots, y_{N'}]$ behaves as the maximum of a set of N' IID random variables.

either short-ranged or long-ranged. In terms of the entries of the time series, the short-ranged interactions would mean that the variables x_i are correlated only over a finite correlation length $\xi \ll N$. We will call this the *weakly correlated* case. In contrast, if the interaction is long-ranged, $\xi \geq N$, it corresponds to the case where the N variables are all correlated with each other. This case will be called *strongly correlated*. We first briefly discuss the weakly correlated case below, followed by an example of a strongly correlated case.

We start with the weakly correlated case. Suppose that the entries $\{x_1, x_2, \ldots, x_N\}$ of our time series are not independent, but correlated such that the connected part of the correlation function decays fast (say exponentially) over a certain finite correlation length $1 \leq \xi \ll N$ (see Fig. 1.8):

$$C_{i,j} = \langle x_i x_j \rangle - \langle x_i \rangle \langle x_j \rangle \sim e^{-|i-j|/\xi}. \tag{1.17}$$

Clearly, when two entries are separated over a length scale larger than ξ, i.e., when $|i - j| \gg \xi$, then they essentially get uncorrelated. Now, weak correlation implies that $\xi \ll N$, where N is the total size of the sample. A simple example is the time series generated by the recursion relation

$$x_i = (1 - \alpha)x_{i-1} + \eta_i, \tag{1.18}$$

where $0 \leq \alpha \leq 1$ and the η_i are IID Gaussian random variables. Such a process appears quite naturally in finance and in queuing theory, where it is called the AR(1) process. It is easy to show that the correlation function behaves as in Eq. (1.17) with $\xi = -1/\ln(1 - \alpha)$ (for a detailed calculation of this model see

Section 8.1). In this weakly correlated case, it is very easy to see heuristically that, in the large-N limit, the distribution of the global maximum reduces to that of the IID case discussed above. This follows from a "block renormalization" type argument that is discussed in detail in Section 4.2. Roughly speaking, one divides the system into $N' = N/\xi$ blocks. While the variables are strongly correlated inside each block, they are essentially uncorrelated across blocks since the distance between them exceeds the correlation length ξ. Let y_i denote the maximum of the ith block. Clearly, the global maximum $X_{\max} = \max[x_1, x_2, \ldots, x_N] = \max[y_1, y_2, \ldots, y_{N'}]$. Since the y_i are IID, this reduces to computing the distribution of the maximum of N' IID variables.

However, this simple argument breaks down completely when the correlation length $\xi \geq N$, i.e., in the strongly correlated case. Computing the extreme value statistics of such strongly correlated time series is thus a major challenging problem, and a lot of research has been devoted to its study in recent years in both physics and mathematics. There is no general result as in the IID case, hence one tries to find solvable models to gain insights on the statistics of the maximum of strongly correlated time series. One simple solvable example of such a strongly correlated time series is a random walk in one dimension, for which it is possible to derive exactly several properties of extreme observables. A large part of this book is indeed devoted to studying this example, which we briefly introduce below.

The RW case. As a simple example of a correlated problem, let us consider a discrete-time random walk on a line. The random walker starts at the origin at step 0, and at each step jumps by a certain random distance. Thus its position x_n at step n evolves as

$$x_n = x_{n-1} + \eta_n, \tag{1.19}$$

where η_n is the jump distance at the nth step. We assume that the jump increments at each step are uncorrelated, i.e., the η_n are IID random variables, each drawn independently from $f(\eta)$. Note that even though the jump increments are uncorrelated, the positions x_n at different n get correlated. To see this, we note from Eq. (1.19) that the position x_n can be expressed as a sum of IID random variables,

$$x_n = \sum_{i=1}^{n} \eta_i. \tag{1.20}$$

Since the positions x_m and x_n at two different times m and n share some common η_is, they are obviously correlated. Using Eq. (1.19) and the fact that the jumps η_i are IID, the joint distribution of the positions x_i can be expressed as

$$P_{\text{joint}}(x_1, x_2, \ldots, x_N) = \prod_{i=1}^{N} f(x_i - x_{i-1}). \tag{1.21}$$

Even though this joint distribution has a factorized form, the factorization is in terms of the difference coordinates $x_i - x_{i-1}$ but not in terms of the x_i themselves.

In that sense, it is different from the IID case in Eq. (1.15) and it manifestly shows that the x_i are *correlated*. Moreover, in Section 2.2 we show in more detail why this is a strongly correlated case, since the correlation length ξ is infinite.

In each of the IID and RW cases we will study in detail the following extreme observables:

- **First-passage probability** We discuss in detail the statistics of the first-passage time to an arbitrary level a in both models. We denote this arbitrary level by a for notational convenience later, though it was denoted by w in Eq. (1.13) and in the right panel of Fig. 1.7.
- **Extreme values** The statistics of $X_{\max}$ and $X_{\min}$ defined in Eqs. (1.1) and (1.2).
- **Time at which an extreme event occurs** The statistics of $t_{\max}$ and $t_{\min}$ denoting respectively the time at which the maximum and the minimum of a time series occurs.
- **Order statistics** After ordering the time series entries $\{x_1, x_2, \ldots, x_N\}$ in decreasing order $\{M_{1,N} > M_{2,N} > \cdots > M_{N,N}\}$ such that $M_{k,N}$ represents the kth maximum, we will study the statistics of $M_{k,N}$ and of the successive gaps $g_{k,N} = M_{k,N} - M_{k+1,N}$. The density of *near-extreme events*, as defined in Eq. (1.7), will also be studied.
- **Record statistics** Several observables associated with record statistics in both models, such as the number of records M up to step N and the statistics of the ages $\{\ell_1, \ell_2, \ldots, \ell_M\}$ of the records.

The remainder the book is organized as follows. In Chapter 2 we discuss some very basic facts and tools from probability theory that are useful in studying the statistics of extreme variables. In Chapter 3, we discuss the first-passage probability for the IID and RW cases. In Chapter 4, we discuss the statistics of $X_{\max}$ and $X_{\min}$ for both models, and also discuss several applications of these results. Chapter 5 is devoted to the computation of the statistics of $t_{\max}$ and $t_{\min}$. In Chapter 6 we discuss the statistics, and in Chapter 7 record statistics. In Chapter 8, we discuss the statistics of extremes in several other models along with applications. Finally, in Chapter 9, we conclude with some general perspectives on extreme value statistics.

2
The Two Principal Models and Some Basic Tools

In this chapter, we introduce the two principal models studied in this book, namely, the IID model and the RW model. The first one corresponds to a time series of size N where the entries are independent random variables, while in the second model the entries correspond to the positions of a discrete-time random walker on a line. In the latter case, the entries of the time series are strongly correlated, as will be shown later. We also discuss some basic tools from probability theory and stochastic processes that will play a crucial role in understanding the extreme statistics in later chapters.

2.1 A single random variable

2.1.1 Probability density function

Imagine tossing a coin and recording the outcome as heads or tails. Let us define $x = +1$ if the outcome is heads and $x = -1$ if it is tails. Thus, the probability distribution of the outcome x can be written as

$$p(x) = \frac{1}{2}\delta(x - 1) + \frac{1}{2}\delta(x + 1). \tag{2.1}$$

This distribution, of course, is not continuous. A variable is called deterministic if it has only one fixed value, say $x = a$; in other words, its distribution is a single delta function, i.e., $p(x) = \delta(x - a)$. Anything different from this single delta function makes the variable random. Generically, we will denote by $p(x)$ the probability distribution function (PDF). In general, $p(x)$ can have any form, either symmetric or asymmetric around 0. A few examples of symmetric distributions are

$$p(x) = \frac{1}{\sqrt{2\pi\ell_0^2}}\,\mathrm{e}^{-x^2/2\ell_0^2} \qquad \text{(Gaussian)}, \tag{2.2}$$

$$p(x) = \frac{1}{2\ell_0}\,\mathrm{e}^{-|x|/\ell_0} \qquad \text{(double exponential)}, \tag{2.3}$$

Statistics of Extremes and Records in Random Sequences. Satya N. Majumdar and Grégory Schehr, Oxford University Press.
© Satya N. Majumdar and Grégory Schehr (2024). DOI: 10.1093/9780191838781.003.0002

$$p(x) = \frac{1}{2\ell_0}[\Theta(x + \ell_0) - \Theta(x - \ell_0)] \quad \text{(uniform in } [-\ell_0, \ell_0]), \tag{2.4}$$

$$p(x) = \frac{1}{\pi} \frac{\ell_0}{x^2 + \ell_0^2} \qquad\qquad \text{(Cauchy distribution)}, \tag{2.5}$$

where $\Theta(z)$ is the Heaviside theta function and ℓ_0 denotes a microscopic length scale associated to these distributions. These distributions are easy to generate numerically. For instance, one can draw a uniform random variable x over $[0, 1]$ very easily using a standard algorithm in available packages. From this uniform random variable, one can very easily generate a random variable y with PDF $p(y)$ using the simple relation

$$x = \int_{-\infty}^{y} p(y') \, \mathrm{d}y'. \tag{2.6}$$

This relation maps any random variable y with PDF $p(y)$ to a uniformly distributed random variable x over $[0, 1]$ (can you prove this?). For instance, if we want to generate a positive exponential random variable $p(y) = \Theta(y)\mathrm{e}^{-y}$, from Eq. (2.6) we get $x = 1 - \mathrm{e}^{-y}$ which implies $y = -\ln(1 - x)$.

Little exercise 2.1 *Generate a random variable with distribution* $p(x) = \frac{1}{2\ell_0}\mathrm{e}^{-|x|/\ell_0}$.

The distributions $p(x)$ are normalized to unity (as in the examples in Eqs. (2.2)–(2.5)),

$$\int_{-\infty}^{\infty} p(x) \, \mathrm{d}x = 1, \tag{2.7}$$

where the integral runs over the support of $p(x)$. In the examples above we have chosen the distributions to be symmetric about 0 but, in principle, they can be asymmetric as well. For example, a Gaussian distribution with a non-zero mean $\langle x \rangle$ and variance ℓ_0^2 reads

$$p(x) = \frac{1}{\sqrt{2\pi\ell_0^2}} \exp\left\{-\frac{(x - \langle x \rangle)^2}{2\ell_0^2}\right\}. \tag{2.8}$$

For a given $p(x)$, it is useful to define its moments,

$$\mu_n = \int_{-\infty}^{\infty} x^n p(x) \, \mathrm{d}x, \tag{2.9}$$

where the integral runs over the support of $p(x)$. The nth moment may not exist for certain classes of $p(x)$. For instance, for the Cauchy distribution defined in Eq. (2.5), all moments for $n \geq 2$ are divergent (the first moment, i.e., the average, is zero by symmetry). Therefore, in this case the distribution has a "fat tail," meaning that large values of $|x|$ can occur with a relatively large probability (compared, for example, to the Gaussian case).

2.1.2 Characteristic function

At this point, it is also convenient to define the Fourier transform of the distribution (sometimes referred to as the characteristic function of $p(x)$):

$$\hat{p}(k) = \int_{-\infty}^{\infty} e^{ikx} p(x)\, dx. \tag{2.10}$$

The moments can be computed from the Fourier transform by expanding Eq. (2.10) in a power series in k:

$$\hat{p}(k) = \sum_{m=0}^{\infty} \frac{(ik)^m}{m!} \mu_m. \tag{2.11}$$

One can also define the cumulants C_m as

$$\ln \hat{p}(k) = \sum_{m=1}^{\infty} \frac{(ik)^m}{m!} C_m. \tag{2.12}$$

By comparing Eqs. (2.11) and (2.12), one can relate the cumulants to the moments. The first few examples are

$$C_1 = \mu_1, \tag{2.13}$$

$$C_2 = \mu_2 - \mu_1^2, \tag{2.14}$$

$$C_3 = \mu_3 - 3\mu_1\mu_2 + 2\mu_1^3, \tag{2.15}$$

$$C_4 = \mu_4 - 4\mu_1\mu_3 - 3\mu_2^2 + 12\mu_1^2\mu_2 - 6\mu_1^4. \tag{2.16}$$

For instance, for the Gaussian distribution in Eq. (2.8) the Fourier transform reads

$$\hat{p}(k) = \int_{-\infty}^{\infty} e^{ikx} \frac{1}{\sqrt{2\pi\ell_0^2}} \exp\left\{-\frac{(x - \langle x \rangle)^2}{2\ell_0^2}\right\} dx = \exp\left\{ik\langle x \rangle - \frac{k^2\ell_0^2}{2}\right\}. \tag{2.17}$$

Hence, in this case, $\ln \hat{p}(k) = ik\langle x \rangle - k^2\ell_0^2/2$ and consequently the cumulants are

$$C_1 = \langle x \rangle, \quad C_2 = \ell_0^2, \quad C_3 = C_4 = C_5 = \cdots = 0. \tag{2.18}$$

All higher cumulants beyond the second are identically zero for a Gaussian variable.

Little exercise 2.2 *For the double exponential in Eq. (2.3), show that all the odd cumulants are 0 while the even cumulants are given by $C_{2m} = \ell_0^{2m}(2m)!/m$ for $m = 1, 2, \ldots$*

2.1.3   Cumulative distribution function

It is also useful to define the cumulative distribution function (CDF) of a random variable as

$$P_<(x) = \int_{-\infty}^{x} p(y)\, \mathrm{d}y, \tag{2.19}$$

which denotes the probability that the random variable takes values less than or equal to x. As $x \to \infty$, the CDF $P_<(x) \to 1$, since the PDF is normalized to unity. In the opposite limit $x \to -\infty$, the CDF $P_<(x) \to 0$. Clearly the PDF is just the derivative of the CDF, i.e.,

$$p(x) = \frac{\mathrm{d}P_<(x)}{\mathrm{d}x}. \tag{2.20}$$

Little exercise 2.3 *For the Gaussian random variable with PDF $p(x)$ given in Eq. (2.8), show that*

$$P_<(x) = 1 - \frac{1}{2}\mathrm{erfc}\left(\frac{x - \langle x \rangle}{\sqrt{2}\ell_0}\right), \tag{2.21}$$

where $\mathrm{erfc}(z) = (2/\sqrt{\pi}) \int_z^\infty \mathrm{e}^{-u^2}\, \mathrm{d}u$ is called the complementary error function.

2.2   Multiple random variables

Often, we actually encounter multiple random variables $\{x_1, x_2, \ldots, x_N\}$. For example, these variables may be the entries in a random sequence. This sequence can represent any possible time series; for instance, it can be the temperature in a given city for N successive days, or it can represent the price of a stock for N different days, and so on. These x_n are random variables. Typically, they are correlated and are distributed according to a joint PDF $P_{\mathrm{joint}}(x_1, x_2, \ldots, x_N)$. In this book we will mostly focus on two representative cases: (1) an uncorrelated sequence of identically distributed random variables, and (2) a correlated sequence where the entry x_n represents the position of a random walker on a line after n steps, starting at $x_0 = 0$. We will see throughout the book that many observables, such as the first-passage probability, extreme statistics, order statistics or record statistics, can be computed exactly for these two solvable sequences. Hence, they will serve as a toy laboratory to understand the effects of correlations on different observables related to extremes and records.

2.2.1   First model: IID sequence

In the special case when the variables are uncorrelated, their joint distribution factorizes,

$$P_{\mathrm{joint}}(x_1, x_2, \ldots, x_N) = \prod_{i=1}^{N} p_i(x_i), \tag{2.22}$$

where $p_i(x_i)$ is called the marginal PDF of the ith variable x_i. In addition, if $p_i(x) = p(x)$ is independent of the label i, each of these variables has the same marginal PDF. In that case,

$$P_{\text{joint}}(x_1, x_2, \ldots, x_N) = \prod_{i=1}^{N} p(x_i). \tag{2.23}$$

As we will see later, the extreme value statistics, as well as other observables such as the first-passage probability, order statistics, record statistics and so on, for this IID model can be studied with just the basic tools discussed in the previous section. One does not require any additional tools.

2.2.2   Second model: Random walk sequence

When the random variables are correlated, the joint PDF $P_{\text{joint}}(x_1, x_2, \ldots, x_N)$ no longer factorizes into marginal distributions. The simplest, and perhaps the most widely studied, example of such a correlated sequence is the celebrated discrete-time random walk sequence. A mathematical description of such random walks can be found in the classical book by Feller (2008*a*,*b*). For a pedagogical introduction to recent developments, see Majumdar (2010*b*).

We consider a random walker on an infinite line starting initially at $x_0 = 0$. At every discrete time step, the walker chooses independently a distance η from a distribution $f(\eta)$ and jumps by this distance. The position x_n of the walker at step n therefore evolves according to the stochastic equation

$$x_n = x_{n-1} + \eta_n, \tag{2.24}$$

where the η_n are IID noise (jump) variables, each drawn independently from $f(\eta)$ (see Fig. 2.1). The jump PDF $f(\eta)$ may be symmetric (no drift) or asymmetric (e.g., in the presence of a constant drift).

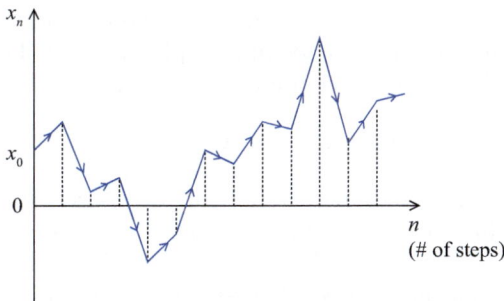

Fig. 2.1 The trajectory of a random walker starting at the initial position x_0 and evolving via Eq. (2.24).

A few examples of symmetric jump distributions are:

$$f(\eta) = \frac{1}{\sqrt{2\pi\ell_0^2}} e^{-\eta^2/2\ell_0^2} \qquad \text{(Gaussian)}, \qquad (2.25)$$

$$f(\eta) = \frac{1}{2\ell_0} e^{-|\eta|/\ell_0} \qquad \text{(double exponential)}, \qquad (2.26)$$

$$f(\eta) = \frac{1}{2\ell_0}[\Theta(\eta + \ell_0) - \Theta(\eta - \ell_0)] \qquad \text{(uniform in } [-\ell_0, \ell_0]), \qquad (2.27)$$

$$f(\eta) \sim |\eta|^{-1-\mu} \text{ for large } |\eta| \text{ with } 0 \le \mu < 2 \qquad \text{(Lévy flights)}, \qquad (2.28)$$

$$f(\eta) = \frac{1}{2}[\delta(\eta - 1) + \delta(\eta + 1)] \qquad \text{(lattice random walk with} \qquad (2.29)$$
$$\text{lattice constant 1)},$$

where ℓ_0 represents a microscopic length scale associated to these distributions. In the first four of these examples, the cumulative jump distribution $P_<(\eta) = \int_{-\infty}^{\eta} f(\eta')\,d\eta'$ is a continuous function. In Eq. (2.29), where the walker is restricted to move on a one-dimensional lattice with unit lattice spacing, the cumulative jump distribution $P_<(\eta)$ is a non-continuous function. We will see later that this continuity property of $P_<(\eta)$ will play an important role. Note further that in Eqs. (2.25)–(2.27) and (2.29), the variance of the step length,

$$\sigma^2 = \int_{-\infty}^{\infty} \eta^2 f(\eta)\,d\eta, \qquad (2.30)$$

is finite. We will see that in such cases the central limit theorem holds. In the Lévy case, Eq. (2.28), where σ^2 is divergent, the central limit theorem breaks down. For a detailed mathematical discussion on the central limit theorem we refer the reader to Feller (2008a,b). For a simple derivation of the central limit theorem see the discussion after Eq. (2.57).

The evolution equation in Eq. (2.24) is Markovian since the position x_n at step n depends only on the position at the previous time step x_{n-1} (and not on the full history before the $(n-1)$th step) and on the current noise, i.e., the noise η_n at step n. This Markovian property makes life simple, as we will see later. As a simple example of a non-Markovian evolution consider the rule

$$x_n = 2x_{n-1} - x_{n-2} + \eta_n, \qquad (2.31)$$

where the η_n are again IID random variables. In Eq. (2.31) the current position x_n, in addition to depending on the previous position x_{n-1} and the current noise η_n, also depends on x_{n-2}—this makes the process non-Markovian in a simple way. In this book we will not discuss non-Markovian processes and focus only on Markov processes such as Eq. (2.24).

Iterating the Markov evolution rule in Eq. (2.24) up to n steps, it follows that the position x_n of the walker after n steps, starting at $x_0 = 0$, is simply the sum of n IID random variables,

$$x_n = \sum_{k=1}^{n} \eta_k. \tag{2.32}$$

In the case when σ^2 is finite, using the independence property of the step lengths η_k, it follows that the mean square displacement of the particle after n steps, for all n, is simply

$$\langle x_n^2 \rangle = n\sigma^2. \tag{2.33}$$

Furthermore, for finite σ^2 the two-point correlation function can also be computed using Eq. (2.32). Using $\langle \eta_i \eta_j \rangle = \sigma^2 \delta_{i,j}$, we get

$$\langle x_m x_n \rangle = \sigma^2 \min(m, n). \tag{2.34}$$

The non-zero correlations emerge from the fact that x_n and x_m, for $n \neq m$, have some noise variables in common, as is evident from Eq. (2.32). Of course, for $m = n$, the result in Eq. (2.34) reduces to Eq. (2.33). When σ^2 is divergent, such as in the case of Lévy flights, the process remains correlated, even though the second and higher moments (including two-point or higher correlation functions) are divergent. Thus, even though the jump variables η_n are uncorrelated, the position variables x_n are correlated and hence the random sequence $\{x_1, x_2, \ldots, x_N\}$ is a *correlated sequence*, unlike the IID case discussed before.

Green's function. Let us get back to our basic discrete-time Markov evolution in Eq. (2.24). In this subsection we compute a basic object, namely the free (bare) Green's function $G(x, x_0, n)$ defined as the probability density of the position of the walker after step n at x, given that it started from x_0 at step 0. Using the Markov property, one can easily write down a recursion relation for the evolution

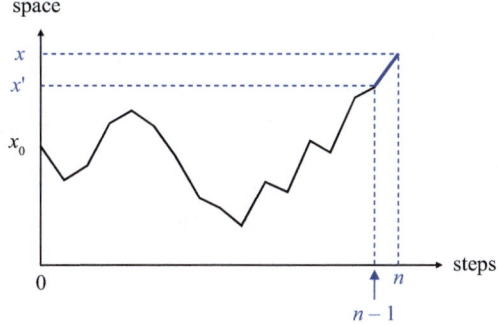

Fig. 2.2 The evolution of the trajectory from position x' at step $n-1$ to position x at the next step n.

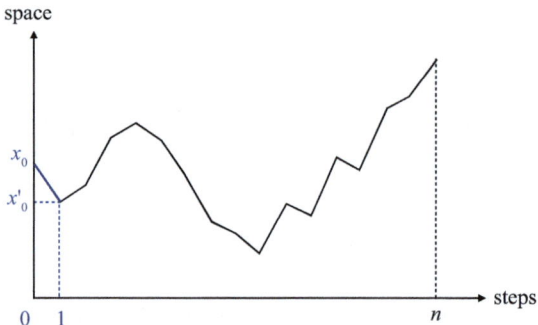

Fig. 2.3 This random walk starting at x_0 makes a jump to x_0' at the first step and the rest of the trajectory evolves from step one onwards, starting from x_0'.

of $G(x, x_0, n)$,

$$G(x, x_0, n) = \int_{-\infty}^{\infty} G(x', x_0, n-1) f(x - x') \, dx', \tag{2.35}$$

which counts the event of the particle jumping from its position x' at step $n-1$ to its position x at step n by an amount $(x - x')$ drawn from the distribution $f(x - x')$ (see Fig. 2.2). The recursion relation in Eq. (2.35) starts from the initial condition

$$G(x, x_0, 0) = \delta(x - x_0) \tag{2.36}$$

and is called the *forward* Kolmogorov equation, since one considers the current position x of the walker as a variable (with the initial position x_0 fixed). Alternatively, one can also write down a *backward* Kolmogorov equation where one considers the starting position of the walker x_0 as a variable (with the final position x kept fixed):

$$G(x, x_0, n) = \int_{-\infty}^{\infty} G(x, x_0', n-1) f(x_0' - x_0) \, dx_0'. \tag{2.37}$$

Here, one considers the displacement of the particle at the first step from x_0 to x_0', and for the subsequent evolution up to $(n-1)$ steps the starting position of the walker is at x_0' (see Fig. 2.3). Both equations are completely equivalent. We will see later, however, that for certain first-passage-related quantities the backward equation is often computationally more advantageous than the forward one.

The integral equations in Eqs. (2.35) or (2.37) can be easily solved using Fourier transforms. For example, for the forward equation, we define

$$\hat{G}(k, x_0, n) = \int_{-\infty}^{\infty} G(x, x_0, n) e^{ikx} \, dx \tag{2.38}$$

and use the convolution form of Eq. (2.35) to get $\hat{G}(k, x_0, n) = \hat{G}(k, x_0, n-1) \hat{f}(k)$, where

$$\hat{f}(k) = \int_{-\infty}^{+\infty} f(x) e^{ikx} \, dx \tag{2.39}$$

is the Fourier transform of $f(x)$. Iterating n times and using the initial condition $(G(x, x_0, 0) = \delta(x - x_0)$ and hence $\hat{G}(k, x_0, 0) = e^{ikx_0})$, one gets

$\hat{G}(k, x_0, n) = [\hat{f}(k)]^n e^{ikx_0}$. Inverting the Fourier transform, one obtains the exact Green's function

$$G(x, x_0, n) = \int_{-\infty}^{\infty} [\hat{f}(k)]^n e^{-ik(x-x_0)} \frac{dk}{2\pi}. \tag{2.40}$$

Let us consider few examples of jump distributions $f(\eta)$ whose Fourier transform can be computed explicitly. For instance, for Eqs. (2.25)–(2.27) one gets

$$\hat{f}(k) = e^{-k^2 \ell_0^2/2} \qquad (f(\eta) \text{ is Gaussian}), \tag{2.41}$$

$$\hat{f}(k) = \frac{1}{1 + k^2 \ell_0^2} \qquad (f(\eta) \text{ is double exponential}), \tag{2.42}$$

$$\hat{f}(k) = \frac{\sin(k\ell_0)}{k\ell_0} \qquad (f(\eta) \text{ is uniform in } [-\ell_0, \ell_0]). \tag{2.43}$$

For Eq. (2.29), i.e., for the lattice walk where $f(\eta) = (1/2)\delta(\eta+1) + (1/2)\delta(\eta-1)$, the Fourier transform is simply

$$\hat{f}(k) = \cos k \qquad (\text{lattice walk}). \tag{2.44}$$

Finally, for Eq. (2.28), i.e., for jump distributions with a power-law tail, $f(\eta) \sim |\eta|^{-1-\mu}$, it is difficult, in general, to write the Fourier transform $\hat{f}(k)$ explicitly. However, there is a special class of jump distributions, known as stable jump distributions, for which the Fourier transform is explicitly given by

$$\hat{f}(k) = e^{-|\ell k|^\mu} \qquad \text{for } 0 < \mu \leq 2, \tag{2.45}$$

where the parameter ℓ characterizes the scale of the jump and is, in fact, proportional to the microscopic length scale ℓ_0 defined at the beginning of this chapter. For $\mu = 1$, this corresponds to the Fourier transform of the Cauchy distribution,

$$f(\eta) = \frac{1}{\pi} \frac{\ell}{\eta^2 + \ell^2}. \tag{2.46}$$

For $\mu = 2$, it corresponds to the Gaussian distribution,

$$f(\eta) = \frac{1}{\sqrt{4\pi\ell^2}} e^{-\eta^2/(4\ell^2)}. \tag{2.47}$$

Exact Green's function for all n in some solvable cases. Given these explicit $\hat{f}(k)$, we need to evaluate the integral over k in Eq. (2.40) to compute the Green's function. However, for finite n, this integral can only be done explicitly in a few cases. The simplest example is the lattice random walk where $\hat{f}(k) = \cos k$. In this case, the integral over k in Eq. (2.40) can be performed explicitly by expanding $(\cos k)^n = 2^{-n}(e^{ikx} + e^{-ikx})^n$ using the binomial formula

$$G(x, x_0, n) = \sum_{m=0}^{n} \frac{1}{2^n} \binom{n}{m} \int_{-\infty}^{\infty} \frac{dk}{2\pi} e^{ikm} (e^{-ik})^{n-m} e^{-ik(x-x_0)}. \tag{2.48}$$

The integral over k simply gives a delta function $\delta(x - x_0 + n - 2m)$. This clearly shows that $x - x_0$ has to be an integer, as expected for a lattice random walk. Hence, finally, the Green's function reads

$$G(x, x_0, n) = \begin{cases} \dfrac{1}{2^n} \begin{pmatrix} n \\ (x - x_0 + n)/2 \end{pmatrix} & \text{if } x - x_0 + n \text{ is even} \\ 0 & \text{if } x - x_0 + n \text{ is odd.} \end{cases} \tag{2.49}$$

The Green's function can also be computed explicitly for the double-exponential jump distribution with $f(\eta) = 1/(2\ell_0)e^{-|\eta|/\ell_0}$. In this case the Fourier transform of the jump distribution is $\hat{f}(k) = 1/(1 + (k\ell_0)^2)$ and the Green's function $G(x, x_0, n)$ in Eq. (2.40) reads, for all $n \geq 1$,

$$G(x, x_0, n) = \frac{1}{\ell_0} \frac{2^{1/2-n}}{\sqrt{\pi}\Gamma(n)} \left(\frac{x - x_0}{\ell_0} \right)^{n-1/2} K_{n-1/2}\left(\frac{x - x_0}{\ell_0} \right), \tag{2.50}$$

where $K_\nu(z)$ is the modified Bessel function of index ν (Gradshteyn and Ryzhik, 2014).

Little exercise 2.4 *Prove the result in Eq. (2.50).*

There is another class of jump distributions, the so-called stable distributions given in Eq. (2.45), for which $G(x, x_0, n)$ can in principle be computed for all n (a discussion about stable distributions can be found in many places in the literature; see, e.g., the review in Bouchaud and Georges (1990)). Substituting $\hat{f}(k) = e^{-|\ell k|^\mu}$ in the integral in Eq. (2.40) and rescaling $\ell n^{1/\mu}k \to k$, the Green's function can be written as

$$G(x, x_0, n) = \frac{1}{\ell n^{1/\mu}} \mathcal{L}_\mu\left(\frac{x - x_0}{\ell n^{1/\mu}} \right) \qquad \text{with } 0 < \mu \leq 2 \tag{2.51}$$

where the function

$$\mathcal{L}_\mu(z) = \int_{-\infty}^{\infty} e^{-|k|^\mu - ikz} \frac{\mathrm{d}k}{2\pi} \tag{2.52}$$

is called the Lévy stable function of index μ. Thus, for this stable class indexed by $0 < \mu \leq 2$, the distribution of the total displacement $x - x_0$ after step n is distributed by the same law as the individual jump distributions, up to a scale factor $n^{1/\mu}$. Thus, the sum of such random variables remains invariant, statistically, up to the scale factor—this is why such jump distributions are called "stable." For the special case $\mu = 2$, this says that the sum of n independent Gaussian random variables is itself a Gaussian with a variance rescaled by a factor $\sqrt{n}$. This scaling function $\mathcal{L}_\mu(z)$ is symmetric in z and it has the asymptotic behaviors

$$\mathcal{L}_\mu(z) \approx \begin{cases} \dfrac{1}{\pi\mu}\left[\Gamma\left(\dfrac{1}{\mu}\right) - \dfrac{z^2}{2}\Gamma\left(\dfrac{3}{\mu}\right) \right] & \text{as } z \to 0, \\ \dfrac{\Gamma(1+\mu)\sin(\pi\mu/2)}{\pi} \dfrac{1}{|z|^{1+\mu}} & \text{as } |z| \to \infty. \end{cases} \tag{2.53}$$

Little exercise 2.5 *Prove these asymtotic behaviors in Eq. (2.53), starting from Eq. (2.52).*

For generic $0 < \mu \leq 2$, it is hard to obtain a simple explicit formula for $\mathcal{L}_\mu(z)$. Only for the two special cases $\mu = 1$ and $\mu = 2$ one can compute it explicitly. For $\mu = 1$ (the Cauchy case), one gets

$$\mathcal{L}_1(z) = \frac{1}{\pi} \frac{1}{1 + z^2}. \tag{2.54}$$

Similarly, for $\mu = 2$, $\mathcal{L}_2(z)$ is simply the Gaussian

$$\mathcal{L}_2(z) = \frac{1}{\sqrt{2\pi}} e^{-z^2/2}. \tag{2.55}$$

Large-n behavior of the Green's function. We have seen that the Green's function $G(x, x_0, n)$ can be explicitly computed for all n in a few specific examples, which include lattice random walks, random walks with a double-exponential jump distribution, and Lévy stable random walks. For other generic jump distributions, it is hard to find an explicit formula valid for all n. However, one can easily derive the asymptotic behavior of the Green's function for large n. In this limit, the integral over k in Eq. (2.40) is dominated by the region of small k. Thus, we need to investigate the small-k behavior of $\hat{f}(k)$. Indeed, in all the examples discussed above, the small-k behavior of $\hat{f}(k)$ is of the form

$$\hat{f}(k) \approx 1 - |\ell k|^\mu \qquad \text{with } 0 < \mu \leq 2, \tag{2.56}$$

where ℓ is the characteristic jump length. For the first three examples above, we have $\mu = 2$ and $\ell = \ell_0/\sqrt{2}$ (Gaussian), $\ell = \ell_0$ (double exponential) and $\ell = \ell_0/\sqrt{3}$ (uniform). For the lattice random walk, $\mu = 2$ and $\ell = 1/\sqrt{2}$. Here, we will restrict ourselves to jump distributions whose Fourier transform has small-k behavior as in Eq. (2.56). Substituting the small-k behavior $\hat{f}(k) \approx 1 - |\ell k|^\mu$ in Eq. (2.40), one gets, to leading order for large n, the same behavior as the stable law of index μ in Eq. (2.51):

$$G(x, x_0, n) \approx \int_{-\infty}^{\infty} e^{-n|\ell k|^\mu} e^{-ik(x-x_0)} \frac{dk}{2\pi} = \frac{1}{\ell n^{1/\mu}} \mathcal{L}_\mu \left(\frac{x - x_0}{\ell n^{1/\mu}} \right), \tag{2.57}$$

with the scaling function $\mathcal{L}_\mu(z)$ given in Eq. (2.52).

Consider first the case $\mu = 2$. In this case, from Eq. (2.56), $\hat{f}(k) \approx 1 - \ell^2 k^2$ as $k \to 0$, where $2\ell^2 = \sigma^2$ is the variance of the jump distribution $f(\eta)$. Thus, for $\mu = 2$, the above result states that the sum of n IID random jump variables converges, for large n, to a Gaussian random variable, irrespective of the jump distribution $f(\eta)$, as long as σ^2 is finite. This is indeed the statement of the celebrated central limit theorem (CLT). For general $0 < \mu < 2$, where the variance is divergent, the above result states that the distribution of the sum of IID random variables converges to a Lévy stable distribution. This is the generalization of the CLT ($\mu = 2$) to the cases where $0 < \mu < 2$.

Continuous-time limit.   In the large-n limit, the discrete-time stochastic sequence $\{x_1, x_2, \ldots, x_n\}$ converges statistically to a continuous-time stochastic process $\{x(\tau), 0 \leq \tau \leq t\}$. This is most easily seen for the case $\mu = 2$, where the standard CLT holds and the associated continuous-time process turns out to be just Brownian motion. To see this, let us define Δt as a small time interval and set $t = n\Delta t$. Thus, the large-n limit corresponds to taking $\Delta t \to 0$ but with t fixed. Then, (2.33) gives

$$\langle x_n^2 \rangle = \frac{\sigma^2}{\Delta t} t. \tag{2.58}$$

If one now takes the limit $\Delta t \to 0$, it follows that $\sigma^2 \to 0$ also in order that $\langle x^2(t) \rangle$ remains finite at finite time t. Thus, to have a meaningful continuous-time limit, the mean square step length $\sigma^2 \to 2D\Delta t$ as $\Delta t \to 0$ with a finite diffusion constant D, leading to the diffusive law of Brownian motion $\langle x^2(t) \rangle = 2Dt$ for all t. In this continuous-time limit, one can also rewrite the Markov evolution rule in Eq. (2.24) as

$$\frac{\Delta x_n}{\Delta t} = \frac{\eta_n}{\Delta t} = \eta(t), \tag{2.59}$$

where $\eta(t)$ is random noise with zero mean that is uncorrelated at two different times, $\langle \eta(t_1)\eta(t_2) \rangle = 0$ for $t_1 \neq t_2$. At the same time instant, however, $\langle \eta^2(t) \rangle = \sigma^2/(\Delta t)^2 = 2D/\Delta t$. Thus, as $\Delta t \to 0$, $\langle \eta^2(t) \rangle$ diverges. A useful physicist's way of writing this correlation function of the noise is $\langle \eta(t_1)\eta(t_2) \rangle = 2D\delta(t_1 - t_2)$. In this limit it is called white noise and one writes Eq. (2.59) as a stochastic Langevin equation,

$$\frac{\mathrm{d}x}{\mathrm{d}t} = \eta(t), \tag{2.60}$$

where $\eta(t)$ is white noise with zero mean and a correlator $\langle \eta(t_1)\eta(t_2) \rangle = 2D\delta(t_1 - t_2)$. Note that, for all practical purposes, such as in numerical simulation, one can interpret the delta function as $\delta(t) = 0$ for $t \neq 0$ and $\delta(0) = 1/\Delta t$.

In this Brownian limit, the integral equations in Eqs. (2.35) or (2.37) reduce to partial differential equations. For example, given the Langevin evolution in Eq. (2.60), the forward Kolmogorov equation in Eq. (2.35) reduces to

$$G(x, x_0, t + \Delta t) = \int_{-\infty}^{\infty} G(x - \eta(t)\Delta t, x_0, t) f(\eta(t)) \, \mathrm{d}\eta(t). \tag{2.61}$$

We now expand the Green's function in the integrand on the right-hand side (RHS) in a Taylor series in powers of Δt. To organize this Taylor series, it is convenient to define the moments of the jump distribution at time t,

$$\mu_k = \left[\int_{-\infty}^{\infty} (\eta(t))^k f(\eta(t)) \, \mathrm{d}\eta(t) \right]. \tag{2.62}$$

Note that, by normalization, $\mu_0 = 1$, and, by symmetry, $\mu_1 = 0$. Furthermore, $\mu_2 = 2D/\Delta t$, as discussed before. Expanding Eq. (2.62) we then get

$$G(x, x_0, t + \Delta t) = \mu_0 G(x, x_0, t) - \mu_1 \frac{\partial G}{\partial x}\Delta t + \frac{1}{2}\mu_2 \frac{\partial^2 G}{\partial x^2}(\Delta t)^2 + \mathcal{O}((\Delta t)^3). \quad (2.63)$$

Note that we expanded up to order $\mathcal{O}((\Delta t)^2)$ since $\mu_2 = 2D/\Delta t$ and hence the second-order term is actually of order $\mathcal{O}(\Delta t)$.

Using the explicit values of the first three moments, and taking the limit $\Delta t \to 0$, we get the well-known diffusion equation for the Green's function,

$$\frac{\partial G}{\partial t} = D\frac{\partial^2 G}{\partial x^2}, \quad (2.64)$$

starting from the initial condition $G(x, x_0, 0) = \delta(x - x_0)$. This equation is just the forward Fokker–Planck equation for one-dimensional Brownian motion.

By taking a similar continuous-time limit of the backward Kolmogorov equation in Eq. (2.37), one can write down a backward diffusion equation with x in Eq. (2.64) replaced by x_0,

$$\frac{\partial G}{\partial t} = D\frac{\partial^2 G}{\partial x_0^2}, \quad (2.65)$$

with the initial condition $G(x, x_0, 0) = \delta(x - x_0)$.

Little exercise 2.6 *Prove Eq. (2.65) starting from Eq. (2.37).*

The solution of the forward (or the backward) diffusion equation can be easily found using Fourier transforms (show this as a little exercise) and one recovers, as expected, the Gaussian behavior

$$G(x, x_0, t) = \frac{1}{\sqrt{4\pi Dt}} \exp\left[-\frac{(x - x_0)^2}{4Dt}\right]. \quad (2.66)$$

We will see later that in the Brownian limit many properties of the random walk, such as its first-passage probability, become much simpler. In contrast, for discrete-time evolution, even though the process is Markov, some of these properties are highly non-trivial.

We next turn to the continuous-time limit of Lévy flights with $0 < \mu < 2$, where σ^2 is infinite and the CLT breaks down. One can still *formally* define a continuous-time limit, and obtain the so-called Lévy fractional diffusion equation [for a review and discussion, see Metzler and Klafter (2000)]. This simply follows by rewriting the basic recursion relation in Eq. (2.35) as

$$G(x, x_0, n) = \int_{-\infty}^{\infty} G(x - \eta_n, x_0, n - 1)f(\eta_n)\,\mathrm{d}\eta_n. \quad (2.67)$$

Next, we write $G(x - \eta_n, x_0, n - 1) = \int_{-\infty}^{\infty} \hat{G}(k, x_0, n - 1)e^{ik(x-\eta_n)} \, dk$ and substitute it into Eq. (2.67). This gives

$$G(x, x_0, n) = \int_{-\infty}^{\infty} dk \, \hat{G}(k, x_0, n - 1)\hat{f}(k). \qquad (2.68)$$

Following similar arguments as in the Brownian case, in the large-n limit one needs to keep only the small-k contribution of $\hat{f}(k) = 1 - |ak|^{\mu}$ in Eq. (2.68). This gives

$$G(x, x_0, n) - G(x, x_0, n - 1) \simeq -a^{\mu} \int_{-\infty}^{\infty} dk \, |k|^{\mu} \hat{G}(k, x_0, n - 1). \qquad (2.69)$$

Now, we need to divide both sides by the time increment Δt and take the limit $\Delta t \to 0$. To obtain a sensible limit, one needs to take the $a \to 0$ limit as well, keeping the ratio $a^{\mu}/\Delta t = K$ fixed. This gives the continuous-time integro-differential equation

$$\frac{\partial G}{\partial t} = -K \int_{-\infty}^{\infty} dk \, \hat{G}(k, x_0, n - 1)|k|^{\mu} = -K(-\partial_x^2)^{\mu/2}G, \qquad (2.70)$$

where the integral in the k space can be *formally* interpreted as a fractional derivative. Note that for $\mu = 2$ one recovers the standard diffusion equation, but for $0 \leq \mu < 2$ one still needs to solve an integral equation even in the continuous-time limit. Thus, for the Lévy flights, even though one can formally write down a continuous-time equation, it is not as useful as the ordinary Brownian case where one has a true differential equation in real space whose solution can be easily obtained. A word of caution to the reader: in the literature there is also a "Lévy walk" model in continuous time, which is different from the continuous-time Lévy flight model discussed in Eq. (2.70) (Zaburdaev et al., 2015). The continuous-time fractional diffusion equation in Eq. (2.70) has been studied extensively in the recent past [for a review, see Metzler and Klafter (2000)] and many interesting results, in particular concerning the first-passage properties, have been derived [see, for instance, Koren et al. (2007) and Zoia *et al.* (2007)]. However, in this book we will not use this approach and will rather stick to the discrete-time evolution as in Eq. (2.24).

To summarize this chapter, we have discussed two models of random sequences: (1) the IID sequence where the entries are uncorrelated, and (2) the random walk sequence where the entries are correlated; we introduced some basic properties of these two models that we will need to use in later chapters. In fact, these are the two toy models for which we will show that various observables associated to extremes can be computed exactly; these include first-passage properties, extreme statistics, order statistics and record statistics.

3
First-Passage Probability

Let us start with our favorite random sequence, $\{x_1, x_2, \ldots, x_N\}$. The first observable that plays a central role in understanding the extreme statistics is the so-called first-passage probability. As a simple example, imagine tossing an unbiased coin N times. At each time, we record the outcome $x_i = +1$ if the ith toss returns a "head" and $x_i = -1$ if the outcome is "tail." We may ask: what is the probability that the *first* head occurs at the nth step? A little note of caution to the reader: we will reserve the notation N for the size of the sequence and use n for a "running time step." Clearly, for this simple example where the x_i are independent, the answer is trivially $(1/2)^{n-1} \times (1/2) = (1/2)^n$. This is because the first $n-1$ entries must be a tail, which happens with probability $(1/2)^{n-1}$, followed by a head at the nth step, which happens with probability $1/2$. This is indeed probably the simplest example of a first-passage probability, which plays an important role for any stochastic process. For example, in the context of finance, one may ask about the probability $F(a, n)$ that the price of a stock exceeds a given value a for the *first time* at step n. Generically, for any arbitrary stochastic sequence, the first-passage probability $F(a, n)$ is precisely the probability that an entry of the sequence exceeds a fixed value a for the first time at step n. More precisely,

$$F(a, n) = \text{Prob}[x_1 \leq a, \, x_2 \leq a, \ldots, x_{n-1} \leq a, \, x_n > a]. \tag{3.1}$$

To compute this probability, it is useful first to define the survival probability, or persistence,

$$Q(a, n) = \text{Prob}[x_1 \leq a, \, x_2 \leq a, \ldots, x_{n-1} \leq a, \, x_n \leq a], \tag{3.2}$$

which is simply the probability that the sequence stays below the level a up to step n. It then follows that

$$F(a, n) = Q(a, n - 1) - Q(a, n). \tag{3.3}$$

Going back to our simple example of coin tossing, the probability that the head occurs for the first time at step n is precisely the first-passage probability $F(0, n)$ that the sequence of x_i exceeds the value $a = 0$ for the first time at step n. In this example, the survival probability $Q(0, n) = (1/2)^n$ is just the probability that all the n first tosses give tails. Clearly, $F(0, n) = Q(0, n - 1) - Q(0, n) = (1/2)^n$. For a generic sequence, the first-passage probability $F(a, n)$—or equivalently the

Statistics of Extremes and Records in Random Sequences. Satya N. Majumdar and Grégory Schehr, Oxford University Press.
© Satya N. Majumdar and Grégory Schehr (2024). DOI: 10.1093/9780191838781.003.0003

survival probability $Q(a, n)$—turns out to be extremely hard to compute. In the rest of this chapter we will focus on our two toy models, (1) the IID sequence and (2) the random walk sequence, and show how the first-passage probability can be computed for these two models.

3.1   IID sequence

In the case where the x_i form an IID sequence, with each entry drawn from a distribution $p(x)$, the survival probability $Q(a, n)$ is actually trivial and is given by

$$Q(a, n) = \left[\int_{-\infty}^{a} p(x') \, dx' \right]^n. \tag{3.4}$$

This simply follows from the fact that the x_i are independent, and hence the probability that the first n of them lie below the level a is simply given by the product of probabilities that each of them is below a. Indeed, note that this can be rewritten in a more suggestive form as

$$Q(a, n) = e^{-\theta(a)n}, \quad \text{where } \theta(a) = -\ln \left(\int_{-\infty}^{a} p(x') \, dx' \right). \tag{3.5}$$

In particular, for large n it decays exponentially with n, with a decay constant $\theta(a)$ which depends on a as well as on the distribution $p(x)$. In that sense, $\theta(a)$ is not universal. From the result in Eq. (3.5) and the relation in Eq. (3.3), we find that the first-passage probability is given by

$$F(a, n) = \left[\int_{-\infty}^{a} p(x') \, dx' \right]^{n-1} - \left[\int_{-\infty}^{a} p(x') \, dx' \right]^n$$

$$= \int_{a}^{\infty} p(x') \, dx' \left[\int_{-\infty}^{a} p(x') \, dx' \right]^{n-1}, \tag{3.6}$$

where we used the fact that $\int_{-\infty}^{a} p(x') \, dx' = 1 - \int_{a}^{+\infty} p(x') \, dx'$, which simply follows from the normalization of $p(x)$. Equation (3.6) is simple to understand: it just denotes the probability of the event where each of the first $(n-1)$ entries is less than a while the nth entry is bigger than a. Clearly, for large n, $F(a, n)$ also decays exponentially with n with the same rate $\theta(a)$ as that of the survival probability $Q(a, n)$ in Eq. (3.5).

Little exercise 3.1 *Calculate the kth moment $\langle n^k \rangle$ of the first-passage time n distributed via Eq. (3.6).*

Thus, we see that for the IID sequence, the survival probability is trivial to compute. However, as we will see below, the same quantity is much harder to compute for the random walk sequence because of the non-trivial correlations between the entries of the random sequence.

3.2   Random walk sequence

We now turn to the random walk sequence where x_n (the position of the walker at step n) evolves via the Markov rule in Eq. (2.24), starting at $x_0 = 0$. We are interested in the first-passage probability $F(a, n)$ at a level $a > 0$, i.e., the probability that the sequence crosses the level a for the first time at step n. The first-passage properties of such a random walk sequence have again been discussed in great detail in the classical books Feller (2008*a*, *b*). However, these rigorous mathematical results are not easy to comprehend/use for a broad audience of physicists. For a discussion more accessible to physicists, we refer the reader to Redner (2001), as well as the extensive complementary review in Bray *et al.* (2013).

In this random walk model it is also convenient to consider the survival probability $Q(a, n)$ defined as in Eq. (3.2), i.e., the probability that the sequence stays below the level a up to step n, *starting at $x_0 = 0$* (see Fig. 3.1):

$$Q(a, n) = \text{Prob}[x_1 \leq a, \, x_2 \leq a, \ldots, x_{n-1} \leq a, \, x_n \leq a \mid x_0 = 0]. \qquad (3.7)$$

As before, the first-passage probability is given by

$$F(a, n) = Q(a, n - 1) - Q(a, n). \qquad (3.8)$$

To compute the survival probability $Q(a, n)$, it is convenient to first make a geometrical transformation as follows. Let us define a new sequence

$$y_n = a - x_n. \qquad (3.9)$$

The sequence y_n is also a random walk sequence that starts at $y_0 = a$ and evolves via the equation

$$y_n = y_{n-1} - \eta_n, \qquad (3.10)$$

which follows from the original evolution equation in Eq. (2.24). The jump increment $-\eta_n$ has the same distribution as $+\eta_n$ since the jump distribution $f(\eta)$ is symmetric. Thus, the survival probability $Q(a, n)$ for the original x-sequence translates into the following probability for the y-sequence (which starts at $y_0 = a$):

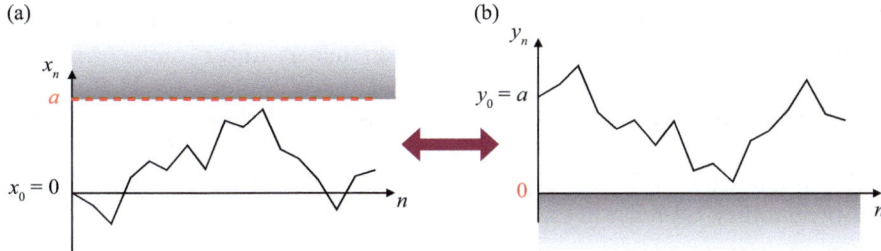

Fig. 3.1 Illustration of the survival probability $Q(a, n)$ in Eq. (3.7) for the random walks (a) x_n and (b) y_n [see Eq. (3.11)]. Note that the two probabilities coincide provided the jump distribution is symmetric, i.e., $f(\eta) = f(-\eta)$.

$$Q(a, n) = \text{Prob}[y_1 > 0, \, y_2 > 0, \ldots, y_n > 0 \mid y_0 = a]. \tag{3.11}$$

This is then just the probability that a random walker, starting at $a > 0$, stays positive up to step n (this is usually called the persistence probability in the physics literature).

To compute this persistence probability $Q(a, n)$, it is convenient to define a restricted Green's function for the random walk sequence. Let us define $G_+(x, x_0, n)$ as the probability (density) that the walker, starting at $x_0 > 0$, reaches the position $x > 0$ at step n *without crossing the origin* in between, i.e., it stays positive at all intermediate steps, and lands at $x > 0$ at the nth step:

$$G_+(x, x_0, n) = \text{Prob}[x_1 > 0, \, x_2 > 0, \ldots, x_{n-1} > 0, \, x_n = x > 0 \mid x_0 > 0]. \tag{3.12}$$

This restricted Green's function turns out to be the basic/key building block: knowing this, one can compute many observables of interest related to first-passage probability, extremes or records. For example, the survival probability $Q(a, n)$ is obtained by integrating over the final position,

$$Q(a, n) = \int_0^\infty G_+(x, a, n) \, \mathrm{d}x. \tag{3.13}$$

The question then boils down to computing this restricted Green's function $G_+(x, x_0, n)$. How should we proceed? Using the Markov property of the evolution, one can again write down the evolution equation for the restricted Green's function, both *forward* and *backward* as in case of the free Green's function in the previous chapter:

$$G_+(x, x_0, n) = \int_0^\infty G_+(x', x_0, n - 1) f(x - x') \, \mathrm{d}x' \qquad \text{(forward)}, \tag{3.14}$$

$$G_+(x, x_0, n) = \int_0^\infty G_+(x, x_0', n - 1) f(x_0' - x_0) \, \mathrm{d}x_0' \qquad \text{(backward)}. \tag{3.15}$$

Both equations are valid for $x \geq 0$ and $x_0 \geq 0$, and they both start from the initial condition $G_+(x, x_0, n) = \delta(x - x_0)$. The interpretation of these two recursion relations is as before. For example, in the forward case, one considers the walker reaching x' at step $(n - 1)$ (staying positive throughout) and then making a final jump $x' \to x$ at step n by drawing a random length $x - x'$ from the jump distribution $f(\eta)$. Similarly, in the backward equation, the particle at step 1 jumps from its initial position x_0 to a new position x_0' and subsequently evolves for $(n - 1)$ steps starting from this new initial position x_0' while staying positive all along. One then integrates over all possible jumps at the first step, but making sure that x_0' is positive.

At this point, there are two different ways to proceed to the calculation of the survival probability $Q(a, n)$: (1) a *forward* method and (2) a *backward* method.

Forward method. The forward method consists of two steps:

- **Step 1**: Solve the integral equation in Eq. (3.14) with the initial condition $G_+(x, x_0, 0) = \delta(x - x_0)$ and obtain $G_+(x, x_0, n)$. This is already very hard.

- **Step 2:** Using this solution, perform the integral $Q(x_0, n) = \int_0^\infty G_+(x, x_0, n)\, dx$. This gives the survival probability $Q(x_0, n)$ that the walker stays positive up to step n, starting at x_0. As mentioned before, setting $x_0 = a$, the probability $Q(a, n)$ is also the probability that the walker, starting at 0, stays below the level a up to step n.

Backward method. In the backward method, we first integrate Eq. (3.15) over $x \in [0, \infty)$ to directly obtain an equation for $Q(x_0, n)$. This gives a backward integral equation,

$$Q(x_0, n) = \int_0^\infty Q(x_0', n - 1) f(x_0' - x_0)\, dx_0', \qquad (3.16)$$

starting from the initial condition $Q(x_0, 0) = 1$ for all $x_0 \geq 0$. This initial condition follows from the fact that the walker definitely (with probability 1) does not cross 0 in 0 steps. Thus, in the backward method, one just has to solve one integral equation and there is no need for an extra integration at the end, as in the forward approach. Note that, in the integral equation in Eq. (3.16), we have treated x_0 as a variable. To connect to the survival probability $Q(a, n)$ in Eq. (3.13), i.e., the probability of staying below the level a in Eq. (3.11), we just have to solve the integral equation in Eq. (3.16) and set $x_0 = a$.

So, to compute the first-passage or the survival probability, we need to solve either the integral equations in Eq. (3.14) (forward approach) or Eq. (3.16) (backward approach)—see, e.g., Majumdar (2010*b*). Note the important differences in these equations compared to the free Green's functions in Eqs. (2.35) and (2.37): they look almost similar, but not quite. In Eqs. (3.14), (3.15) or (3.16), the limit of integration on the RHS is from 0 to ∞, as opposed to $-\infty$ to ∞ in the free Green's function in Eqs. (2.35) and (2.37). This makes a huge difference! The reason is that even though Eq. (3.16) apparently seems to have a convolution form, the limit of integration is only over the half-space $[0, \infty)$ and not the full space $(-\infty, \infty)$. If the limits were over the full space, as in the case of free Green's functions, one can simply use the Fourier transform methods. But for the half-space problem, unfortunately, one cannot use the simple Fourier transform technique. In fact, such half-space integral equations have been well studied in the mathematics literature and are known as Wiener–Hopf integral equations (Pollaczek, 1952; Morse and Feshbach, 1953; Spitzer, 1956, 1957; Comtet and Majumdar, 2005; Feller, 2008*a,b*). For a general kernel $f(x - x')$, they are notoriously difficult to solve! However, for the particular case where the kernel $f(x - x')$ has the interpretation of a probability density function (i.e., a non-negative and normalizable function), one can obtain an explicit solution (Comtet and Majumdar, 2005), as discussed later.

The discussion above makes clear the technical reason as to why computing the first-passage properties of even a simple random walker (but with arbitrary jump distribution $f(\eta)$) is non-trivial. Before we present the solution of Eq. (3.14) for arbitrary jump distribution $f(\eta)$, it is instructive to present in detail the explicit solutions for three different cases:

- the continuous-time Brownian limit;
- double-exponential jump distribution;
- lattice random walk.

For each of these three cases we will compute the restricted Green's function $G_+(x, x_0, n)$ by the forward method (by solving Eq. (3.14)), and we will also compute the survival probability $Q(x_0, n)$ by the backward method (by solving Eq. (3.16)).

3.2.1 Brownian limit

In the Brownian limit (obtained by taking the continuous-time limit of the discrete-time random walks whose jump distributions have a finite variance σ^2), we illustrate how the restricted Green's function and the survival probability can be computed by both the forward and backward methods discussed above.

Forward method. Let us start with the forward method. The starting point is the integral equation in Eq. (3.14). The main idea is to show that, in the Brownian limit, this integral equation can be reduced to a partial differential equation (the diffusion equation) in the half-space $x \geq 0$. The procedure is exactly the same as in the case of the Brownian limit of the free Green's function in the previous chapter [see Eq. (2.61)]. Proceeding as in the case of the free Green's function, it is easy to show that the restricted Green's function satisfies the same diffusion equation as Eq. (2.64),

$$\frac{\partial G_+}{\partial t} = D \frac{\partial^2 G_+}{\partial x^2}, \qquad x \geq 0. \qquad (3.17)$$

The only difference from Eq. (2.64) is that we need to solve this diffusion equation only in the half-space $x \geq 0$. Thus, we have to provide a boundary condition at $x = 0$, in addition to the $x \to \infty$ limit. Clearly, in the $x \to \infty$ limit $G_+(x, x_0, t) \to 0$, since the walker cannot reach infinity in a finite time starting from a finite $x_0 \geq 0$. The boundary condition at $x = 0$ turns out to be *absorbing*, i.e.,

$$G_+(x = 0, x_0, t) = 0 \qquad \text{for all} \ \ t > 0. \qquad (3.18)$$

This condition comes from the fact that, in the continuous-time limit, the particle cannot be at $x = 0$ at time t while staying positive at all previous times. Hence, the probability that the particle is at $x = 0$ at any time t is identically 0. We also have to provide the initial condition, which is simply $G_+(x, x_0, t = 0) = \delta(x - x_0)$, since the particle starts at $x = x_0 > 0$.

We thus have to solve the diffusion equation in Eq. (3.17) with the absorbing boundary condition at the origin, Eq. (3.18). This can be done by various methods, for instance by the standard image method (Redner, 2001; Feller, 2008a,b;

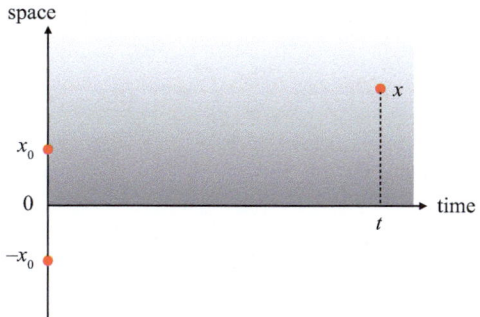

Fig. 3.2 Illustration of the image method in Eq. (3.20). There is an absorbing wall at the origin and we are interested in the propagator $G_+(x, x_0, t)$ from x_0 at $t = 0$ to $x > 0$ at time t. The process stays in the upper half-space $x > 0$ shown by the shaded area. This restricted propagator can be computed from the full-space propagator by ignoring the absorbing wall, but putting an image at $-x_0$ and taking the linear combination of the free propagators with opposite signs from these two "sources" starting at x_0 and $-x_0$.

Bray *et al.*, 2013). We first consider the free propagator from x_0 to x, without the boundary at the origin, which is simply given by Eq. (2.66):

$$G(x, x_0, t) = \frac{1}{\sqrt{4\pi Dt}} \exp\left\{-\frac{(x - x_0)^2}{4Dt}\right\}. \tag{3.19}$$

We now consider the image propagator $G(x, -x_0, t)$ where the paths propagate from the image at $-x_0$ to x at time t. This image propagator also satisfies the same diffusion equation. To find a solution that satisfies the absorbing boundary condition at $x = 0$, we then take a linear superposition of the free propagator and its image, with a minus sign (see Fig. 3.2):

$$G_+(x, x_0, t) = G(x, x_0, t) - G(x, -x_0, t)$$

$$= \frac{1}{\sqrt{4\pi Dt}} \left(\exp\left\{-\frac{(x - x_0)^2}{4Dt}\right\} - \exp\left\{-\frac{(x + x_0)^2}{4Dt}\right\} \right). \tag{3.20}$$

This solution is valid everywhere in the half-space $x \geq 0$. By construction, it satisfies the absorbing boundary condition at $x = 0$ and it is also easy to check that it satisfies the initial condition. Indeed, as $t \to 0$, one gets, from Eq. (3.20),

$$G_+(x, x_0, t \to 0) = \delta(x - x_0) - \delta(x + x_0). \tag{3.21}$$

Note, however, that since $x > 0$ and $x_0 > 0$ the second delta function is always 0 and hence it satisfies the correct initial condition $G_+(x, x_0, t \to 0) = \delta(x - x_0)$.

To compute the survival probability, we integrate the restricted Green's function in Eq. (3.20) over the final position x as in Eq. (3.13):

$$Q(x_0, t) = \int_0^\infty G_+(x, x_0, t)\, dx$$

$$= \frac{1}{\sqrt{4\pi Dt}} \int_0^\infty \left(\exp\left\{-\frac{(x - x_0)^2}{4Dt}\right\} - \exp\left\{-\frac{(x + x_0)^2}{4Dt}\right\} \right) dx. \tag{3.22}$$

Each of these integrals can be computed in terms of the error function, defined as

$$\operatorname{erf}(z) = \frac{2}{\sqrt{\pi}} \int_0^z e^{-u^2} \, \mathrm{d}u, \tag{3.23}$$

and this finally gives, for $x_0 \geq 0$,

$$Q(x_0, t) = \operatorname{erf}\left(\frac{x_0}{\sqrt{4Dt}} \right). \tag{3.24}$$

Thus, the probability of staying below the level a, starting at the origin, is given by $Q(x_0 = a, t)$, as explained earlier.

Little exercise 3.2 *Prove the result in Eq. (3.24) from Eq. (3.22).*

Backward method. In this method, we take the Brownian limit of the backward integral equation in Eq. (3.16) for the survival probability and again reduce it to a partial differential equation. Here we directly consider the survival probability $Q(x_0, t + \Delta t)$ up to time $t + \Delta t$. Let us break the interval $[0, t + \Delta t]$ into two, $[0, \Delta t]$ and $[\Delta t, t + \Delta t]$. In the first small interval the particle evolves from its initial position x_0 to a new random position $x_0 + \xi(0)\Delta t$, where $\xi(0)$ is the initial noise in the Langevin equation, Eq. (2.60). Subsequently, the particle evolves in the interval $[\Delta t, t + \Delta t]$ starting from its new initial position $x_0 + \xi(0)\Delta t$. Thus, the analogue of Eq. (3.16) is

$$Q(x_0, t + \Delta t) = \int_0^\infty Q(x_0 + \xi(0)\Delta t, t) f(\xi(0)) \, \mathrm{d}(\xi(0)). \tag{3.25}$$

Expanding in a Taylor series as in the case of the free Green's function and using the properties of white noise, one then gets the backward Fokker–Planck equation for the survival probability,

$$\frac{\partial Q}{\partial t} = D \frac{\partial^2 Q}{\partial x_0^2}, \tag{3.26}$$

valid for all $x_0 \geq 0$ and to be solved with the boundary conditions (i) $Q(x_0 = 0, t) = 0$ for all t and (ii) $Q(x_0 \to \infty, t) = 1$ for all t, and subject to the initial condition $Q(x_0, 0) = 1$ for all $x_0 > 0$. Thus, in the Brownian limit we are able to reduce the Wiener–Hopf integral equation into a partial differential equation (PDE): that's already a big simplification!

Little exercise 3.3 *Prove Eq. (3.26) by expanding Eq. (3.25) in a Taylor series and taking the limit $\Delta t \to 0$.*

The solution to the PDE in Eq. (3.26) can be obtained conveniently by using the Laplace transform with respect to time. Let us define

$$\tilde{Q}(x_0, s) = \int_0^\infty e^{-st} Q(x_0, t) \, dt. \tag{3.27}$$

We take the Laplace transform on both sides of Eq. (3.26). On the left-hand side (LHS), we perform integration by parts and use the initial condition $Q(x_0, t = 0) = 1$ for all $x_0 > 0$. This gives an ordinary differential equation for $\tilde{Q}(x_0, s)$:

$$D \frac{\partial^2 \tilde{Q}(x_0, s)}{\partial x_0^2} = -1 + s\tilde{Q}(x_0, s). \tag{3.28}$$

The general solution of this differential equation is given by

$$\tilde{Q}(x_0, s) = \frac{1}{s} + A e^{-\sqrt{s/D}\, x_0} + B e^{+\sqrt{s/D}\, x_0}, \tag{3.29}$$

where A and B are two arbitrary constants. Using the boundary condition that $\tilde{Q}(x_0 \to \infty, s)$ remains finite, one has $B = 0$. Furthermore, the absorbing boundary condition at the origin $\tilde{Q}(x_0 = 0, s) = 0$ fixes the constant $A = -1/s$. Hence, we get

$$\tilde{Q}(x_0, s) = \frac{1}{s} \left(1 - e^{-\sqrt{s/D}\, x_0} \right). \tag{3.30}$$

It remains to invert this Laplace transform with respect to s. Fortunately, this can be done exactly, leading to

$$Q(x_0, t) = L^{-1}_{s \to t} \tilde{Q}(x_0, s) = \mathrm{erf}\left(\frac{x_0}{\sqrt{4Dt}} \right). \tag{3.31}$$

Little exercise 3.4 *Show that by taking the Laplace transform of Eq. (3.31) with respect to t you get the result in Eq. (3.30).*

Thus, both the forward, Eq. (3.24), and the backward, Eq. (3.31), methods give the same answer for the survival probability, $Q(x_0, t) = \mathrm{erf}(x_0/\sqrt{4Dt})$, as expected. In particular, the survival probability decays to 0 as $t \to \infty$, indicating that the walker eventually crosses the origin with probability one. One can also ask how this survival probability decays for large time t. From this exact solution, using the small-z behavior of the error function, $\mathrm{erf}(z) \approx (2/\sqrt{\pi})z$, one gets a power-law decay of this survival probability at late times:

$$Q(x_0, t) \simeq \frac{x_0}{\sqrt{\pi D t}}. \tag{3.32}$$

The fact that $Q(x_0, t)$ vanishes as $t \to \infty$ indicates that the walker certainly (with probability one) crosses the origin at some time. This reflects the fact that the Brownian motion is *recurrent* in one dimension, i.e., it definitely passes through any point with probability one.

The first-passage probability is given by $F(x_0, t) = -\partial Q(x_0, t)/\partial t$, which is just the continuous-time limit of the relation in Eq. (3.8). Using Eq. (3.31), one then gets, for arbitrary t,

$$F(x_0, t) = \frac{x_0}{\sqrt{4\pi D t^3}} e^{-x_0^2/4Dt}. \tag{3.33}$$

Note that $F(x_0, t)\, dt$ in Eq. (3.33) represents the PDF of the random variable t_f denoting the first time the process, starting at x_0, crosses the origin, i.e.,

$$\mathrm{Prob}[t_f = t \mid x_0] = F(x_0, t). \tag{3.34}$$

As explained before, $F(x_0, t)\, dt$ also represents the probability that the walker, starting at 0, crosses the level x_0 for the first time in the time interval $[t, t+dt]$. For large t and fixed x_0, the first-passage probability density in Eq. (3.33) decays as

$$F(x_0, t) \simeq \frac{x_0}{\sqrt{4\pi D}} t^{-3/2}, \tag{3.35}$$

thus recovering the famous first-passage exponent 3/2 (Lévy, 1940; Chandrasekhar, 1943; Redner, 2001; Feller, 2008a; Bray et al., 2013). Thus, the mean first-passage time diverges,

$$\langle t_f \rangle = \int_0^\infty t F(x_0, t)\, dt = +\infty. \tag{3.36}$$

This result may appear contradictory to the recurrence property discussed above, namely that the walker definitely crosses the origin with probability one. Actually, there is really no contradiction due to the fact that $F(x_0, t)$ is normalized to unity: $\int_0^\infty F(x_0, t)\, dt = 1$. However, the moments of this PDF $F(x_0, t)$, including its average, diverge due to the power-law tail $t^{-3/2}$. This is because the algebraic tail indicates that the trajectories that cross the origin at very late times occur with a relatively high probability and hence contribute dominantly to the mean, leading to its divergence.

3.2.2 Double exponential jump distribution

Forward method. Here we compute the restricted Green's function by solving the forward integral equation in Eq. (3.14) for the special case of a double-exponential jump distribution,

$$f(\eta) = \frac{1}{2\ell_0} e^{-|\eta|/\ell_0}, \tag{3.37}$$

where ℓ_0 is a characteristic jump length. The mean of this distribution is zero, while the variance is given by

$$\sigma^2 = \frac{1}{\ell_0} \int_{-\infty}^\infty \eta^2 e^{-|\eta|/\ell_0}\, d\eta = 2\ell_0^2. \tag{3.38}$$

We will see that this is a very special jump distribution, for which the integral equation in Eq. (3.14) can be reduced to a differential equation which, subsequently, can be solved explicitly. It is always very useful to have a solvable toy example, and this double-exponential jump distribution plays precisely this role. We will see later that, not just for the first-passage probability but even for extreme, order and record statistics, the exact solution obtained for this toy example will play a crucial role in understanding several properties for random walks with generic jump distributions $f(\eta)$.

Our goal is to calculate the restricted Green's function $G_+(x, x_0, n)$ from Eq. (3.14), which reads, in this case,

$$G_+(x, x_0, n) = \frac{1}{2\ell_0} \int_0^\infty G_+(x', x_0, n-1) e^{-|x-x'|/\ell_0}\, dx' \quad \text{for } n \geq 1, \qquad (3.39)$$

starting from the initial condition $G_+(x, x_0, n=0) = \delta(x-x_0)$. To proceed, we first define the generating function

$$\tilde{G}_+(x, x_0, s) = \sum_{n=1}^\infty G_+(x, x_0, n) s^n. \qquad (3.40)$$

Multiplying Eq. (3.39) by s^n and summing over n from $n = 1$ to $+\infty$, we obtain

$$\tilde{G}_+(x, x_0, s) = \frac{s}{2\ell_0} \int_0^\infty [\tilde{G}_+(x', x_0, s) + G_+(x', x_0, n=0)] e^{-|x-x'|/\ell_0}\, dx', \qquad (3.41)$$

where $G_+(x', x_0, n=0) = \delta(x'-x_0)$ corresponds to the initial condition. Hence, we get

$$\tilde{G}_+(x, x_0, s) = \frac{s}{2\ell_0} \int_0^\infty \tilde{G}_+(x', x_0, s) e^{-|x-x'|/\ell_0}\, dx' + \frac{s}{2\ell_0} e^{-|x-x_0|/\ell_0}. \qquad (3.42)$$

This integral equation can be reduced to a differential equation by making use of the following identity:

$$\frac{\partial^2}{\partial x^2}\left[\frac{1}{2\ell_0} e^{-|x-x'|/\ell_0}\right] = \frac{1}{2\ell_0^3} e^{-|x-x'|/\ell_0} - \frac{1}{\ell_0^2}\delta(x-x'). \qquad (3.43)$$

Little exercise 3.5 *Prove the identity in Eq. (3.43). As a hint, you can use the fact that*

$$\frac{d|x|}{dx} = \text{sgn}(x) = 2\Theta(x) - 1, \qquad (3.44)$$

where $\Theta(x)$ is the Heaviside step function and $\text{sgn}(x)$ gives the sign of x. You can also use the fact that $d\Theta(x)/dx = \delta(x)$.

To proceed, we differentiate Eq. (3.42) twice with respect to x and use the identity in Eq. (3.43) to obtain

$$\frac{\partial^2 \tilde{G}_+}{\partial x^2} = \frac{1}{\ell_0^2}(1-s)\tilde{G}_+(x, x_0, s) - \frac{s}{\ell_0^2}\delta(x-x_0), \qquad x \geq 0. \qquad (3.45)$$

To solve this differential equation, we need appropriate boundary conditions. One boundary condition is obvious: when $x \to \infty$, clearly $G_+(x, x_0, n) \to 0$ and consequently $\tilde{G}_+(x, x_0, s) \to 0$. What about the boundary condition at $x = 0$? In fact, while we have gained by reducing the integral equation to a differential equation, the price we have to pay is that we do not have any a priori information on the boundary condition at $x = 0$. Indeed, from the integral equation in Eq. (3.42), we see that if we put $x = 0$ on the RHS, it gives a finite non-zero value for $\tilde{G}_+(x = 0, x_0, s)$ which needs to be determined self-consistently. However, this is possible since we have lost some information in going from the integral equation to the differential equation. The integral equation contains more information than the differential equation. The idea is thus to solve the differential equation, find the most general solution compatible with the boundary condition at $x \to \infty$ and then inject it back into the integral equation in Eq. (3.42). As we will see, this determines the complete solution without a priori knowledge of the boundary condition at $x = 0$.

Due to the presence of the delta function on the RHS of Eq. (3.45), we need to solve it in two different regions of space, (i) $x \geq x_0$ and (ii) $0 \leq x \leq x_0$, and then match the solutions at $x = x_0$. The solution must be continuous at $x = x_0$ and its derivative must have a jump discontinuity

$$\left.\frac{\partial \tilde{G}_+}{\partial x}\right|_{x=x_0^+} - \left.\frac{\partial \tilde{G}_+}{\partial x}\right|_{x=x_0^-} = -\frac{s}{\ell_0^2}, \tag{3.46}$$

which follows by integrating Eq. (3.45) over x on a small region around x_0.

In the first region, $x \geq x_0$ the most general solution reads

$$\tilde{G}_+(x, x_0, s) = A_1 \exp\left\{-\frac{\sqrt{1-s}}{\ell_0}(x - x_0)\right\} + B_1 \exp\left\{+\frac{\sqrt{1-s}}{\ell_0}(x - x_0)\right\}. \tag{3.47}$$

Since the solution cannot diverge as $x \to +\infty$, we must have $B_1 = 0$ and hence, for $x \geq x_0$, we have

$$\tilde{G}_+(x, x_0, s) = A_1 \exp\left\{-\frac{\sqrt{1-s}}{\ell_0}(x - x_0)\right\}. \tag{3.48}$$

In the second region, $0 \leq x \leq x_0$, the most general solution reads

$$\tilde{G}_+(x, x_0, s) = A_2 \exp\left\{-\frac{\sqrt{1-s}}{\ell_0}(x - x_0)\right\} + B_2 \exp\left\{+\frac{\sqrt{1-s}}{\ell_0}(x - x_0)\right\}. \tag{3.49}$$

The continuity at x_0 gives

$$A_1 = A_2 + B_2. \tag{3.50}$$

The jump discontinuity of the derivative in Eq. (3.46) gives another relation,

$$A_2 - A_1 - B_2 = -\frac{s}{\ell_0\sqrt{1-s}}. \tag{3.51}$$

We then have three unknown constants A_1, A_2 and B_2, but only two relations, Eqs. (3.50) and (3.51), connecting them. We thus need a third relation and this is obtained by injecting the general solution in Eqs. (3.48) and (3.49) back into the original integral equation, Eq. (3.42). One can verify that, indeed, this solution satisfies the integral equation in Eq. (3.42) provided the following relation holds (please check this as a little exercise):

$$\frac{A_2}{s_1}(1 - e^{-s_1 x_0}) + \frac{B_2}{s_2}(1 - e^{-s_2 x_0}) - \frac{A_1}{s_1} + 1 = 0, \tag{3.52}$$

where

$$s_1 = \frac{1 - \sqrt{1 - s}}{\ell_0}, \qquad s_2 = \frac{1 + \sqrt{1 - s}}{\ell_0}. \tag{3.53}$$

We thus now have three linear equations, Eqs. (3.50), (3.51) and (3.52), for three unknowns, A_1, A_2 and B_2. We can solve them easily and obtain

$$A_1 = \frac{s}{2\ell_0\sqrt{1-s}}\left[1 - \frac{1 - \sqrt{1-s}}{1 + \sqrt{1-s}}\exp\left\{-2\frac{\sqrt{1-s}}{\ell_0}x_0\right\}\right],$$

$$A_2 = -\frac{s(1 - \sqrt{1-s})}{2\ell_0\sqrt{1-s}(1 + \sqrt{1-s})}\exp\left\{-2\frac{\sqrt{1-s}}{\ell_0}x_0\right\}, \tag{3.54}$$

$$B_2 = \frac{s}{2\ell_0\sqrt{1-s}}.$$

With these constants fully determined, we then have the full solution of $\tilde{G}_+(x, x_0, s)$:

$$\tilde{G}_+(x, x_0, s) = \begin{cases} A_1 \exp\left\{-\frac{\sqrt{1-s}}{\ell_0}(x - x_0)\right\}, & x \geq x_0, \\ A_2 \exp\left\{-\frac{\sqrt{1-s}}{\ell_0}(x - x_0)\right\} + B_2 \exp\left\{\frac{\sqrt{1-s}}{\ell_0}(x - x_0)\right\}, & x \leq x_0. \end{cases} \tag{3.55}$$

The restricted propagator $G_+(x, x_0, n)$ in the time domain can then be formally obtained, using Eq. (3.40), by expanding $\tilde{G}_+(x, x_0, s)$ in Eq. (3.55) in powers of s and identifying the coefficient of s^n. For finite n, it is difficult to obtain an explicit formula for $G_+(x, x_0, n)$. However, one can make progress in the scaling limit where $n \to \infty$ and $x, x_0 \to \infty$ but the ratios $x/\sqrt{n}$ (and $x_0/\sqrt{n}$) are kept fixed. To investigate this scaling limit, we first write $s = e^{-p}$ on the RHS of Eq. (3.40), which then becomes

$$\tilde{G}_+(x, x_0, s = e^{-p}) = \sum_{n=1}^{\infty} G_+(x, x_0, n)e^{-pn}. \tag{3.56}$$

To extract the large-n behavior of $G_+(x, x_0, n)$, it is clear that we need to examine the limit $p \to 0$ of the LHS. Indeed, in this $p \to 0$ limit, one can replace the sum over n by an integral,

$$\tilde{G}_+(x, x_0, s = e^{-p}) \approx \int_0^{\infty} G_+(x, x_0, \tau)e^{-p\tau}\, d\tau. \tag{3.57}$$

Thus, in this limit, the generating function reduces to a Laplace transform and one can invert this Laplace transform using the standard Bromwich integral in the complex p-plane. To proceed, we consider the exact expression in Eq. (3.55) with $s = e^{-p}$ and expand for small p but large x and x_0 with $x\sqrt{p}$ (as well as $x_0\sqrt{p}$) fixed. The leading and the subleading behaviors, in the scaling limit, are given by

$$\tilde{G}_+(x, x_0, s = e^{-p}) \approx \frac{1}{2\ell_0\sqrt{p}}\left[\exp\left\{-\frac{\sqrt{p}}{\ell_0}|x - x_0|\right\} - \exp\left\{-\frac{\sqrt{p}}{\ell_0}(x + x_0)\right\}\right]$$

$$+ \frac{1}{\ell_0}\exp\left\{-\frac{\sqrt{p}}{\ell_0}(x + x_0)\right\}. \tag{3.58}$$

We now invert the Laplace transform in Eq. (3.58) using the identities

$$L_{p\to\tau}^{-1}\left[\frac{1}{\sqrt{p}}e^{-b\sqrt{p}}\right] = \frac{e^{-b^2/4\tau}}{\sqrt{\pi\tau}}, \qquad L_{p\to\tau}^{-1}[e^{-b\sqrt{p}}] = \frac{b}{\sqrt{4\pi\tau^3}}e^{-b^2/4\tau}. \tag{3.59}$$

Little exercise 3.6 *Prove the two identities in Eq. (3.59).*

Identifying τ with n, for large n, we then get, in the scaling limit, the following two leading contributions:

$$G_+(x, x_0, n) \approx \frac{1}{\sqrt{4\pi\ell_0^2 n}}\left[\exp\left\{-\frac{(x - x_0)^2}{4\ell_0^2 n}\right\} - \exp\left\{-\frac{(x + x_0)^2}{4\ell_0^2 n}\right\}\right]$$

$$+ \frac{x + x_0}{\ell_0^2\sqrt{4\pi n^3}}\exp\left\{-\frac{(x + x_0)^2}{4\ell_0^2 n}\right\} + \cdots. \tag{3.60}$$

Noting that $\sigma^2 = 2\ell_0^2$ for the double-exponential jump distribution, let us express $G_+(x, x_0, n)$ in terms of σ^2:

$$G_+(x, x_0, n) \approx \frac{1}{\sqrt{2\pi\sigma^2 n}}\left[\exp\left\{-\frac{(x - x_0)^2}{2\sigma^2 n}\right\} - \exp\left\{-\frac{(x + x_0)^2}{2\sigma^2 n}\right\}\right]$$

$$+ \frac{x + x_0}{\sigma^2\sqrt{\pi n^3}}\exp\left\{-\frac{(x + x_0)^2}{2\sigma^2 n}\right\} + \cdots. \tag{3.61}$$

From this formula, we see that in the Brownian limit $\sigma \to 0$, $n \to \infty$ with $\sigma^2 n = 2Dt$ fixed, the first term in Eq. (3.61) reduces to the Brownian result given in Eq. (3.20), obtained by solving the diffusion equation with an absorbing boundary at $x = 0$ using the image method. Note, however, that the subleading term does not vanish as $x \to 0$. In general, the solution for $G_+(x, x_0, n)$ for any n does not satisfy the absorbing boundary condition at $x = 0$, i.e., $G_+(x = 0, x_0, n) \neq 0$. This is in contrast to the Brownian limit where, indeed, the Green's function does satisfy the boundary condition $G_+(x = 0, x_0, t) = 0$ which enabled us to obtain the solution using the method of images. Thus, the method of images does not work for a generic discrete-time random walk with a continuous jump distribution.

Finally, upon integrating $\tilde{G}_+(x, x_0, n)$ over x for $x \in [0, +\infty)$, we get the survival probability $Q(x_0, n) = \int_0^\infty G_+(x, x_0, n)\, dx$. In fact, its generating function is defined as

$$\tilde{Q}(x_0, s) = \sum_{n=1}^{\infty} Q(x_0, n)s^n = \int_0^{\infty} \tilde{G}_+(x, x_0, s)\,\mathrm{d}x. \tag{3.62}$$

Substituting the exact solution in Eq. (3.55) then gives

$$\tilde{Q}(x_0, s) = \frac{s}{1-s} - \frac{(1 - \sqrt{1-s})}{1-s} \exp\left\{-\frac{\sqrt{1-s}}{\ell_0}x_0\right\}, \qquad x_0 \geq 0. \tag{3.63}$$

We will see that we could obtain this result for $\tilde{Q}(x_0, s)$ directly by the simpler backward method. Let us remark that the generating function in Eq. (3.62) starts from $n = 1$. For future purposes, it is also useful to compute the generating function that starts at $n = 0$. Indeed, using $Q(x_0, 0) = 1$ we get, from Eq. (3.63),

$$\sum_{n=0}^{\infty} Q(x_0, n)s^n = \frac{1}{1-s}\left(1 - \frac{(1 - \sqrt{1-s})}{1-s} \exp\left\{-\frac{\sqrt{1-s}}{\ell_0}x_0\right\}\right). \tag{3.64}$$

Backward method. We start with the backward equation in Eq. (3.16) with $f(\eta)$ given in Eq. (3.37) and the initial condition $Q(x_0, n = 0) = 1$ for $x_0 \geq 0$. We first take the generating function of this equation, which gives (Comtet and Majumdar, 2005)

$$\tilde{Q}(x_0, s) = \sum_{n=1}^{\infty} Q(x_0, n)s^n$$

$$= \int_0^{\infty} [s\tilde{Q}(x', s) + 1]f(x' - x_0)\,\mathrm{d}x'$$

$$= s\int_0^{\infty} \tilde{Q}(x', s)f(x' - x_0)\,\mathrm{d}x + s\int_0^{\infty} f(x' - x_0)\,\mathrm{d}x'. \tag{3.65}$$

Differentiating twice with respect to x_0, and using the identity in Eq. (3.43), we obtain

$$\frac{\partial^2 \tilde{Q}}{\partial x_0^2} = \frac{1}{\ell_0^2}(1 - s)\tilde{Q}(x_0, s) - \frac{s}{\ell_0^2}, \tag{3.66}$$

with the boundary condition $\tilde{Q}(x_0 \to \infty, s) = s/(1 - s)$. Where does this condition come from? We note that, in the time domain, we have the condition

$$Q(x_0 \to \infty, n) = 1, \tag{3.67}$$

which follows from the fact that, if the particle starts at $x_0 \to \infty$, it will definitely stay positive (with probability 1) for any finite n. We now take the generating function of this condition in Eq. (3.67), which simply gives the boundary condition $\tilde{Q}(x_0 \to \infty, s) = s/(1 - s)$. As in the forward method, we do not have any a priori information about the boundary condition at $x_0 = 0$. In fact, this information is already contained in the integral equation (3.65).

The most general solution of Eq. (3.66) is

$$\tilde{Q}(x_0, s) = \frac{s}{(1-s)} + C_1 \exp\left\{-\frac{\sqrt{1-s}}{\ell_0}x_0\right\} + C_2 \exp\left\{+\frac{\sqrt{1-s}}{\ell_0}x_0\right\}, \qquad (3.68)$$

where the first term is obtained by a constant shift in $\tilde{Q}$ that renders the differential equation in Eq. (3.66) homogeneous. The boundary condition at $x_0 \to \infty$ imposes $C_2 = 0$. Hence, the solution is given by

$$\tilde{Q}(x_0, s) = \frac{s}{(1-s)} + C_1 \exp\left\{-\frac{\sqrt{1-s}}{\ell_0}x_0\right\}. \qquad (3.69)$$

To determine the unknown constant C_1, we substitute this solution back into the integral equation in Eq. (3.65) which gives $C_1 = -(1-\sqrt{1-s})/(1-s)$. Hence, the backward method gives the same solution $\tilde{Q}(x_0, s)$ as the forward method in Eq. (3.63).

As in the case of the restricted Green's function, it is hard to invert the generating function in Eq. (3.63) to get $Q(x_0, n)$ for arbitrary x_0 and n. However, a great simplification occurs if the walker starts at the origin $x_0 = 0$. In this case, Eq. (3.63) gives

$$\tilde{Q}(x_0 = 0, s) = \frac{1}{\sqrt{1-s}} - 1. \qquad (3.70)$$

Expanding in powers of s, it is easy to see that, for $n \geq 1$,

$$Q(0, n) = \frac{1}{2^{2n}}\binom{2n}{n}. \qquad (3.71)$$

Note that this formula is also valid for $n = 0$, since $Q(0, 0) = 1$. In particular, note that for large n, $Q(0, n)$ has the asymptotic behavior

$$Q(0, n) \approx \frac{1}{\sqrt{\pi n}}. \qquad (3.72)$$

Little exercise 3.7 *Show the asymptotic behavior in Eq. (3.72) starting from Eq. (3.71) and using Stirling's formula, $n! \sim \sqrt{2\pi n}e^{n\ln n - n}$.*

There are two important remarks to be made about the result in Eq. (3.71):

- Even if the particle starts at the origin, it can actually survive, i.e., it stays on the positive side $x \geq 0$, with a finite probability. This is in stark contrast with the continuous-time Brownian motion result in Eq. (3.31), which is identically 0, for $x_0 = 0$, at all time $t \geq 0$.
- This result is completely independent of the characteristic length scale ℓ_0 that parameterizes the jump distribution. The parameter ℓ_0 has completely disappeared from the formula in Eq. (3.71)! We will come back to this "universality" later when we discuss arbitrary jump distributions. We will see that

this universality is a consequence of the so-called Sparre Andersen theorem (Sparre Andersen, 1954).

For general $x_0 > 0$, as remarked earlier, it is hard to invert the generating function in Eq. (3.63) to obtain $Q(x_0, n)$ for any finite n. However, it is possible to extract the result in the scaling limit $x_0 \to \infty$ and $n \to \infty$, but keeping the ratio $x_0/\sqrt{n}$ fixed, as we did before for $G_+(x, x_0, n)$ in Eq. (3.60). As before, to obtain the large-n limit, we set $s = e^{-p}$ in Eq. (3.62) and replace the sum over n by an integral over τ:

$$\tilde{Q}(x_0, s = e^{-p}) = \sum_{n=1}^{\infty} e^{-pn} Q(x_0, n) \approx \int_0^{\infty} Q(x_0, \tau) e^{-p\tau} \, d\tau. \tag{3.73}$$

We now expand the RHS of Eq. (3.63) for small-p expansion, keeping $x_0\sqrt{p}$ fixed. The first two leading terms are given by

$$\tilde{Q}(x_0, s = e^{-p}) \approx \frac{1}{p}\left(1 - \exp\left\{-\frac{\sqrt{p}}{\ell_0}x_0\right\}\right) + \frac{1}{\sqrt{p}}\exp\left\{-\frac{\sqrt{p}}{\ell_0}x_0\right\} + \cdots. \tag{3.74}$$

Now, this Laplace transform can be inverted using the identities

$$L_{p\to\tau}^{-1}\left[\frac{1}{p}(1 - e^{-b\sqrt{p}})\right] = \text{erf}\left(\frac{b}{\sqrt{4\tau}}\right), \quad L_{p\to\tau}^{-1}\left[\frac{1}{\sqrt{p}}e^{-b\sqrt{p}}\right] = \frac{b}{\sqrt{\pi\tau}}e^{-b^2/4\tau}. \tag{3.75}$$

Identifying τ with n, for large n, we then get, in the scaling limit, the following two leading contributions:

$$Q(x_0, n) \approx \text{erf}\left(\frac{x_0}{\sqrt{4\ell_0^2 n}}\right) + \frac{1}{\sqrt{\pi n}}e^{-x_0^2/4\ell_0^2 n} + \cdots. \tag{3.76}$$

Note that by integrating over $x \in [0, +\infty)$ the two leading terms in the large-n behavior of $G_+(x, x_0, n)$ in Eq. (3.60), we obtain the same result as in Eq. (3.76), as expected.

Why did we bother to keep the subleading term in the scaling limit in Eqs. (3.60) and (3.76)? Notice that in Eq. (3.76) the leading term in the scaling limit is just the Brownian result in Eq. (3.31), upon identifying $\ell_0^2 n \to Dt$. The subleading term in the scaling limit is smaller by a factor $1/\sqrt{n}$ compared to the leading term. However, if we just keep the leading term and then set $x_0 = 0$, we get $Q(0, n) = 0$. This is clearly in contradiction with the exact result in Eq. (3.71), which says that $Q(0, n) > 0$. However, if we keep the subleading term in Eq. (3.76) we see that

$$Q(0, n) \approx \frac{1}{\sqrt{\pi n}}, \tag{3.77}$$

which is completely consistent with the exact asymptotic behavior of $Q(0, n)$ in Eq. (3.72). Note that this non-zero contribution to $Q(0, n)$ comes entirely from the subleading term in Eq. (3.76). In other words, the region $x_0 = O(1)$ is not part of the Brownian scaling limit which corresponds to $x_0 = O(\sqrt{n})$ for large n. Thus, the two leading terms in Eq. (3.76) clearly demonstrate that the two limits, (i) where we first take the scaling limit $n \to \infty$, $x_0 \to \infty$ with the ratio $x_0/\sqrt{n}$ fixed and after taking this scaling limit we set $x_0 = 0$, and (ii) we set $x_0 = 0$ first and then take the large-n limit, do not commute (Comtet and Majumdar, 2005). This is an interesting fact which actually holds for arbitrary jump distributions and we will discuss this in more detail later when we present the results for general jump distributions.

3.2.3   Lattice random walk

This is another solvable case, where the jump distribution is given by

$$f(\eta) = \frac{1}{2}\delta(\eta - 1) + \frac{1}{2}\delta(\eta + 1). \tag{3.78}$$

The cumulative distribution of this jump distribution is discontinuous. This corresponds to a random walker jumping by ± 1 on a line. Hence, if the initial position x_0 is an integer, the walker always moves on the set of integer points—hence the name "lattice random walk." Our first aim is to compute the restricted Green's function $G_+(x, x_0, n)$ from the forward integral equation in Eq. (3.14). Substituting the jump distribution $f(\eta)$, Eq. (3.78), on the RHS of Eq. (3.14), we get

$$G_+(x, x_0, n) = \frac{1}{2}[G_+(x + 1, x_0, n - 1) + G_+(x - 1, x_0, n - 1)]. \tag{3.79}$$

This equation holds for $x \geq 0$, $x_0 \geq 0$ and $n \geq 1$ with the convention that

$$G_+(x = -1, x_0, n) = 0 \text{ for all } x_0 \geq 0, \ n \geq 1. \tag{3.80}$$

This last condition follows from the fact that the particle can be at the origin at step n only if it was at $x = 1$ at step $n - 1$ (since it is not allowed to cross the origin, there is no jump contribution from $x = -1$). Hence, if Eq. (3.79) holds for $x = 0$, we have to impose the convention in Eq. (3.80). In addition, as $x \to \infty$ we have another boundary condition,

$$G_+(x \to \infty, x_0, n) = 0, \tag{3.81}$$

since the particle cannot reach infinity in a finite number of steps starting at a finite x_0. Finally, the initial condition for this equation is

$$G_+(x, x_0, n = 0) = \delta_{x, x_0}, \qquad x, x_0 \geq 0, \tag{3.82}$$

where $\delta_{x, x_0} = 1$ if $x = x_0$ and zero otherwise (the Kronecker delta function). To solve this set of equations, one can again employ the method of images, as for the continuous-time Brownian motion case. In terms of the unrestricted Green's function $G(x, x_0, n)$ given in Eq. (2.49) and valid for all x, x_0 integers (both positive and negative), it is easy to see that the solution of this set of equations satisfying the correct initial and boundary conditions is given by

$$G_+(x, x_0, n) = G(x, x_0, n) - G(-x - 2, x_0, n). \tag{3.83}$$

Note that this method of images works for both continuous-time Brownian motion and the lattice random walk at arbitrary time. However, it does not work for the discrete-time random walk with any continuous jump distribution, including the double-exponential jump distribution as demonstrated in the previous subsection. We will see later that, for an arbitrary continuous jump distribution, there is a general formula for the generating function: $\sum_{n=0}^{\infty} G_+(x, x_0, n)s^n = \tilde{G}_+(x, x_0, s)$. It is natural to ask what the corresponding formula for this generating function for the lattice random walk is. However, computing $\tilde{G}_+(x, x_0, s)$ for the lattice random walk directly from the image solution in Eq. (3.83) is not completely straightforward. Below, we show that it can be derived by an alternative approach similar to the double-exponential case discussed in Section 3.2.2.

To proceed, we consider directly the generating function, as in the double-exponential case,

$$\tilde{G}_+(x, x_0, s) = \sum_{n=0}^{\infty} G_+(x, x_0, n)s^n. \tag{3.84}$$

Note that, here, for convenience, we have taken the sum over n in the generating function from $n = 0$ to ∞, at variance with the previous double-exponential case in Eq. (3.40) where the sum starts at $n = 1$. Taking the generating function of Eq. (3.79), we get

$$\tilde{G}_+(x, x_0, s) - \delta_{x,x_0} = \frac{s}{2}[\tilde{G}_+(x+1, x_0, s) + \tilde{G}_+(x-1, x_0, s)], \tag{3.85}$$

valid for $x \geq 0$ and $x_0 \geq 0$. Note that the Kronecker delta function in Eq. (3.85) comes from the initial condition in Eq. (3.82). To solve this equation, we now introduce another generating function, now with respect to the positive integer variable x:

$$\mathcal{G}_+(x_0, w, s) = \sum_{x=0}^{\infty} \tilde{G}_+(x, x_0, s)w^x. \tag{3.86}$$

Multiplying Eq. (3.85) by w^x, summing over x and simplifying, we get (verify this!)

$$\mathcal{G}_+(x_0, w, s) = \frac{\tilde{G}_+(0, x_0, s) - \frac{2}{s}w^{x_0+1}}{w^2 - \frac{2}{s}w + 1}. \tag{3.87}$$

The term $\tilde{G}_+(0, x_0, s)$ comes from the $x = 0$ term when one performs the sum over x on the RHS of Eq. (3.85). This quantity $\tilde{G}_+(0, x_0, s)$ has to be determined self-consistently using the appropriate boundary conditions.

To determine this unknown quantity $\tilde{G}_+(0, x_0, s)$ we proceed as follows. We first formally invert Eq. (3.86) using Cauchy's formula,

$$\tilde{G}_+(x, x_0, s) = \oint \frac{dw}{2\pi i} \frac{1}{w^{x+1}} \frac{\tilde{G}_+(0, x_0, s) - (2/s)w^{x_0+1}}{w^2 - (2/s)w + 1}, \tag{3.88}$$

where the contour circles around the origin in the complex w-plane. Next, we rewrite the denominator in the integrand of Eq. (3.88) as

$$w^2 - \frac{2}{s}w + 1 = (w - w_+(s))(w - w_-(s)), \quad \text{where } w_\pm(s) = \frac{1}{s}(1 \pm \sqrt{1 - s^2}). \quad (3.89)$$

Thus, the integrand in Eq. (3.88) has two poles, $w = w_\pm(s)$, in the complex w-plane, in addition to the multiple pole at $w = 0$. Of these two non-zero poles, it is easy to see that $w_-(s) < 1$ while $w_+(s) > 1$ (their product is 1). The residue of the integrand in Eq. (3.88) at the pole $w = w_-(s) < 1$ is given by

$$Res(w_-(s)) = \frac{1}{(w_+ - w_-)} \frac{\tilde{G}_+(0, x_0, s) - (2/s)w_-^{x_0+1}}{w_-^{x+1}}. \quad (3.90)$$

Since $w_-(s) < 1$, the contribution of this residue to $G_+(x_0, w, s)$ in Eq. (3.87) diverges as $x \to \infty$. However, the boundary condition says that $\tilde{G}_+(x, x_0, s)$ must vanish as $x \to \infty$. Hence, to satisfy this boundary condition, the coefficient of w_-^{-x-1} in the first term must be identically 0, leading to the condition

$$\tilde{G}_+(0, x_0, s) = \frac{2}{s}[w_-(s)]^{x_0+1} = \frac{2}{s}\left[\frac{1}{s}(1 - \sqrt{1 - s^2})\right]^{x_0+1}. \quad (3.91)$$

This then fixes the unknown quantity $\tilde{G}_+(0, x_0, s)$. Let us make a remark here: one frequently encounters this problem in determining an unknown constant when one takes a generating function with respect to indices which run only over a half-space. The method presented here, called the "pole cancellation" method, is a very powerful and elegant way to fix this unknown quantity (Rajesh and Majumdar, 2000a,b, 2001). Substituting the result from Eq. (3.91) in Eq. (3.87) determines completely the restricted Green's function,

$$\sum_{x=0}^{\infty} \tilde{G}_+(x, x_0, s)w^x = \frac{2}{s}\frac{[w_-(s)]^{x_0+1} - w^{x_0+1}}{(w - w_+(s))(w - w_-(s))}, \quad \text{where } w_\pm(s) = \frac{1}{s}(1 \pm \sqrt{1 - s^2}).$$
$$(3.92)$$

From this expression we can evaluate the generating function of the survival probability $Q(x_0, n) = \sum_{x=0}^{\infty} G_+(x, x_0, n)$ that denotes the probability that the walker, starting at x_0, does not enter the negative side up to step n. Setting $w = 1$ in Eq. (3.92), one obtains

$$\sum_{n=0}^{\infty} Q(x_0, n)s^n = \frac{1 - [w_-(s)]^{x_0+1}}{1 - s}$$

$$= \frac{1}{1-s}\left(1 - \frac{1}{s}(1 - \sqrt{1 - s^2})\exp\left\{x_0 \ln\left(\frac{1 - \sqrt{1 - s^2}}{s}\right)\right\}\right). \quad (3.93)$$

One can compare this expression with the one obtained for the double-exponential jump distribution in Eq. (3.64). They have a similar structure but, obviously, are not identical.

Finally, as in the double-exponential case, one can extract the scaling behavior of the survival probability in the limit when $x_0 \to \infty$ and $n \to \infty$ but keeping the ratio $x_0/\sqrt{n}$ fixed. Performing a similar analysis to the double-exponential case [see Eqs. (3.73)–(3.76)], we get

$$Q(x_0, n) \approx \mathrm{erf}\left(\frac{x_0}{\sqrt{2n}}\right). \tag{3.94}$$

Thus, as expected, the survival probability for the lattice random walk converges to the result for the Brownian motion in Eq. (3.24) with the identification $n = 2Dt$.

Little exercise 3.8 *Show the asymptotic behavior in Eq.* (3.94) *starting from the generating function in Eq.* (3.93).

3.2.4 Pollaczek–Spitzer formula

Let us now go back to the basic Wiener–Hopf integral equation in Eq. (3.16) that describes the evolution of the survival probability $Q(x_0, n)$ for an arbitrary jump distribution $f(\eta)$. As mentioned before, the solution of this integral equation is not easy to find for a general $f(\eta)$. However, when the cumulative distribution $P_<(x) = \int_{-\infty}^{x} f(\xi)\,\mathrm{d}\xi$ is a continuous function such as in Eqs. (2.25)–(2.28) (but not for the lattice random walk, Eq. (2.29), where $P_<(x)$ is a discontinuous function), a solution was first found by Pollaczek (1952) and later independently by Spitzer (1956, 1957) in a slightly different context. Pollaczek was interested in finding the distribution of the ordered partial sums of a set of IID variables, whereas Spitzer was interested in finding the distribution of the maximum of the set of partial sums, which is related (see later) to the survival probability. Spitzer's derivation was more combinatorial. The same integral equation also appeared previously in a variety of half-space transport problems in physics and astrophysics (see Ivanov (1994) and references therein), and several other derivations of the solution of this equation, mostly algebraic in nature, are known (Frisch and Frisch, 1995). Unfortunately, all these derivations, the combinatorial as well as the algebraic ones, are highly technical in nature and there is no easy way! Here we will avoid these technical steps and instead just state the final result and discuss its implications and applications.

The solution of Eq. (3.16), with the initial condition $Q(x_0, 0) = 1$ for all $x_0 > 0$, is in terms of a double Laplace transform of $Q(x_0, n)$,

$$\int_0^{\infty} \left[\sum_{n=0}^{\infty} Q(x_0, n)s^n\right] \mathrm{e}^{-p_0 x_0}\,\mathrm{d}x_0 = \frac{\varphi(s, p_0)}{p_0\sqrt{1-s}}, \tag{3.95}$$

where

$$\varphi(s, p_0) = \exp\left[-\frac{p_0}{\pi}\int_0^{\infty} \frac{\ln(1 - s\hat{f}(k))}{p_0^2 + k^2}\,\mathrm{d}k\right], \tag{3.96}$$

$$\hat{f}(k) = \int_{-\infty}^{\infty} f(\eta)\mathrm{e}^{\mathrm{i}k\eta}\,\mathrm{d}\eta. \tag{3.97}$$

We will refer to the solution in Eq. (3.95) as the Pollaczek–Spitzer formula. Note that the dependence on the jump distribution $f(\eta)$ appears in this Pollaczek–Spitzer formula only through the Fourier transform $\hat{f}(k)$ in the expression of $\varphi(s, p_0)$ in Eq. (3.96). While the function $\varphi(s, p_0)$ looks formidable, its asymptotic behaviors for small and large p are quite simple. To extract the small-p_0 behavior, we make a change of variable $k = p_0 z$ in the integrand that appears inside the argument of the exponential in Eq. (3.96). This gives

$$\frac{p_0}{\pi} \int_0^\infty \frac{\ln(1 - s\hat{f}(k))}{p_0^2 + k^2} \, dk = \frac{1}{\pi} \int_0^\infty \frac{\ln(1 - s\hat{f}(p_0 z))}{1 + z^2} \, dz \xrightarrow[p_0 \to 0]{} \frac{1}{2} \ln(1 - s), \qquad (3.98)$$

where we used $\hat{f}(0) = 1$, since the jump distribution $f(\eta)$ is normalized to unity, and the identity $\int_0^\infty dz/(1 + z^2) = \pi/2$. This leads to the small-p_0 behavior for $\varphi(s, p_0 \to 0) = 1/\sqrt{1 - s}$. In the opposite limit, $p_0 \to \infty$, it is much easier, since the integrand inside the exponential in Eq. (3.96) just vanishes in this limit, leading to $\varphi(s, p_0 \to \infty) = 1$. Summarizing the two limiting behaviors, one gets

$$\varphi(s, p_0) = \begin{cases} \dfrac{1}{\sqrt{1 - s}}, & p_0 \to 0, \\ 1, & p_0 \to \infty. \end{cases} \qquad (3.99)$$

Let us now discuss some consequences of these asymptotic limits.

3.2.5   Sparre Andersen theorem

Although the survival probability $Q(x_0, n)$ for arbitrary x_0 depends explicitly on the jump length distribution $f(\eta)$ as evident in Eq. (3.95), it turns out that $Q(0, n)$ (the survival probability of the particle up to n steps starting at the origin) becomes, somewhat miraculously, *independent* of the distribution $f(\eta)$ as long as it is a continuous function. To see this, let us take the $p_0 \to \infty$ limit in Eq. (3.95). Making a change of variable $p_0 x_0 = y$ on the LHS of Eq. (3.95) and taking the $p_0 \to \infty$ limit, the LHS reduces, to leading order, to $(1/p_0) \sum_{n=0}^\infty Q(0, n) s^n$. On the RHS, taking the $p_0 \to \infty$ limit and using the result in the second line of Eq. (3.99) gives $1/(p_0 \sqrt{1 - s})$. Equating the leading-order terms (of $O(1/p_0)$ for large p_0) on both sides gives the identity, for all s,

$$\sum_{n=0}^\infty Q(0, n) s^n = \frac{1}{\sqrt{1 - s}}. \qquad (3.100)$$

Equating powers of s one gets the Sparre Andersen theorem (Sparre Andersen, 1954)

$$q(n) = Q(0, n) = \binom{2n}{n} 2^{-2n}, \qquad (3.101)$$

where we have used, for convenience, the shorthand notation $q(n)$ for $Q(0, n)$. Thus, quite amazingly, the survival probability $q(n) = Q(0, n)$ (starting from the origin) is completely *universal* and *for all* n (and not just for large n). No matter whether the jump length distribution is exponential, Gaussian or uniform, $q(n)$ is the same and is given by the simple formula in Eq. (3.101). Sparre Andersen originally derived this formula using a rather involved combinatorial approach (Sparre Andersen, 1954). This simple-looking formula is, however, somewhat deceptive and led several authors to try to derive it in a "simple" way! Unfortunately, all attempts led to equally complicated derivation (see Frisch and Frisch (1995) and Bauer *et al.* (1999)). However, recently, a rather simple derivation was found (Majumdar *et al.*, 2021*c*) using the statistics of the time of the maximum, and this derivation is detailed in Chapter 5 [see the discussion after Eq. (5.31)]. Note that deriving this formula as a special case of the Pollaczek–Spitzer solution is instructive as it shows that the role of the starting point $x_0 = 0$ is important for this *universality*. One loses this universality the moment x_0 is non-zero.

Let us also note another interesting fact. In the limit of large n, the survival probability $q(n)$ in Eq. (3.101) decays, to leading order, as

$$q(n) = Q(0, n) \simeq \frac{1}{\sqrt{\pi n}}. \tag{3.102}$$

Let us emphasize again that this result holds for an arbitrary continuous jump distribution $f(\eta)$, even including the Lévy flights! One may "naively" remark that this $n^{-1/2}$ asymptotic decay is equivalent to the $t^{-1/2}$ decay of the survival probability in the Brownian limit derived in Eq. (3.32). However, this is not correct and is actually rather subtle, as was shown in Majumdar *et al.* (2006). Consider first a continuous and symmetric jump distribution with a finite second moment $\sigma^2 = \int_{-\infty}^{\infty} \eta^2 f(\eta) \, d\eta$. To derive the Brownian limit from the Pollaczek–Spitzer formula in Eq. (3.95), one first considers the scaling limit $x_0 \to \infty$ and $n \to \infty$ but keeping the ratio $x_0/\sqrt{n}$ fixed. A careful asymptotic analysis of Eq. (3.95) shows that in this limit the first two leading terms for large n are given by (Majumdar *et al.*, 2006)

$$Q(x_0, n) \simeq \text{erf}\left(\frac{x_0}{\sqrt{2\sigma^2 n}}\right) + \frac{1}{\sqrt{\pi n}} e^{-x_0^2/2\sigma^2 n}. \tag{3.103}$$

For the special case of the double-exponential jump distribution $f(\eta) = 1/(2\ell_0)e^{-|\eta|/\ell_0}$, this result was derived explicitly in Eq. (3.76). However, the result in Eq. (3.103) is more general and holds for arbitrary jump distributions with finite σ^2 (Majumdar *et al.*, 2017). If one now takes the limit $x_0 \ll \sqrt{n}$, one recovers the universal Sparre Andersen result in Eq. (3.24) from the second term on the RHS of Eq. (3.103). On the other hand, if one keeps the scaling ratio $x_0/\sqrt{n}$ fixed and takes the strict $n \to \infty$ limit, the second term in Eq. (3.103) becomes subleading and the first term on the RHS (which remains non-universal in this limit as it contains σ^2 explicitly) becomes the leading term that provides the Brownian result in Eq. (3.24) upon identifying $\sigma^2 n = 2Dt$. Thus, the $n^{-1/2}$ universal decay of the survival probability (for $x_0 = 0$) is not quite related to the Brownian result $t^{-1/2}$: they originate from two different terms in Eq. (3.103).

A natural question is what happens to the survival probability $Q(x_0, n)$ for random walks where the variance of the jump distribution σ^2 is not finite, such as in

Lévy flights. We have seen above that for $x_0 = 0$, the universal Sparre Andersen result in Eq. (3.101) holds. In particular, for large n, $Q(0, n) \simeq 1/\sqrt{\pi n}$ for arbitrary symmetric and continuous jump distributions. On the other hand, we expect quite different results in the scaling limit where $x_0 = O(n^{1/\mu})$ for Lévy flights with $0 < \mu < 2$, as in the Brownian limit for random walks with a finite σ^2. How do the two limiting behaviors match with each other as one varies the starting position x_0 from $x_0 = 0$ to $x_0 = O(n^{1/\mu})$? Indeed, by exploiting the Pollaczeck–Spitzer formula in Eq. (3.95), the survival probability $Q(x_0, n)$, as a function of x_0 for fixed but large n, has been computed (Majumdar *et al.*, 2017) for arbitrary jump distributions including Lévy flights, which clearly demonstrates this interpolation between the Sparre Andersen result and the result in the scaling limit $x_0 = O(n^{1/\mu})$. We do not give the detailed derivation of these results here, but just summarize the main results. For a detailed derivation, see Majumdar *et al.* (2017). One finds, to leading order for large n,

$$
Q(x_0, n) \sim
\begin{cases}
\dfrac{1}{\sqrt{n}} U(x_0) & \text{for } n \to +\infty \text{ and } x_0 = O(1), \\[3mm]
V_\mu\left(\dfrac{x_0}{n^{1/\mu}}\right) & \text{for } n \to +\infty \text{ and } x_0 = O(n^{1/\mu}).
\end{cases}
\tag{3.104}
$$

The Laplace transform of the function $U(x_0)$ reads

$$
\int_0^\infty U(x_0) e^{-\lambda x_0} \, dx_0 = \frac{1}{\lambda\sqrt{\pi}} \exp\left[-\frac{\lambda}{\pi} \int_0^\infty \frac{dk}{\lambda^2 + k^2} \ln(1 - \hat{f}(k)) \right],
\tag{3.105}
$$

where $\hat{f}(k)$ is the Fourier transform of the jump PDF $f(\eta)$. Thus, $U(x_0)$ depends on the full $\hat{f}(k)$ and not just on its small-k expansion. The scaling function $V_\mu(z)$, by contrast, depends only on the small-k behavior of $\hat{f}(k)$ in Eq. (2.56) and hence can be labelled just by the index μ. (Note that $U(x_0)$ cannot be labelled just by μ as it depends on the full $\hat{f}(k)$ and not just its small-k behavior). The scaling function $V_\mu(z)$ is a little complicated and is given by the solution of the integral equation

$$
\int_0^\infty dy \, e^{-y} y^{1/\mu} \int_0^\infty dz \, V_\mu(z) e^{-wy^{1/\mu} z} = \frac{1}{w} J_\mu(w),
$$

$$
\text{where } J_\mu(w) = \exp\left[-\frac{1}{\pi} \int_0^\infty \frac{du}{1 + u^2} \ln(1 + (\ell w u)^\mu) \right]
\tag{3.106}
$$

and ℓ is the characteristic length of the jump distribution as defined in Eq. (2.56). In the special case $\mu = 2$, one can show that the scaling function reduces to $V_2(z) = \text{erf}(z/(2\ell))$, thus recovering the first term in Eq. (3.103).

In general, it is hard to derive the functions $U(x_0)$ and $V_\mu(z)$ explicitly for arbitrary jump distributions. However, one can extract the asymptotic behaviors of both functions for small and large arguments (Majumdar *et al.*, 2017). For instance, one finds $U(0) = 1/\sqrt{\pi}$ irrespective of the jump distributions, thus recovering the Sparre Andersen result for large n in Eq. (3.102). Similarly, for large x_0, $U(x_0) \sim x_0^{\mu/2}$, which matches with the small-argument behavior of the scaling function $V_\mu(z)$.

Generalization to asymmetric jump distributions. Actually, there exists a generalized Sparre Andersen theorem (Sparre Andersen, 1954) which holds for non-symmetric (but still continuous) jump length distributions $f(\eta)$. Unlike in the symmetric case, for an asymmetric jump distribution of a random walk starting at $x_0 = 0$, the probability that the walker is on the positive side up to n steps is different from the probability that it is on the negative side up to n steps. Thus, one needs to define two different survival probabilities:

$$q_+(n) = \text{Prob}[x_n \geq 0, x_{n-1} \geq 0, \ldots, x_1 \geq 0 \mid x_0 = 0], \tag{3.107}$$

$$q_-(n) = \text{Prob}[x_n \leq 0, x_{n-1} \leq 0, \ldots, x_1 \leq 0 \mid x_0 = 0]. \tag{3.108}$$

For a symmetric jump distribution, $q_+(n) = q_-(n) = q(n)$. In the asymmetric case, the generalized Sparre Andersen theorem reads

$$\tilde{q}_+(s) = \sum_{n=0}^{\infty} q_+(n)s^n = \exp\left[\sum_{n=1}^{\infty} \frac{p_n^+}{n} s^n\right], \tag{3.109}$$

$$\tilde{q}_-(s) = \sum_{n=0}^{\infty} q_-(n)s^n = \exp\left[\sum_{n=1}^{\infty} \frac{p_n^-}{n} s^n\right], \tag{3.110}$$

where

$$p_n^+ = \text{Prob}(x_n \geq 0) = \int_0^{\infty} G(x, x_0, n)\, \mathrm{d}x, \tag{3.111}$$

$$p_n^- = \text{Prob}(x_n \leq 0) = \int_{-\infty}^0 G(x, x_0, n)\, \mathrm{d}x \tag{3.112}$$

are just the probabilities that exactly at the nth step the particle position is positive or negative respectively. For the symmetric (zero bias) case, $p_n^+ = p_n^- = 1/2$ (by symmetry) and then both Eqs. (3.109) and (3.110) reduce to Eq. (3.100). We remark that the results in Eqs. (3.109) and (3.110) are quite remarkable and non-trivial as they relate the history-dependent probabilities $q_\pm(n)$ to local (in time) quantities $p_\pm(n)$.

Let us mention here a special case with a drift (Bauer *et al.*, 1999; Le Doussal and Wiese, 2009; Majumdar, 2010*b*; Majumdar *et al.*, 2012) that is explicitly solvable and that gives rise to a power-law decay of the survival probability with a continuously dependent exponent. Consider the evolution

$$x_n = x_{n-1} + \mu + \eta_n, \tag{3.113}$$

with $x_0 = 0$. Here, μ represents a drift and the η_n are IID noise variables each drawn from a symmetric Cauchy distribution

$$f(\eta) = \frac{1}{\pi(\eta^2 + 1)}. \tag{3.114}$$

In this case, the variable $y_n = x_n - \mu n$ undergoes a symmetric random walk, $y_n = y_{n-1} + \eta_n$. Hence, the probability distribution of y_n at step n, starting from $y_0 = 0$, can be easily computed from the free Green's function discussed in Chapter 2. In fact, the Cauchy distribution corresponds to the Lévy laws in Eq. (2.51) with index $\mu = 1$. Hence,

$$G(y, 0, n) = \frac{1}{n}\mathcal{L}_1\left(\frac{y}{n}\right) = \frac{n}{\pi(y^2 + n^2)}. \tag{3.115}$$

Thus,

$$p_n^+ = \mathrm{Prob}(x_n \geq 0) = \mathrm{Prob}(y_n \geq -\mu n) = \int_{-\mu n}^{\infty} \frac{an}{\pi(y^2 + n^2)}\, dy = \frac{1}{2} + \frac{1}{\pi}\tan^{-1}(\mu), \tag{3.116}$$

$$p_n^- = \mathrm{Prob}(x_n \leq 0) = \mathrm{Prob}(y_n \leq -\mu n) = \int_{-\infty}^{-\mu n} \frac{an}{\pi(y^2 + n^2)}\, dy = \frac{1}{2} - \frac{1}{\pi}\tan^{-1}(\mu). \tag{3.117}$$

Substituting these results into Eqs. (3.109) and (3.110), one gets

$$\tilde{q}_\pm(s) = \sum_{n=0}^{\infty} q_\pm(n)s^n = \frac{1}{(1-s)^{\zeta_\pm}}; \qquad \zeta_\pm = \frac{1}{2} \pm \frac{1}{\pi}\tan^{-1}(\mu). \tag{3.118}$$

Inverting the generating function, one then finds that, for large n,

$$q_\pm(n) \simeq \frac{1}{\Gamma(\zeta_\pm)}\frac{1}{n^{\theta_\pm(\mu)}}; \qquad \theta_\pm(\mu) = 1 - \zeta_\pm = \frac{1}{2} \mp \frac{1}{\pi}\tan^{-1}(\mu). \tag{3.119}$$

Thus, the persistence exponents $\theta_\pm(\mu)$ are non-trivial and vary continuously with the drift μ. For example, as $\mu \to \infty$ (drift away from the origin), $\theta_+ \to 0$ (the particle always remains on the positive side), and as $\mu \to -\infty$ (drift towards the origin), $\theta_+ \to 1$ leading to a faster decay than the driftless ($\mu = 0$) case where $\theta_\pm(0) = 1/2$.

Little exercise 3.9 *Show the result in Eq. (3.119).*

3.2.6   A more general result due to Ivanov

Going beyond the Pollaczeck–Spitzer formula for the survival probability $Q(x_0, n)$ in Eq. (3.95), one may ask if it is possible to obtain a result for the full restricted Green's function $G_+(x, x_0, n)$ defined in Eq. (3.12). This restricted Green's function satisfies the Wiener–Hopf integral equation in Eq. (3.14) and carries information about both the survival probability $Q(x_0, n)$ and the final position x of the walker at step n. Indeed, by integrating over the final position x, one can recover the survival probability from the restricted Green's function, $Q(x_0, n) = \int_0^\infty dx\, G_+(x, x_0, n)$. It turns out that there indeed exists a formula for $G_+(x, x_0, n)$, valid for an arbitrary symmetric and continuous jump distribution $f(\eta)$. This can be seen as a generalization of the Pollaczek–Spitzer formula and is known as the Ivanov formula (Ivanov, 1994). It reads

$$\sum_{n=0}^{\infty} \left[\int_0^{\infty} dx\, e^{-px} \int_0^{\infty} dx_0\, e^{-p_0 x_0} G_+(x, x_0, n) \right] s^n = \frac{\varphi(s, p)\varphi(s, p_0)}{p + p_0}, \tag{3.120}$$

where $\varphi(s, p)$ is given in Eq. (3.96). Note that, in the limit $p \to 0$, the LHS of Eq. (3.120) reduces to the LHS of the Pollaczeck–Spitzer formula in Eq. (3.95), using $Q(x_0, n) = \int_0^{\infty} dx\, G_+(x, x_0, n)$. The RHS of Eq. (3.120), using $\varphi(s, p \to 0) = 1/\sqrt{1-s}$ from the first line of Eq. (3.99), reduces to the RHS of the Pollaczeck–Spitzer formula in Eq. (3.95). Thus, the Pollaczeck–Spitzer formula, as well as the Sparre Andersen formula, is actually contained in this more general formula in Eq. (3.120). While this Ivanov formula holds for continuous and symmetric jump distributions, it can be generalized to the case where there is a non-zero drift (Mounaix *et al.*, 2018).

As a nice application of this Ivanov formula, let us again consider our canonical example of the double-exponential jump distribution $f(\eta) = 1/(2\ell_0)e^{-|\eta|/\ell_0}$ in Eq. (3.37). In this case, the Fourier transform $\hat{f}(k)$ is given explicitly by

$$\hat{f}(k) = \frac{1}{1 + (k\ell_0)^2}. \tag{3.121}$$

We insert this expression in $\varphi(s, p)$ in Eq. (3.96) and use the identity

$$\int_0^{\infty} dx\, \frac{\ln(a^2 + b^2 x^2)}{c^2 + d^2 x^2} = \frac{\pi}{cd} \ln\left(\frac{ad + bc}{d}\right), \qquad a, b, c, d > 0, \tag{3.122}$$

to obtain

$$\varphi(s, p) = \frac{1 + \ell_0 p}{\sqrt{1 - s} + \ell_0 p}. \tag{3.123}$$

The Ivanov formula in this case thus reads

$$\sum_{n=0}^{\infty} \left[\int_0^{\infty} dx\, e^{-px} \int_0^{\infty} dx_0\, e^{-p_0 x_0} G_+(x, x_0, n) \right] s^n$$

$$= \frac{1 + \ell_0 p}{(\sqrt{1-s} + \ell_0 p)} \frac{1 + \ell_0 p_0}{(\sqrt{1-s} + \ell_0 p_0)} \frac{1}{(p + p_0)}. \tag{3.124}$$

We now have to invert the double Laplace transform with respect to p and p_0. We can first invert it with respect to p, for fixed p_0, followed by an inversion with respect to p_0. In the first inversion with respect to p, we note that the RHS of Eq. (3.124) can be written as a sum of partial fractions and then each term can be inverted simply. As an example, consider inverting the Laplace transform

$$\mathcal{L}_{p \to x}^{-1} \left[\frac{1}{(p+a)(p+b)} \right] = \mathcal{L}_{p \to x}^{-1} \left[\left(\frac{1}{p+a} - \frac{1}{p+b} \right) \frac{1}{b-a} \right] = \frac{1}{b-a}(e^{-ax} - e^{-bx}), \tag{3.125}$$

where in the last equality we have just used the identity $\mathcal{L}_{p \to x}^{-1} 1/(p+a) = e^{-ax}$ that simply follows from the fact that there is a simple pole at $p = -a$ in the complex p-plane. Using the result in Eq. (3.125), one can then invert the double Laplace transform with respect to both p and p_0, and one recovers the exact solution obtained by a different method detailed in Eq. (3.55). We invite the enthusiastic reader to verify this!

We conclude this chapter by pointing out that these Pollaczeck–Spitzer/Ivanov results for discrete-time random walks / Lévy flights in the semi-infinite geometry have been heavily used in several applications, including radiative transfer of photons (Frisch and Frisch, 1982; Frisch, 1988; Ivanov, 1994), the Smoluchowski flux problem in chemical processes (Majumdar *et al.*, 2006, 2019*a*; Ziff *et al.*, 2007, 2009), survival and first-passage properties in semi-infinite (Zumofen and Klafter, 1995; Majumdar *et al.*, 2017) and finite domains (Buldyrev et al., 2001*a,b*; Klinger et al., 2022), the statistics of ordered maxima of random walks (Comtet and Majumdar, 2005; Schehr and Majumdar, 2012; Majumdar *et al.*, 2014; Mounaix *et al.*, 2018; De Bruyne *et al.*, 2021, 2023), statistics of records (Majumdar and Ziff, 2008; Majumdar *et al.*, 2012, 2021*b,c*; Wergen *et al.*, 2012; Godrèche et al., 2016, 2017), etc. Some of these topics will be discussed in the later chapters.

4

Extreme Statistics

Extreme means either the largest (maximum) or the smallest (minimum) of a set of variables. When the variables are random, their extremes (both maximum and minimum) also become random variables. Our goal in this chapter is to understand the statistics of the maximum and the minimum in our two principal models: (i) IID random variables, and (ii) the random walk model. We start with the IID model and discuss the random walk model later.

4.1 The statistics of extremes for IID random variables: Maximum and minimum

Of course, both the maximum and the minimum are examples of extremes. In general, the maximum and the minimum are typically correlated. We start with just the maximum for IID variables and later discuss the statistics of the minimum, as well as the joint statistics of both the maximum and the minimum. Before deriving the distribution of the maximum formally for the IID case, it is useful to start with an amusing numerical experiment.

4.1.1 A little experiment on the maximum

Let us start with a simple numerical experiment that you can easily do on your laptop. Suppose we draw a set of N (say $N = 100$) random numbers independently from a common distribution $p(x)$. For instance, $p(x)$ can be the normal (Gaussian) distribution, or the uniform distribution over $[0, 1]$, etc. For a fixed choice of this distribution $p(x)$, we then look at our set of $N = 100$ random numbers and denote them by $\{x_1^{(1)}, x_2^{(1)}, \ldots, x_N^{(1)}\}$, where the superscript (1) refers to this first batch of N random numbers. We now identify the maximum from this set and denote it by $X_{\max}^{(1)} = \max\{x_1^{(1)}, x_2^{(1)}, \ldots, x_N^{(1)}\}$. Now let us repeat the experiment and choose another set of $N = 100$ random numbers drawn from the same $p(x)$. Let $X_{\max}^{(2)}$ denote the maximum obtained from this second batch. Imagine repeating this experiment a large number of times, N_{samp}. Then we will have a set $\{X_{\max}^{(1)}, X_{\max}^{(2)}, \ldots\}$ and we can construct a histogram of this set. In the limit when the number of samples N_{samp} becomes very large, we expect that this histogram will converge to a probability distribution $P_{\max}(x, N) = \mathrm{Prob}(X_{\max} = x, N)$ (normalized to unity) for a given

Statistics of Extremes and Records in Random Sequences. Satya N. Majumdar and Grégory Schehr, Oxford University Press.
© Satya N. Majumdar and Grégory Schehr (2024). DOI: 10.1093/9780191838781.003.0004

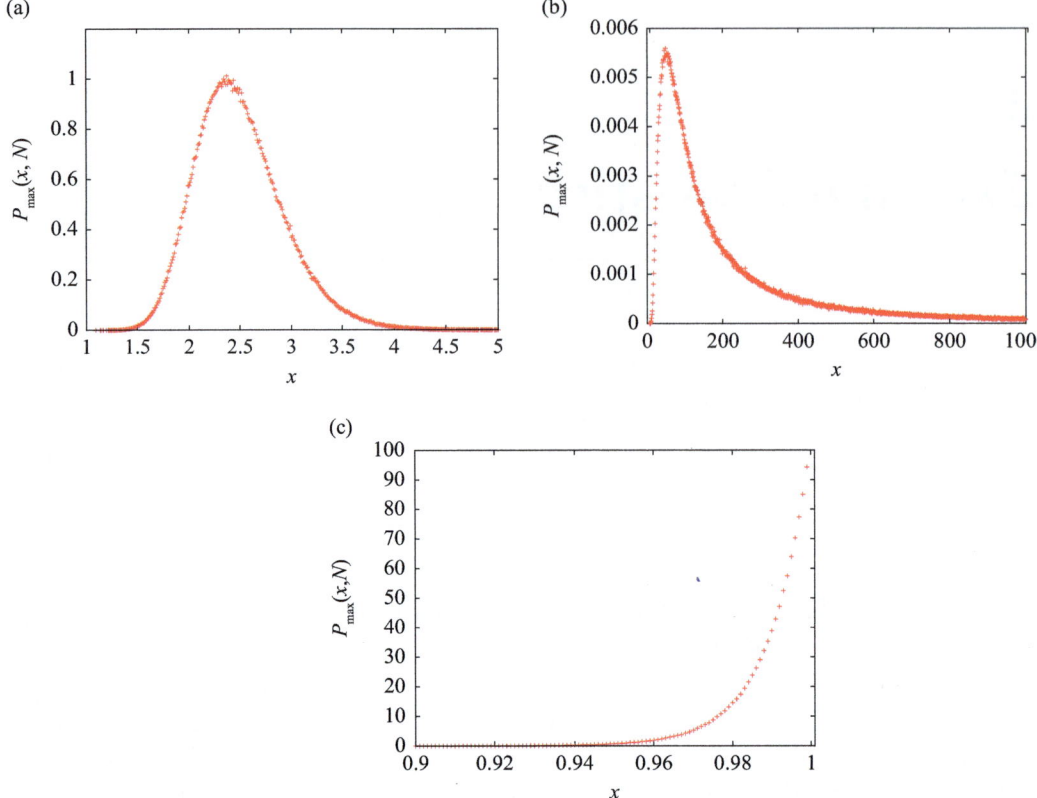

Fig. 4.1 Plot of $P_{\max}(x, N) = \mathrm{Prob}(X_{\max} = x, N)$ vs. x for $N = 100$ (numerically generated) for three different choices of $p(x)$: (a) the normal centered Gaussian distribution, (ii) a Pareto distribution with $1/x^2$ tail and (iii) the uniform distribution over $[0, 1]$.

fixed N. In Fig. 4.1 we plot this histogram $P_{\max}(x, N)$ vs. x for $N = 100$, $N_{\mathrm{samp}} = 10^4$ and for three different choices of $p(x)$:

(a) the normal centered Gaussian distribution $p(x) = \mathrm{e}^{-x^2/2}/\sqrt{2\pi}$;
(b) a Pareto distribution which has a power-law tail $p(x) = 1/x^2$ for $x \in [1, +\infty)$;
(c) a uniform distribution over $[0, 1]$.

We can see from these three plots that $P_{\max}(x, N)$ looks rather different in the three cases, and it seems to depend very much on the choice of $p(x)$. But are there some aspects of the extreme statistics that are universal, i.e., independent of the choice of $p(x)$? Why should we be interested in such a concept of universality? It turns out that the universality—when the details of a system become irrelevant for certain large-scale properties—is a key concept that has always fascinated physicists. Our goal in this chapter will be to understand and explore the universal emerging behavior of extremes in this simple setting of IID random variables. Indeed, this is a classical subject in probability theory with many applications (see, e.g., Fréchet, 1927; Fisher and Tippett, 1928; Gnedenko, 1943; Weibull, 1951; Bouchaud and Mézard, 1997; Gumbel, 2004; Bertin and Clusel, 2006; Leadbetter et al., 2012). For a review of recent developments and applications, we refer the reader to Majumdar *et al.* (2020).

To understand if there is any hidden universality, we need to first compute the distribution of $X_{\max}$. But how does one proceed? It is actually very simple. Let us imagine that we have a set of N random variables $\{x_1, x_2, \ldots, x_N\}$. In the most general setting, we assume that these random variables are drawn from a joint distribution $P_{\text{joint}}(x_1, x_2, \ldots, x_N)$. As discussed in the introduction, if the random variables are IID, this joint distribution factorizes as

$$P_{\text{joint}}(x_1, x_2, \ldots, x_N) = p(x_1)p(x_2) \cdots p(x_N), \tag{4.1}$$

where $p(x)$ is the marginal distribution of any single random variable. In cases where the random variables $\{x_1, x_2, \ldots, x_N\}$ are correlated, the joint distribution $P_{\text{joint}}(x_1, x_2, \ldots, x_N)$ does not factorize. It turns out that in the IID case where Eq. (4.1) is valid, one can compute the distribution of $X_{\max}$ quite easily, as we will demonstrate.

4.1.2 The statistics of the maximum

We therefore start with Eq. (4.1). Our goal is to compute the distribution of $X_{\max} = \max\{x_1, x_2, \ldots, x_N\}$. It turns out to be convenient to compute the cumulative distribution of $X_{\max}$ defined as

$$Q_{\max}(x, N) = \text{Prob}(X_{\max} \leq x). \tag{4.2}$$

Can you guess why we use the cumulative distribution? This is because the event $\{X_{\max} \leq x\}$ is exactly the same event as $\{x_1 \leq x, x_2 \leq x, \ldots, x_N \leq x\}$. Indeed, if the maximum of a set is $\leq x$, it necessarily implies that all elements of this set are $\leq x$. Similarly, if all elements of a set are $\leq x$, this necessarily means that their maximum $X_{\max} \leq x$. But the probability that all of them are $\leq x$ can be very simply written in terms of the joint distribution:

$$Q_{\max}(x, N) = \text{Prob}[x_1 \leq x, x_2 \leq x, \ldots, x_N \leq x]$$
$$= \int_{-\infty}^{x} dx_1 \int_{-\infty}^{x} dx_2 \cdots \int_{-\infty}^{x} dx_N \, P_{\text{joint}}(x_1, x_2, \ldots, x_N). \tag{4.3}$$

Note that Eq. (4.3) is actually valid in the most general case, even when the variables x_i are correlated, i.e., when the joint distribution does not factorize. In the general case, when the variables are correlated, computing $Q_{\max}(x, N)$ using the above formula is very hard because we quite often do not know how to carry out this N-fold multiple integral—this really is at the heart of the difficulties of computing the extreme value distribution $Q_{\max}(x, N)$ for correlated variables. However, for IID random variables a neat simplification occurs due to the factorization property in Eq. (4.1). Using this property in Eq. (4.3) one immediately gets a rather explicit result:

$$Q_{\max}(x, N) = \left(\int_{-\infty}^{x} p(x')\, \mathrm{d}x' \right)^{N} = \left(1 - \int_{x}^{+\infty} p(x')\, \mathrm{d}x' \right)^{N}, \qquad (4.4)$$

where we used the fact that $p(x)$ is normalized to unity. The function $Q_{\max}(x, N)$ typically has a sigmoid shape, with the limiting behaviors $Q_{\max}(x \to \infty, N) \to 1$ and $Q_{\max}(x \to -\infty, N) = 0$. Knowing the cumulative distribution $Q_{\max}(x, N)$, how can one compute the PDF $P_{\max}(x, N)$? This is simple because, by definition, the cumulative distribution is

$$Q_{\max}(x, N) = \int_{-\infty}^{x} P_{\max}(y, N)\, \mathrm{d}y. \qquad (4.5)$$

Taking the derivative with respect to x, one finds

$$P_{\max}(x, N) = \frac{\mathrm{d}Q_{\max}(x, N)}{\mathrm{d}x}. \qquad (4.6)$$

Remember that the PDF $P_{\max}(x, N)$ is normalized to one, $\int_{-\infty}^{\infty} P_{\max}(x, N)\, \mathrm{d}x = 1$. Consequently, taking the derivative in Eq. (4.4), one finds

$$P_{\max}(x, N) = N p(x) \left(\int_{-\infty}^{x} p(x')\, \mathrm{d}x' \right)^{N-1}. \qquad (4.7)$$

Little exercise 4.1 *Verify that $P_{\max}(x, N)$ in Eq. (4.7) is normalized to unity.*

Little exercise 4.2 *Evaluate Eq. (4.7) for the three distributions (a), (b) and (c) with $N = 100$ in the experiment done at the beginning of the chapter and check that the formula reproduces the experimental curves shown in Fig. 4.1.*

The search for universality. For any finite N, we then have the exact formula for the cumulative distribution $Q_{\max}(x, N)$ in Eq. (4.5), or equivalently for the PDF $P_{\max}(x, N)$ in Eq. (4.7). These formulae manifestly depend on $p(x)$ and one would naively conclude that there is no universality. Is this true? It turns out that, while the universality is not manifest for any finite N, it actually emerges only in the large-N limit. But that is not all. In addition to taking the large-N limit, we need to introduce the notion of "centering and scaling." Centering means that we zoom in to the vicinity of the region where $P_{\max}(x, N)$ has its peak. Scaling means that, after zooming in to the peak, we rescale the axes (both x and y). If we do this right, we may get a limiting curve that may have the universal properties. You may be confused by this, so let us give a couple of simple examples to illustrate how it works.

Example 4.1 Suppose that the $x_i \geq 0$ are positive random variables, each drawn independently from the exponential distribution

$$p(x) = \mathrm{e}^{-x}, \quad x \geq 0. \qquad (4.8)$$

Substituting into Eq. (4.5) (with the lower limit $-\infty$ replaced by 0 in this case), we get

$$Q_{\max}(x, N) = \left(\int_0^x e^{-x'} \, dx' \right)^N = (1 - e^{-x})^N. \tag{4.9}$$

If we now take the limit $N \to \infty$ for any fixed $x \geq 0$, it is clear that $Q_{\max}(x, N) \to 0$ as $N \to \infty$, except at $x = +\infty$ where $Q_{\max}(x \to \infty, N) = 1$. So, we get nothing interesting, and need to do something else. The idea is to re-express the result in Eq. (4.9) as

$$Q_{\max}(x, N) = e^{N \ln(1 - e^{-x})} \tag{4.10}$$

and then simultaneously take the limits $N \to \infty$ and $x \to \infty$ such that the product Ne^{-x} is finite. Expanding the logarithm in a Taylor series for large x and keeping only the first term of this expansion, we get

$$Q_{\max}(x, N) \approx e^{-Ne^{-x}}, \tag{4.11}$$

which is now of order $O(1)$ in this scaling limit. We can further rewrite this as

$$Q_{\max}(x, N) \approx e^{-Ne^{-x}} = e^{-e^{-(x - \ln N)}}. \tag{4.12}$$

Thus, finally, we can express $Q_{\max}(x, N)$ in the scaling form

$$Q_{\max}(x, N) \approx F_{\mathrm{I}}\left(\frac{x - a_N}{b_N} \right), \quad \text{where } a_N = \ln N, \ b_N = 1, \tag{4.13}$$

and the scaling function $F_{\mathrm{I}}(z)$ is given by

$$F_{\mathrm{I}}(z) = e^{-e^{-z}}. \tag{4.14}$$

This scaling function is called the Gumbel distribution (Gumbel, 2004). Note that the scaling form in Eq. (4.13) holds only in the scaling limit when $N \to \infty$, $x \to \infty$ but keeping the scaled variable $z = (x - a_N)/b_N$ fixed. This is equivalent to zooming in on the vicinity of $x = a_N$ and then rescaling the centered distance $x - a_N$ by a factor b_N: this centering and scaling procedure then reduces the distribution $Q_{\max}(x, N)$, which is a function of two variables x and N, into a function of only one variable $z = (x - a_N)/b_N$.

While this example illustrates the idea of scaling for large N, to illustrate universality we need to consider other examples of distributions $p(x)$ to see if some universal features emerge in this scaling limit.

Example 4.2 Let us now consider another example where

$$p(x) = \frac{1}{\sqrt{2\pi}} e^{-x^2/2}, \quad x \in \mathbb{R}. \tag{4.15}$$

As in the previous example, substituting this $p(x)$ into Eq. (4.4) we obtain

$$Q_{\max}(x, N) = \left(1 - \frac{1}{2}\mathrm{erfc}\left(\frac{x}{\sqrt{2}}\right)\right)^N = \exp\left[N \ln\left(1 - \frac{1}{2}\mathrm{erfc}\left(\frac{x}{\sqrt{2}}\right)\right)\right], \qquad (4.16)$$

where $\mathrm{erfc}(x) = (2/\sqrt{\pi})\int_x^\infty e^{-y^2}\,dy$ is the complementary error function. Expanding the logarithm in a Taylor series for large x, we get the approximate formula

$$Q_{\max}(x, N) \approx \exp\left(-\frac{N}{2}\mathrm{erfc}\left(\frac{x}{\sqrt{2}}\right)\right). \qquad (4.17)$$

We only want to retain the leading behavior for large x. This can be done by using the large-x behavior of $\mathrm{erfc}(x)$,

$$\mathrm{erfc}(x) \approx \frac{e^{-x^2}}{x\sqrt{\pi}}, \qquad x \to \infty. \qquad (4.18)$$

Little exercise 4.3 *Starting from the definition* $\mathrm{erfc}(x) = (2/\sqrt{\pi})\int_x^\infty e^{-y^2}\,dy$, *show Eq. (4.18). (Hint: use integration by parts.)*

Finally, injecting this asymptotic behavior into Eq. (4.17), we get, for both N and x large (we have not yet taken the scaling limit) the leading behavior

$$Q_{\max}(x, N) \approx \exp\left(-\frac{N}{\sqrt{2\pi}}\frac{e^{-x^2/2}}{x}\right). \qquad (4.19)$$

Now the search for a_N and b_N starts! The idea is to find appropriate a_N and b_N such that $(x - a_N)/b_N = z$ is fixed to an order $O(1)$ value as $N \to \infty$ (just as in the previous example). For this, we set $x = a_N + b_N z$ in Eq. (4.19) to get

$$Q_{\max}(x = a_N + b_N z, N) \approx \exp\left(-\frac{N}{\sqrt{2\pi}}\frac{1}{(a_N + b_N z)}e^{-\frac{1}{2}(a_N^2 + 2a_N b_N z + b_N^2 z^2)}\right). \qquad (4.20)$$

Let us anticipate (and verify a posteriori) that $b_N \ll a_N$ such that in the denominator of the argument of the exponential we can replace $(a_N + b_N z) \approx a_N$ to leading order. Furthermore, we can drop the term $b_N^2 z^2$ inside the exponential, since we expect it to be small. We then choose a_N and b_N such that

$$\frac{N}{\sqrt{2\pi}a_N}e^{-a_N^2/2} = 1, \qquad a_N b_N = 1. \qquad (4.21)$$

With this choice of a_N and b_N we then have

$$Q_{\max}(x = a_N + b_N z, N) \approx e^{-e^{-z}} = F_{\mathrm{I}}(z). \qquad (4.22)$$

The centering and the scaling constants a_N and b_N can be solved for large N from Eq. (4.21). To leading order, one gets

$$a_N \approx \sqrt{2\ln N} - \frac{\ln(\ln N)}{2\sqrt{2\ln N}} + \cdots, \qquad b_N = \frac{1}{a_N}. \qquad (4.23)$$

Now we see the first glimpse of universality! Even though the finite-N formula of $Q_{\max}(x, N)$ for the two distributions $p(x) = e^{-x}$ (with $x \geq 0$) and $p(x) = (1/\sqrt{2\pi})e^{-x^2/2}$ look rather different from each other, once we center and scale by appropriate a_N and b_N, which differ in the two cases, the scaling function for the two cases is exactly identical and is given by the Gumbel law $F_{\mathrm{I}}(z) = e^{-e^{-z}}$.

Little exercise 4.4. (Slightly harder) *Take $p(x)$ such that, for large x, $p(x) \sim ce^{-x^\delta}$ with $\delta > 0$ and $c > 0$. Show that the scale factors a_N and b_N are given (to leading order for large N) by*

$$a_N \approx (\ln N)^{1/\delta}, \qquad b_N \approx \frac{1}{\delta}(\ln N)^{(1/\delta)-1}. \qquad (4.24)$$

Show also that $Q_{\max}(x = a_N + b_N z N)$ converges in the large-N limit to the Gumbel form $F_{\mathrm{I}}(z) = e^{-e^{-z}}$.

From the result in Eq. (4.24) we see that for $\delta > 1$ (when the tail is faster than exponential), the width $b_N \to 0$. This means that the distribution of $X_{\max}$ becomes narrower and narrower as $N \to \infty$ and gets concentrated around a_N. In contrast, for $0 < \delta < 1$ (when the tail is slower than exponential), the width $b_N \to \infty$ as $N \to \infty$. This means that the scale of the typical fluctuations of $X_{\max}$ around a_N becomes broader with increasing N. The case $\delta = 1$ (purely exponential decay) is "marginal" where the width $b_N \to b$, a constant independent of N.

Actually, it turns out (see later) that for any $p(x)$ that decays at large x faster than any power law, the cumulative distribution of $Q_{\max}(x, N)$ approaches, for large N, the scaling form

$$Q_{\max}(x, N) \sim F_{\mathrm{I}}\left(\frac{x - a_N}{b_N}\right), \qquad F_{\mathrm{I}}(z) = e^{-e^{-z}}, \qquad (4.25)$$

where the constants a_N and b_N are non-universal, i.e., they depend on the details of the distribution $p(x)$. However, the scaling function $F_{\mathrm{I}}(z)$ is universal and is called the Gumbel distribution. Its derivative is called the Gumbel PDF and is given by

$$f_{\mathrm{I}}(z) = F_{\mathrm{I}}'(z) = e^{-z-e^{-z}}, \qquad z \in (-\infty, +\infty). \qquad (4.26)$$

One can easily check that this PDF is normalized to unity: $\int_{-\infty}^{\infty} f_{\mathrm{I}}(z)\,dz = 1$. One can also calculate all the moments of the Gumbel PDF. For example, the first two moments are given by

$$\mu_1 = \int_{-\infty}^{\infty} z f_{\mathrm{I}}(z)\,dz = \gamma_{\mathrm{E}}, \qquad \mu_2 = \int_{-\infty}^{\infty} z^2 f_{\mathrm{I}}(z)\,dz = \frac{\pi^2}{6} + \gamma_{\mathrm{E}}^2, \qquad (4.27)$$

where $\gamma_{\mathrm{E}} = 0.577\,216$ is the Euler constant.

Little exercise 4.5 *Calculate the nth moment of $f_{\mathrm{I}}(z)$.*

What happens when $p(x)$ decays as a power law? We have just said that the Gumbel form in Eq. (4.25) will not hold, as is also evident from the middle panel of Fig. 4.1. In this case, we will now see that we get yet another class of extreme value distribution.

Example 4.3 Let us start with experiment (b) in the beginning of the chapter, i.e., the Pareto distribution $p(x) = 1/x^2$ for $x \geq 1$. From the general result in Eq. (4.4), the cumulative distribution of the maximum $X_{\max}$ reads

$$Q_{\max}(x, N) = \left(\int_1^x \frac{\mathrm{d}x'}{x'^2} \right)^N = \left(1 - \frac{1}{x} \right)^N, \quad x \geq 1. \tag{4.28}$$

To obtain the centered and scaled distribution for large N we set $x = Nz$ and take the $N \to \infty$ limit but keeping z fixed:

$$Q_{\max}(x = Nz, N) = \mathrm{e}^{N \ln(1-(1/Nz))} \xrightarrow[N\to\infty]{} \mathrm{e}^{-1/z}, \quad z \geq 0. \tag{4.29}$$

Clearly, we see that this scaling function $\mathrm{e}^{-1/z}$ is different from the Gumbel form $F_{\mathrm{I}}(z) = \mathrm{e}^{-\mathrm{e}^{-z}}$ in Eq. (4.25).

Little exercise 4.6 *Consider $p(x) \approx A/x^{1+\alpha}$ for large x, with $\alpha > 0$. Set $x = a_N + b_N z$, with*

$$a_N = 0, \qquad b_N = (AN/\alpha)^{1/\alpha}, \tag{4.30}$$

and show that, as $N \to \infty$,

$$Q_{\max}(x = b_N z, N) \xrightarrow[N\to\infty]{} F_{\mathrm{II}}(z), \quad \text{with } F_{\mathrm{II}}(z) = \mathrm{e}^{-1/z^\alpha}, \ z \geq 0. \tag{4.31}$$

$F_{\mathrm{II}}(z)$ is called the Fréchet distribution (Fréchet, 1927). Note that this distribution is parameterized by α. However, we suppressed the explicit α-dependence in $F_{\mathrm{II}}(z)$ in order to keep the notation simple.

Remark 4.1 *Note that in this power-law tail case, the scale b_N of the typical fluctuations of $X_{\max}$ increases with increasing N, as in the case of $p(x) \sim \mathrm{e}^{-x^\delta}$ with $0 < \delta < 1$. In fact, it turns out, more generally, that for all $p(x)$ that decay slower than an exponential for large x, the width b_N increases with increasing N.*

So far we have found two different extreme value distributions, Gumbel and Fréchet, depending on the tail of $p(x)$. Notice that in these two cases, the support of $p(x)$ extends all the way up to $+\infty$. Do these examples exhaust all possibilities?

What about the experiment in Fig. 4.1(c), where $p(x)$ is the uniform distribution over $[0, 1]$? Obviously, that one looks very different from (a) and (b).

Example 4.4 Let us now consider the case $p(x) = 1$ for $x \in [0, 1]$ and $p(x) = 0$ otherwise. In other words, the random variables x_i are uniformly (and independently) distributed. From Eq. (4.4), we find that

$$Q_{\max}(x, N) = \begin{cases} x^N, & x \in [0, 1], \\ 1, & x \geq 1. \end{cases} \tag{4.32}$$

The big difference, in this case, compared to the previous examples (the Gumbel and Fréchet cases), is that the distribution of $X_{\max}$ seems to get localized near its maximum allowed value 1 as N increases. In Fig. 4.1(c) we show the experiment for $N = 100$, but please verify this statement by repeating the experiment for increasing values of N. Thus, in this case, we expect the universality to emerge if we zoom in near the upper limit $x = 1$. Indeed, in this case, if we set $x = a_N + b_N z$ with the choice $a_N = 1$ and $b_N = 1/N$, and take the large-N limit, it is easy to see that $Q_{\max}(x = a_N + b_N z, N)$ converges to

$$Q_{\max}(a_N + b_N z, N) \xrightarrow[N \to \infty]{} \begin{cases} e^{-|z|}, & z \leq 0, \\ 1, & z \geq 0. \end{cases} \tag{4.33}$$

Clearly this looks very different from the Gumbel, Eq. (4.25), and Fréchet, Eq. (4.31), cases.

Little exercise 4.7 *Take $p(x)$ such that $p(x) = 0$ for $x \geq a$ and $p(x) \approx B(a - x)^{\gamma - 1}$ as $x \to a$ from below, with $\gamma > 0$. Using Eq. (4.4), and setting*

$$a_N = a, \qquad b_N = (\gamma/NB)^{1/\gamma}, \tag{4.34}$$

show that $Q_{\max}(x = a_N + b_N z, N)$ approaches the scaling form

$$Q_{\max}(x = a_N + b_N z, N) \xrightarrow[N \to \infty]{} F_{\mathrm{III}}(z), \quad \text{with } F_{\mathrm{III}}(z) = \begin{cases} e^{-|z|^{\gamma}}, & z \leq 0, \\ 1, & z > 0. \end{cases} \tag{4.35}$$

This new extreme value distribution is known as the Weibull distribution (Weibull, 1951), which is parameterized by the exponent γ. However, as in the Fréchet case, we again suppress the explicit γ-dependence in $F_{\mathrm{III}}(z)$ in order to keep the notation simple.

Remark 4.2 *For this bounded distribution (which vanishes faster than exponentially for large x), we see that the width b_N of the typical fluctuations of $X_{\max}$ decreases with increasing N. Recall that in the example $p(x) \sim e^{-x^{\delta}}$ with $\delta > 1$ (which also decays faster than exponentially for large x), we also found, in Eq. (4.24), that $b_N \to 0$ as $N \to \infty$. In fact, it turns out that, more generally, for any $p(x)$ that decays faster than exponentially, the width $b_N \to 0$ with increasing N.*

Table 4.1 The three possible large-N limiting scaling forms of the CDF $F_\rho(z)$ for the maximum of N IID random variables: the Gumbel, the Fréchet and the Weibull distributions.

	Gumbel	Fréchet	Weibull
	$\rho = \text{I}$	$\rho = \text{II}$	$\rho = \text{III}$
$F_\rho(z)$	$e^{-e^{-z}}$	$\begin{cases} 0, & z < 0 \\ e^{-z^{-\alpha}}, & z \geq 0 \end{cases}$	$\begin{cases} 1, & z \geq 0 \\ e^{-|z|^\gamma}, & z < 0 \end{cases}$

Three universality classes: Gumbel, Fréchet and Weibull. Have we finally exhausted all the possible extreme value limiting distributions? This time, it seems that we have! More precisely, from the exact finite-N result in Eq. (4.4), Gnedenko (1943) showed that, by appropriately choosing the centering factor a_N and the scale factor b_N, the cumulative distribution approaches a scaling form in the large-N limit:

$$Q_{\max}(x = a_N + b_N z, N) \xrightarrow[N \to \infty]{} F_\rho(z). \tag{4.36}$$

While a_N and b_N are non-universal and depend on the tails of the parent distribution $p(x)$, it turns out that the scaling function $F_\rho(z)$ (with $\rho = \text{I}, \text{II}, \text{III}$) can only be one of the three types depending on the large-x behavior of $p(x)$, namely the Gumbel, Fréchet and Weibull distributions (see Table 4.1). For a generic parent distribution $p(x)$, except for some "pathological" cases, $Q_{\max}(x = a_N + b_N z, N)$ approaches, in the large-N limit, one of these three scaling functions (Gnedenko, 1943). The criteria on $p(x)$ that select which of the three universal scaling functions are the following:

I Gumbel scaling function: $F_\text{I}(z)$ in Eq. (4.25), when $p(x)$ decays, for large x, faster than any power law in an unbounded domain. In the case where the support has a finite upper bound a, but $p(x)$ decays extremely fast (faster than a power law) as $x \to a$, e.g., $p(x) \sim e^{-(a-x)^{-\nu}}$ with $\nu > 0$, one again recovers the Gumbel distribution for $Q_{\max}(x, N)$. For a plot of the PDF $F'_\text{I}(z)$, associated with this centered and scaled extreme distribution, see Fig. 4.2.

II Fréchet scaling function: $F_\text{II}(z)$ in Eq. (4.31), when $p(x) \sim A/x^{1+\alpha}$ decays as a power law for large x in an unbounded domain. For a plot of the PDF $F'_\text{II}(z)$, see Fig. 4.2.

III Weibull scaling function: $F_\text{III}(z)$ in Eq. (4.35), when $p(x)$ has an upper bounded support $x \leq a$, and $p(x) \sim B(a - x)^{\gamma-1}$ with $\gamma > 0$. For a plot of the PDF $F'_\text{III}(z)$, see Fig. 4.2.

Recipes for finding. a_N and b_N

I *The Gumbel class with $p(x)$ having an unbounded support.* In this case, a_N, roughly speaking, is the location of the maximum. To estimate a_N, consider the original parent PDF $p(x)$ of the random variable x. Then, the integral $\int_{a_N}^{\infty} p(x') \, dx'$ (which is just the area under the curve $p(x')$ to the right of a_N)

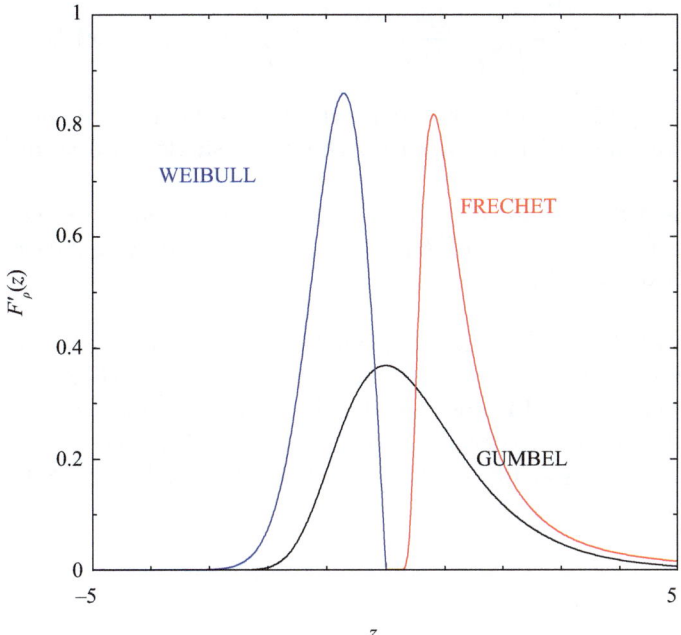

Fig. 4.2 The three limiting scaling functions $F'_\rho(z)$ for the PDF of the maximum of N IID random variables. These are just the derivatives of the functions detailed in Table 4.1.

represents the probability of the event "$x \geq a_N$." Now let us consider N independent variables, each drawn from $p(x)$, and ask what fraction of these N variables have values bigger than a_N. For large N, this fraction is well approximated by $\int_{a_N}^{\infty} p(x')\,\mathrm{d}x'$—this is often referred to as the "self-averaging property." If now a_N happens to be the location of the maximum, we expect that this fraction is just $1/N$, since there is only one variable (out of N) in the range $[a_N, +\infty)$. Hence, we can estimate a_N from the relation

$$\int_{a_N}^{\infty} p(x)\,\mathrm{d}x = \frac{1}{N}. \tag{4.37}$$

From the exact relation in Eq. (4.4), we see that, for large N,

$$Q_{\max}(x, N) = \mathrm{Prob}(X_{\max} \leq x) \approx \exp\left\{ -N \int_{x}^{\infty} p(x')\,\mathrm{d}x' \right\}. \tag{4.38}$$

Setting $x = a_N$ in Eq. (4.38), we see that the condition in Eq. (4.37) corresponds to setting the argument of the exponential to be of order $O(1)$. This value corresponds to the typical value of x at which the sigmoid-shaped $Q_{\max}(x, N)$ crosses over from 0 (for $x \leq a_N$) to 1 (for $x \geq a_N$).

Having determined a_N, we now turn to determining the scaling factor b_N. Since b_N represents the typical scale of fluctuations of $X_{\max}$ around its typical value a_N, one can estimate b_N by the following formula:

$$b_N = \frac{\int_{a_N}^{\infty}(x - a_N)p(x)\,\mathrm{d}x}{\int_{a_N}^{\infty}p(x)\,\mathrm{d}x} = N\int_{a_N}^{\infty}(x - a_N)p(x)\,\mathrm{d}x, \tag{4.39}$$

where we used Eq. (4.37). This can be interpreted as the average fluctuation of $X_{\max}$ around a_N, conditioned on the fact that there is a single variable in $[a_N, +\infty)$.

Little exercise 4.8 *Evaluate a_N and b_N for Examples 4.1 and 4.2 using the formulae for a_N in Eq. (4.37) and b_N in Eq. (4.39). Verify that you recover the previous expressions for a_N and b_N.*

II *The Fréchet class.* In this case, since $p(x)$ decays as a power law, it does not have any natural scale, i.e., $p(x)$ is "scale free." This gets reflected in the extreme distribution as well. In other words, the extreme is also scale free, leading to $a_N = 0$. The scale of typical fluctuations b_N of $X_{\max}$ is thus the same as the typical value of $X_{\max}$ itself. Hence, as in the Gumbel case, we can estimate b_N from the formula

$$\int_{b_N}^{\infty}p(x)\,\mathrm{d}x = \frac{1}{N}. \tag{4.40}$$

Little exercise 4.9 *Check that the formula in Eq. (4.40) gives the correct b_N for Example 4.3.*

III *The Weibull class.* In this case, the support of $p(x)$ has an upper bound at $x = a$. As N becomes large, the typical value of $X_{\max}$ is pushed towards this upper bound a. Hence, we expect that the typical scale of $X_{\max}$ is $a_N = a$. To estimate the scale of the typical fluctuations of $X_{\max}$ around this typical value $a_N = a$, we can use a similar argument to the two previous cases and set

$$\int_{a-b_N}^{a}p(x)\,\mathrm{d}x = \frac{1}{N}. \tag{4.41}$$

Little exercise 4.10 *Check that the formula in Eq. (4.41) gives the correct b_N for Example 4.4.*

So far, we have discussed the limiting distribution of $X_{\max}$ in the limit of large N, and the main conclusion is that the random variable $X_{\max}$ can be expressed as

$$X_{\max} \approx a_N + b_N z, \tag{4.42}$$

where a_N and b_N are non-universal scale factors, while the scaled typical fluctuation z is an N-independent random variable in the large-N limit with a distribution given by $F_{\rho}'(z)$ with $\rho = \mathrm{I}, \mathrm{II}, \mathrm{III}$ depending on the tail of $p(x)$. The physicists prefer to express the relation in Eq. (4.42) in a slightly different language, namely that the PDF of $X_{\max}$ converges, for large N, to the scaling form

$$P_{\max}(x, N) \to \frac{1}{b_N} F'_\rho\left(\frac{x - a_N}{b_N}\right). \tag{4.43}$$

However, this scaling form is valid for $(x - a_N) \sim b_N$, i.e., $z \sim O(1)$, describing the typical fluctuations. One can wonder how atypically large fluctuations of $X_{\max}$ around a_N behave, i.e., when $|x - a_N| \gg b_N$. This is a question of large deviation and it can be analyzed by carefully examining the exact result for $P_{\max}(x, N)$ in Eq. (4.7). We refer the reader to Vivo (2015) for details.

Finite-size effects. We argued, in the previous section, that in the limit of large N, the cumulative distribution $Q_{\max}(x, N)$, appropriately centered and scaled, converges to one of the three extreme value distributions, namely the Gumbel, Fréchet and Weibull distributions. A natural question is: how big does N need to be to see these limiting distributions? Indeed, you can try a numerical experiment yourself and you will find that the convergence is extremely slow. For instance, for the Gaussian $p(x)$ (our Example 4.2), the first correction to the Gumbel result decreases extremely slowly with N, as $\sim 1/\ln N$,

$$Q_{\max}(x = a_N + b_N z, N) \xrightarrow[N \to \infty]{} F_{\mathrm{I}}(z) + \frac{1}{\ln N} F_{\mathrm{S}}(z) + \cdots, \tag{4.44}$$

where $F_{\mathrm{S}}(z)$ is some finite-size scaling function. One can derive this result by expanding the exact finite-N result in Eq. (4.4) for large N keeping $z = (x - a_N)/b_N$ fixed.

Little exercise 4.11 *For the exponential case $p(x) = \mathrm{e}^{-x}$, for $x \geq 0$ (Example 4.1), compute the first correction to the Gumbel law, starting from Eq. (4.9), and show it decays faster than $1/\ln N$ as in the Gaussian case in Eq. (4.44).*

These limiting distributions, as well as the associated finite-size correction results, for the extreme statistics of IID random variables have also been studied in the physics literature. Physicists typically use a technique called the "renormalization group" (RG) method, which was originally introduced to understand the scaling and universality associated with phase transitions and critical phenomena. This method has been adapted to study extreme value distributions (Györgyi et al., 2008, 2010; Bertin and Györgyi, 2010). For IID variables the extreme value distribution is known exactly [as in Eq. (4.4)], so one may ask why the RG machinery is needed. It turns out that the RG method provides a nice physical insight as to why only three limiting distributions emerge for large N (in the RG language, this corresponds to three different "fixed points"). The RG method is also useful to understand the finite-size corrections, as in the case of phase transitions and critical phenomena (Györgyi et al., 2008, 2010; Bertin and Györgyi, 2010; Amir, 2020a,b).

4.1.3 Statistics of the minimum

So far, we have been discussing the distribution of $X_{\max}$ of a set of N IID random variables, each drawn from $p(x)$. What about the statistics of their minimum, $X_{\min} = \min\{x_1, x_2, \ldots, x_N\}$? Let us note a trivial fact: to compute the statistics of the minimum, you may consider the new set of variables $\{-x_1, -x_2, \ldots, -x_N\}$ and consider their maximum. Then

$$X_{\min} = -\max\{-x_1, -x_2, \ldots, -x_N\}. \tag{4.45}$$

Therefore, the statistics of $X_{\min}$ can be easily inferred from the statistics of $X_{\max}$ (which we have already discussed in detail). Indeed, as in the case of the maximum in Eq. (4.4), one can derive an equivalent exact finite-N formula for the cumulative distribution of $X_{\min}$. Due to the identity in Eq. (4.45), the cumulative probability distribution has to be defined in a slightly different manner than for $X_{\max}$. In this case, it is more convenient to define

$$Q_{\min}(x, N) = \text{Prob}(X_{\min} \geq x). \tag{4.46}$$

Once again, using the independence of the x_i, one can write

$$Q_{\min}(x, N) = \text{Prob}(x_1 \geq x, \, x_2 \geq x, \ldots, x_N \geq x) = \left[\int_x^\infty p(x')\, \mathrm{d}x'\right]^N. \tag{4.47}$$

The large-N analysis of Eq. (4.47) can be done in exactly the same way as we analyzed Eq. (4.4) for the maximum. We will not repeat this computation here, but you can verify that the same three universality classes emerge for the minimum as well.

One important application of the statistics of the minimum of a set of N random variables is found in the celebrated random energy model (REM) of Derrida (Derrida, 1981, 1985; Derrida and Gardner, 1986). This is a toy model of disordered systems (Bouchaud and Mézard, 1997). In a disordered system, like a spin glass, the energies associated with different spin configurations are actually random variables as they fluctuate from one realization of disorder to another. In general, these energy levels are correlated and it is rather hard to compute the disorder-averaged free energy of the system. In the REM, one makes the simple assumption that the energy levels of different spin configurations are uncorrelated. Thus, the set of energy levels $\{E_1, E_2, \ldots, E_N\}$ can be considered as IID random variables, each drawn from some underlying distribution $p(E)$. Clearly, the ground-state energy $E_0 = \min\{E_1, E_2, \ldots, E_N\}$ is also a random variable and its distribution plays a paramount role in understanding the behavior of the disordered system at low temperature. Hence, the results on the distribution of the minimum of N IID random variables plays an important role in understanding the physics of disordered systems (Bouchaud and Mézard, 1997; Biroli *et al.*, 2007*a*).

4.1.4 Joint statistics of the maximum and minimum

By now, we know how to compute the distribution of the maximum $X_{\max}$ and, separately, that of the minimum $X_{\min}$ for a set of N IID variables, each drawn from a PDF $p(x)$. But are $X_{\max}$ and $X_{\min}$ correlated or independent? Generically, for finite N, they are indeed correlated. This can be seen by computing the joint cumulative distribution of $X_{\max}$ and $X_{\min}$. As you can guess, the most convenient definition of a joint cumulative distribution, in the present case, is

$$J_N(x, y) = \text{Prob}(X_{\max} \le x, X_{\min} \ge y). \tag{4.48}$$

For any finite N, using the independence of the variables, we see that

$$J_N(x, y) = \text{Prob}(y \le x_1 \le x, y \le x_2 \le x, \ldots, y \le x_N \le x)$$

$$= \left(\int_y^x p(x')\, dx' \right)^N \qquad \text{for } x \ge y. \tag{4.49}$$

Note that, for $x \le y$, $J_N(x, y) = 0$ by definition. Clearly, for any finite N,

$$J_N(x, y) \ne Q_{\max}(x, N) Q_{\min}(y, N), \tag{4.50}$$

where $Q_{\max}(x, N)$ in Eq. (4.4) and $Q_{\min}(y, N)$ in Eq. (4.47) are the separate cumulative distributions of $X_{\max}$ and $X_{\min}$ respectively. Clearly, Eq. (4.50) then shows that $X_{\min}$ and $X_{\max}$ are indeed correlated. However, as N increases, the correlation between $X_{\max}$ and $X_{\min}$ decreases. Let us demonstrate this with the simple example of a positive exponential distribution $p(x) = e^{-x}$ for $x \ge 0$, as in Example 4.1. In this case, Eq. (4.49) reduces to

$$J_N(x, y) = (e^{-y} - e^{-x})^N \quad \text{with } x \ge y. \tag{4.51}$$

In the scaling limit, we set $x = a_N + b_N z$ where $a_N = \ln N$ and $b_N = 1$ (recall Example 4.1 for the maximum). Similarly, for y we set $y = a'_N + b'_N z'$ with $a'_N = 0$ (this corresponds to the Weibull case for the minimum since the support of $p(x)$ has a lower bound at $x = 0$) and $b'_N = 1/N$. Substituting this into Eq. (4.51), and taking the $N \to \infty$ limit, one gets

$$J_N(x = a_N + b_N z, y = a'_N + b'_N z') \xrightarrow[N \to \infty]{} e^{-z'} e^{-e^{-z}} = F_{\text{III}}(-z') F_{\text{I}}(z), \tag{4.52}$$

where $F_{\text{III}}(-z')$ corresponds to the Weibull distribution in Eq. (4.35) with $\gamma = 1$, and $F_{\text{I}}(z)$ is the Gumbel distribution in Eq. (4.25). Thus, we see that, as $N \to \infty$, the joint cumulative distribution of $X_{\max}$ and $X_{\min}$ factorizes and they become statistically independent from each other. Even though we have demonstrated this independence for large N only for a specific example (the exponential case), one can check (please do!) that this is true for other $p(x)$ as well.

As a nice application of this joint distribution of $X_{\max}$ and $X_{\min}$, we consider the observable called the "span,"

$$S = X_{\max} - X_{\min}. \tag{4.53}$$

Clearly, $S \geq 0$ is the range over which the N random variables take their values in a given sample. Hence, S is a random variable as it fluctuates from one sample to another. One example is the temperature range in a given day, where $X_{\max}$ corresponds to the highest temperature, while $X_{\min}$ corresponds to the lowest one.

Another nice observable is the asymmetry between $X_{\min}$ and $X_{\max}$, measured by

$$\tilde{S} = X_{\max} + X_{\min}. \tag{4.54}$$

We will refer to it as the "imbalance" (Biroli *et al.*, 2022). Of course, its mean value is zero for a symmetric $p(x)$, since in that case the marginal distributions of $X_{\max}$ and $-X_{\min}$ are identical. However, this observable has non-zero fluctuations around its mean, and its variance provides a measure of the fluctuations of this asymmetry between $X_{\max}$ and $-X_{\min}$.

To compute the PDF of S and $\tilde{S}$, we need the joint PDF of $X_{\min}$ and $X_{\max}$, i.e.,

$$P_{\mathrm{joint}}(x, y, N) = \mathrm{Prob}[X_{\max} = x, X_{\min} = y, N]. \tag{4.55}$$

Clearly, the cumulative distribution $J_N(x, y)$ defined in Eq. (4.48) can be expressed as

$$J_N(x, y) = \int_{-\infty}^{x} \mathrm{d}x' \int_{y}^{\infty} \mathrm{d}y' \, P_{\mathrm{joint}}(x', y', N). \tag{4.56}$$

Taking derivatives with respect to x and y, we get

$$P_{\mathrm{joint}}(x, y, N) = -\frac{\partial^2 J_N(x, y)}{\partial x \, \partial y}. \tag{4.57}$$

Using the expression for $J_N(x, y)$ in Eq. (4.49), we obtain an exact formula for the joint PDF of $X_{\max}$ and $X_{\min}$,

$$P_{\mathrm{joint}}(x, y, N) = N(N-1)p(x)p(y)\left(\int_{y}^{x} p(x')\,\mathrm{d}x'\right)^{N-2} \quad \text{with } x \geq y, \tag{4.58}$$

and $P_{\mathrm{joint}}(x, y, N) = 0$ if $x \leq y$. This result can be understood very easily. We can pick in $N(N-1)$ ways any two of the N IID random variables with values x and y as the maximum and the minimum respectively—this thus occurs with probability $N(N-1)p(x)p(y)$. Given this, the rest of the $(N-2)$ IID variables have to lie inside the interval $[y, x]$—this occurs with probability $\left(\int_{y}^{x} p(x')\,\mathrm{d}x'\right)^{N-2}$.

Consequently, the PDF of the span $P_{\text{span}}(S, N)$, with $S \geq 0$, is given by

$$
\begin{aligned}
P_{\text{span}}(S, N) &= \int_{-\infty}^{\infty} \mathrm{d}x \int_{-\infty}^{\infty} \mathrm{d}y \, P_{\text{joint}}(x, y, N)\delta(x - y - S) \\
&= \int_{-\infty}^{\infty} \mathrm{d}x \, P_{\text{joint}}(x, x - S, N) \\
&= N(N-1) \int_{-\infty}^{\infty} \mathrm{d}x \, p(x)p(x - S)\left(\int_{x-S}^{x} p(x') \, \mathrm{d}x' \right)^{N-2}.
\end{aligned}
\tag{4.59}
$$

Similarly, the PDF of the imbalance $\tilde{S}$ is given by

$$
\begin{aligned}
P_{\text{imb}}(\tilde{S}, N) &= \int_{-\infty}^{\infty} \mathrm{d}x \int_{-\infty}^{\infty} \mathrm{d}y \, P_{\text{joint}}(x, y, N)\delta(x + y - \tilde{S}) \\
&= \int_{-\infty}^{\infty} \mathrm{d}x \, P_{\text{joint}}(x, x - S, N) \\
&= N(N-1) \int_{-\infty}^{\infty} \mathrm{d}x \, p(x)p(\tilde{S} - x)\left(\int_{\tilde{S}-x}^{x} p(x') \, \mathrm{d}x' \right)^{N-2},
\end{aligned}
\tag{4.60}
$$

which is valid for all $\tilde{S} \in (-\infty, +\infty)$. The subscript "imb" in $P_{\text{imb}}(\tilde{S}, N)$ stands for "imbalance." Equations (4.59) and (4.60) are exact for all N, though they are not particularly illuminating! However, in the limit of large N, things simplify as usual. In that limit, the joint PDF of X_{max} and X_{min}, i.e., $P_{\text{joint}}(x, y, N)$, factorizes, and we can obtain a scaling law for $P_{\text{span}}(S, N)$ and $P_{\text{imp}}(\tilde{S}, N)$ rather simply using the large-N scaling laws for X_{max} and X_{min}. Let us demonstrate this with a simple example.

For example, let us consider the symmetric exponential distribution $p(x) = \frac{1}{2}e^{-|x|}$, which is clearly symmetric around $x = 0$. In this case, one can easily check that, for large N, one needs to scale the maximum and minimum as follows:

$$
X_{\text{max}} = \ln\left(\frac{N}{2}\right) + z, \qquad X_{\text{min}} = -\ln\left(\frac{N}{2}\right) - z',
\tag{4.61}
$$

where z and z' are both of order $O(1)$, and they are each independent random variables distributed via a Gumbel PDF $F_I'(z) = e^{-z - e^{-z}}$ [see Eq. (4.14)]. From Eq. (4.61) we can then compute the limiting distribution of the span S and the imbalance $\tilde{S}$ in the large-N limit as follows.

For the span $S = X_{\text{max}} - X_{\text{min}}$, it is easy to see from Eq. (4.61) that

$$
S = 2\ln\left(\frac{N}{2}\right) + u,
\tag{4.62}
$$

where $u = z + z'$ is of order $O(1)$ and has range $(-\infty, +\infty)$. Since z and z' are independent Gumbel variables, their sum is distributed via the PDF

$$
f_{\text{span}}(u) = \int_{-\infty}^{\infty} \mathrm{d}z \, F_I'(z)F_I'(u - z).
\tag{4.63}
$$

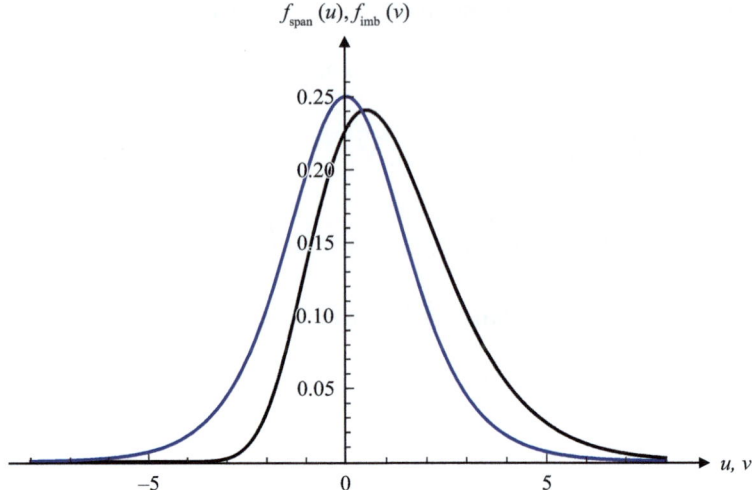

Fig. 4.3 Plot of $f_{\text{span}}(u)$ given in Eq. (4.66) (black line) and $f_{\text{imb}}(v)$ given in Eq. (4.71) (blue line).

Consequently, the distribution of the span S, using Eq. (4.62), approaches, for large N, the scaling form

$$P_{\text{span}}(S, N) \to f_{\text{span}}\left(S - 2\ln\left(\frac{N}{2}\right) = u\right), \tag{4.64}$$

where the scaling function $f_{\text{span}}(u)$ is given in Eq. (4.63). The integral in Eq. (4.63) can be explicitly computed as

$$f_{\text{span}}(u) = e^{-u}\int_0^\infty dz\, \exp\{-e^{-z} - e^{z-u}\} = e^{-u}\int_0^\infty \frac{dy}{y}\, \exp\left\{-\frac{1}{y} - ye^{-u}\right\}, \tag{4.65}$$

where we made a change of variable $y = e^{-z}$ to get the second equality. This last integral can be expressed in terms of the modified Bessel function $K_0(z)$ and one gets a nice explicit expression (Kundu *et al.*, 2013),

$$f_{\text{span}}(u) = 2e^{-u}K_0(2e^{-u/2}), \quad u \in (-\infty, +\infty). \tag{4.66}$$

One can check that $f_{\text{span}}(u)$ is normalized to unity, i.e., $\int_{-\infty}^\infty f_{\text{span}}(u)\, du = 1$. For a plot of this distribution see Fig. 4.3.

Little exercise 4.12 *Show that the mean and the variance of u are given by*

$$\langle u \rangle = 2\gamma_{\text{E}}, \qquad \text{Var}(u) = \langle u^2 \rangle - \langle u \rangle^2 = \frac{\pi^2}{3}. \tag{4.67}$$

Hint: Use the fact that $u = z + z'$, where z and z' are two independent Gumbel random variables, each distributed via the PDF $f_1(z)$ given in Eq. (4.26).

For the imbalance $\tilde{S} = X_{max} + X_{min}$, we find from Eq. (4.61) that in the large-N limit the leading $\ln(N/2)$ terms cancel out, leading to an N-independent variable for the imbalance,

$$\tilde{S} \to v = z - z', \tag{4.68}$$

where z and z' are both independent random variables of order $O(1)$, each distributed via the Gumbel PDF $F'_I(z) = e^{-z-e^{-z}}$. Consequently, the PDF $P_{imb}(\tilde{S}, N)$ becomes independent of N, for large N,

$$P_{imb}(\tilde{S}, N) \to f_{imb}(v), \tag{4.69}$$

where the limiting scaling function $f_{imb}(v)$ can be expressed as

$$f_{imb}(v) = \int_{-\infty}^{\infty} dz \, F'_I(z) F'_I(z - v). \tag{4.70}$$

Substituting $F'_I(z) = e^{-z-e^{-z}}$ and carrying out the integral over z in Eq. (4.70) we get

$$f_{imb}(v) = \frac{e^v}{(1 + e^v)^2}, \quad v \in (-\infty, +\infty). \tag{4.71}$$

One can check that $f_{imb}(v)$ is normalized to unity: $\int_{-\infty}^{\infty} f_{imb}(v) \, dv = 1$. This distribution is known as the "logistic distribution" and is just the derivative of the Fermi function (Balakrishnan, 2013). For a plot of the distribution $f_{imb}(v)$, see Fig. 4.3. From Eq. (4.71), one can compute the moments of v. For example, the first moment $\langle v \rangle = 0$ and the variance is given by

$$\mathrm{Var}(v) = \langle v^2 \rangle - \langle v \rangle^2 = \int_{-\infty}^{\infty} v^2 \frac{e^v}{(1 + e^v)^2} \, dv = \frac{\pi^2}{3}. \tag{4.72}$$

We note that the variance of v coincides with that of u in Eq. (4.67).

Universality. Note that we have so far calculated the distribution of the span and the imbalance in the large-N limit, as given in Eqs. (4.66) and (4.71), for the specific example of the double-exponential distribution. One may naturally wonder how universal they are. Note that the crucial fact we have used is that X_{max} and X_{min} get essentially uncorrelated in the large-N limit. In that limit, we have already shown that the marginal distributions of X_{max} and X_{min} (appropriately centered and scaled) are given by one of the three limiting distributions, namely the Gumbel, Fréchet and Weibull depending on the tail of $p(x)$. Therefore, after appropriate centering and scaling, one can show that these two limiting results in Eqs. (4.66) and (4.71) will hold for any $p(x)$ that is symmetric, unbounded and has a tail that decays faster than any power law (an example being the Gaussian distribution).

In general, given the symmetry and the tails of $p(x)$, one would first compute the marginal distributions of X_{max} and X_{min}, after proper centering and scaling in the large-N limit. From these limiting distributions, one would then calculate the distributions of $S = X_{\max} - X_{\min}$ and $\tilde{S} = X_{\max} + X_{\min}$ by assuming that $X_{\max}$ and $X_{\min}$ are statistically independent for large N. Depending on the limiting distributions of $X_{\max}$ and $X_{\min}$, different limiting scaling forms will emerge for the span S and the imbalance $\tilde{S}$. For example, in the case of the positive exponential $p(x) = e^{-x}$ with $x \geq 0$ (Example 4.1), $X_{\max}$ has a Gumbel scaling form, while $X_{\min}$ has a Weibull scaling form. In this case, the limiting distribution of the span variable, appropriately centered and scaled, will be given by the convolution of the Gumbel and the Weibull PDFs. Similarly, one can calculate the distribution of the imbalance $\tilde{S}$.

Exercise 4.1 Consider N *independent* Brownian particles on a line. They all start initially (at time $t = 0$) at the origin $x = 0$. Each particle subsequently diffuses with time t, independently of the others, with a diffusion constant D.

(i) At time t, write down the expression for the joint PDF $P_{\text{joint}}(x_1, x_2, \ldots, x_N, t)$ of the particle positions.

(ii) What is the probability distribution of the position of the rightmost particle, $X_{\max}(t) = \max\{x_1(t), x_2(t), \ldots, x_N(t)\}$? Likewise, what is the distribution of the position of the leftmost particle, $X_{\min}(t) = \min\{x_1(t), x_2(t), \ldots, x_N(t)\}$?

(iii) Estimate, for large N, the mean distance between the rightmost and leftmost particles. Argue that, for large N, the distance between the rightmost and the leftmost particles becomes deterministic, i.e., the fluctuation around the mean decreases to zero as $N \to \infty$. Estimate roughly how the fluctuation around the mean decreases with N for large N.

(iv) Calculate the joint cumulative probability distribution of the rightmost and leftmost particles, i.e., $Q_{\text{joint}}(x, y, t) = \text{Prob}[X_{\max}(t) \leq x, X_{\min}(t) \geq y]$. From this expression, argue that, for large N, the rightmost and leftmost particles become uncorrelated.

Exercise 4.2 Throw N random points uniformly distributed within a disk of radius R. Compute the distribution of the farthest point from the center of the disk. (The same question in a higher dimension.)

Exercise 4.3 Compute the distribution of the span S and the imbalance $\bar{S}$ for the Fréchet and Weibull classes.

4.2 Statistics of extremes of weakly correlated random variables

Suppose now that the entries $\{x_1, x_2, \ldots, x_N\}$ of our time series are not independent, but correlated such that the connected part of the correlation function decays faster than any power law (e.g., exponentially) over a certain finite correlation length ξ (see Fig. 1.8),

$$C_{i,j} = \langle x_i x_j \rangle - \langle x_i \rangle \langle x_j \rangle \sim e^{-|i-j|/\xi}. \tag{4.73}$$

Clearly, when two entries are separated over a length scale larger than ξ, i.e., when $|i-j| \gg \xi$, then they essentially become uncorrelated. Now weak correlation implies that $\xi \ll N$, where N is the total size of the sample.

For such weakly correlated time series one can construct a heuristic real-space renormalization group argument to study the extreme statistics (Majumdar and Comtet, 2005; Majumdar *et al.*, 2020), as discussed briefly in the introduction and recalled here. Consider $N' = \xi \ll N$ and break the system into identical blocks, each of size ξ (see Fig. 1.8). There are thus N/ξ blocks. While the entries inside each box are still strongly correlated, the entries that belong to different boxes are approximately uncorrelated. So, we can approximate these boxes as non-interacting. Now, for each box i, let y_i denote the "local maximum," i.e., the maximum of all the x variables belonging to the ith block, where $i = 1, 2, \ldots, N' = N/\xi$ (see Fig. 1.8).

By our approximation, the variables y_i are thus essentially *uncorrelated*. Hence, the global maximum $X_{\max}$ can be expressed as

$$X_{\max} = \max[x_1, x_2, \ldots, x_N] = \max[y_1, y_2, \ldots, y_{N'}]. \tag{4.74}$$

Therefore, in principle, if one knows the PDF $p(y)$ of the y_i, then this problem is essentially reduced to calculating the maximum of N' uncorrelated random variables $\{y_1, y_2, \ldots, y_{N'}\}$, the problem that we just discussed in detail in the previous section. So, we know that, depending on the tail of $p(y)$, the limiting distribution of $X_{\max}$ of N weakly correlated variables will, for sure, belong to one of the three (Gumbel, Fréchet or Weibull) limiting extreme distributions of IID random variables. To decide the tail of $p(y)$, of course one needs to solve a *strongly* correlated problem since inside each block the entries are strongly correlated. However, one can often guess the tail of $p(y)$ without really solving for the full PDF of $p(y)$ (or even measure it numerically) and then one knows, from the EVS for IID variables, to which class the distribution of the maximum belongs. As a concrete example of this procedure for weakly correlated variables, we discuss in Chapter 8 the Ornstein–Uhlenbeck stochastic process where one can compute the EVS exactly and demonstrate that indeed this heuristic renormalization group argument works very well. Note also that, in the case of Gaussian random variables x_i with stationary correlations, i.e., $C_{i,j} = c(|i-j|)$ in Eq. (4.73), it is known in the mathematics literature—as Berman's theorem (Berman, 1964)—that if $c(x)$ decays faster than $1/\ln(x)$ for large x, then the distribution of $X_{\max}$, properly centered and scaled, is still given by a Gumbel distribution.

To summarize, the problem of EVS of weakly correlated random variables basically reduces to IID variables with an effective number $N' = N/\xi$ where ξ is the correlation length. So, the real challenge is to compute the EVS of strongly correlated variables where $\xi \geq \mathcal{O}(N)$. An example of such a strongly correlated time series is indeed the random walk, which we discuss now.

4.3 Statistics of extremes for a random walk: Maximum and minimum

We now turn to the random walk sequence $x_n = x_{n-1} + \eta_n$, starting from $x_0 = 0$, where the η_n are the IID jumps, each drawn from a symmetric distribution $f(\eta)$.

Unlike the IID random variables in the previous section, here the positions x_n's (which are the entries of the time series) are strongly correlated, even though the increments η_n's are uncorrelated. As mentioned earlier, this can be seen from the explicit joint distribution of the entries in Eq. (1.21). As before, the two main extreme observables of interest are respectively the maximum and the minimum of the sequence of size N,

$$X_{\max} = \max\{x_0 = 0, x_1, \ldots, x_N\}, \qquad X_{\min} = \min\{x_0 = 0, x_1, \ldots, x_N\}. \qquad (4.75)$$

When the jump distribution $f(\eta)$ is symmetric, one expects that the distributions $P_{\max}(x, N)$ and $P_{\min}(x, N)$ are identical. However, if $f(\eta)$ is asymmetric (for example in the presence of a constant drift), then they will surely differ. However, for simplicity we restrict ourselves here only to the symmetric case and refer the reader to relevant literature when $f(\eta)$ is asymmetric (Mounaix *et al.*, 2018).

When $f(\eta)$ is symmetric, $P_{\max}(x, N)$ and $P_{\min}(x, N)$ defined above are actually the marginal distributions of $X_{\max}$ and $X_{\min}$. However, it turns out that $X_{\max}$ and $X_{\min}$ remain strongly correlated in the random walk problem, even in the limit of large N. This is different from the IID sequence discussed in the previous section where, for large N, the two random variables $X_{\max}$ and $X_{\min}$ do become uncorrelated. Hence, it is important to compute their joint distribution,

$$P_{\text{joint}}(x, y, N) = \text{Prob}[X_{\max} = x, X_{\min} = y, N]. \qquad (4.76)$$

Indeed, for any arbitrary N, one can show that

$$P_{\text{joint}}(x, y, N) \neq P_{\max}(x, N) P_{\max}(y, N), \qquad (4.77)$$

indicating that the maximum and the minimum are strongly correlated. In the following two subsections we discuss in detail how to compute the marginal distributions $P_{\max}(x, N)$ and $P_{\min}(x, N)$ for a discrete-time random walk with arbitrary symmetric jump distribution $f(\eta)$. However, computing the joint distribution $P_{\text{joint}}(x, y, N)$ for arbitrary symmetric $f(\eta)$ turns out to be extremely hard (except for very special jump distributions such as the double-exponential case $f(\eta) = (1/2)e^{-|\eta|}$). Hence, for the joint distribution, we will focus only on the Brownian limit, where it can be computed explicitly, as we will see in Section 4.3.3. Indeed, for any arbitrary symmetric jump distribution with finite variance σ^2, we know, by virtue of the central limit theorem, that the process converges to Brownian motion when N becomes large. Hence, the Brownian result actually describes the asymptotic behavior of the statistics of $X_{\max}$ and $X_{\min}$ for a rather large class of discrete-time random walks with finite σ^2.

4.3.1 Statistics of the maximum

We consider our random walk sequence, $x_n = x_{n-1} + \eta_n$, starting from $x_0 = 0$, where the η_n's are the IID jumps, each drawn from a symmetric distribution $f(\eta)$. Here, we are interested in the distribution of the maximum of this sequence up to N steps,

$$X_{\max} = \max\{x_0 = 0, x_1, \ldots, x_N\}. \tag{4.78}$$

We will denote its cumulative distribution by

$$Q_{\max}(x, N) = \text{Prob}[X_{\max} \leq x, N]. \tag{4.79}$$

Since the walker starts at the origin, $X_{\max}$ is necessarily non-negative. Hence, $Q_{\max}(x, N) = 0$ for $x < 0$. The PDF of the maximum is then given by its derivative,

$$P_{\max}(x, N) = \frac{\mathrm{d}Q_{\max}(x, N)}{\mathrm{d}x}. \tag{4.80}$$

Clearly then the PDF is supported only over $x \geq 0$ and normalized to unity, i.e., $\int_0^\infty P_{\max}(x, N)\,\mathrm{d}x = 1$. Before computing this distribution of $X_{\max}$ explicitly, let us first derive a simple and useful relation between the CDF $Q_{\max}(x, N)$ and the survival probability $Q(x, N)$, where x refers to the starting position of the random walk, that we discussed in Section 3.2.

Relation between the CDF of the maximum and the survival probability. The events that contribute to the CDF $Q_{\max}(x, N)$ correspond to those trajectories of the random walk that start at the origin and stay below the level $x > 0$ up to step N (see Fig. 4.4). If we shift the origin of space to the level x, these trajectories precisely correspond to the event that the random walker *starting* at $-x < 0$ stays below the origin up to step N. But using the symmetry of the random walk, this in turn is exactly equal to the probability that the walker, starting at $x > 0$, stays positive up to step N, which is precisely the survival probability $Q(x, N)$ defined in Section 3.2. Hence, we can simply lift the results on survival probability from Section 3.2 to compute the distribution of the maximum. Note, however, that in Section 3.2 we used the notation $Q(x, n)$ to denote the survival probability up to step n and here we set $n = N$, since we have a sequence of size N in mind.

Let us compute this distribution of $X_{\max}$ explicitly for the three examples of random walks discussed in Section 3.2, namely the continuous-time Brownian limit, the random walk with a double-exponential jump distribution and the lattice random walk.

Brownian limit. In this case, using Eq. (3.24), we get the cumulative distribution of the maximum up to time T as (Redner, 2001; Bray *et al.*, 2013)

$$Q_{\max}(x, T) = \mathrm{erf}\left(\frac{x}{\sqrt{4DT}}\right), \quad x \geq 0. \tag{4.81}$$

Consequently, the PDF of the maximum, using Eq. (4.80), is given by

$$P_{\max}(x, T) = \frac{1}{\sqrt{\pi DT}}\mathrm{e}^{-x^2/4DT}, \quad x \geq 0. \tag{4.82}$$

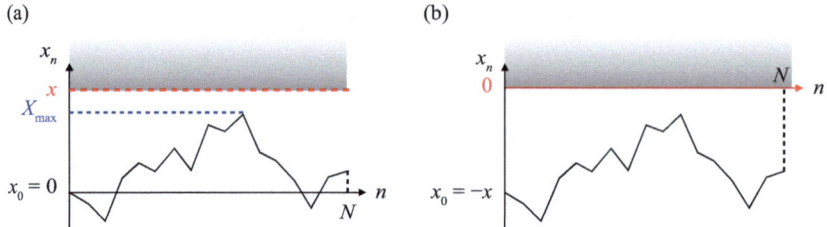

Fig. 4.4 (a) A trajectory of a random walk contributing to the probability $Q_{\max}(x, N) = \text{Prob}(X_{\max} \leq x)$ in Eq. (4.79). The gray shaded area indicates that the walker is constrained to stay below x up to step N. The value of $X_{\max}$ for this trajectory is also marked in blue. (b) By shifting the horizontal axis from 0 to x, we see that this trajectory contributes to the probability that the walker starting at $-x < 0$ stays below the origin. By reversing the vertical axis (assuming that the jump distribution is symmetric), we see that such a trajectory contributes to the survival probability $Q(x, N)$—see Fig. 3.1.

Little exercise 4.13 *Show that $P_{\max}(x, t)$ is normalized to unity when integrated over $[0, +\infty)$. Show also that the kth moment of the maximum is given by*

$$\langle X^k_{\max} \rangle = \frac{1}{\sqrt{\pi}} \Gamma\left(\frac{k+1}{2}\right) (4DT)^{k/2}, \qquad (4.83)$$

where $\Gamma(z) = \int_0^\infty t^{z-1} e^{-t}\, dt$ is the standard gamma function.

Double exponential jump distribution. In this case, we consider a jump distribution of the form $f(\eta) = (1/2\ell_0) e^{-|\eta|/\ell_0}$ for which we obtained an explicit expression for the generating function of the survival probability $\tilde{Q}(x, s) = \sum_{N=1}^\infty Q(x, N) s^N$ in Eq. (3.63). Hence, using the mapping between the CDF of $X_{\max}$ and the survival probability, we therefore have (Comtet and Majumdar, 2005)

$$\tilde{Q}_{\max}(x, s) = \sum_{N=1}^\infty Q_{\max}(x, N) s^N$$

$$= \frac{s}{1-s} - \frac{1 - \sqrt{1-s}}{1-s} \exp\left\{-\frac{\sqrt{1-s}}{\ell_0} x\right\}, \quad x \geq 0. \qquad (4.84)$$

Inverting this generating function explicitly for all N looks tricky (an enthusiastic reader may still give it a try!). It is, however, possible to compute, for example, the first moment $\langle X_{\max} \rangle(N)$ explicitly. We first note that the PDF of $X_{\max}$ is supported over $[0, +\infty)$. This is because the particle starts at the origin. Hence, the maximum is necessarily non-negative. The expected maximum is then given by

$$\langle X_{\max} \rangle(N) = \int_0^\infty x P_{\max}(x, N)\, dx = \int_0^\infty x \frac{d}{dx} Q_{\max}(x, N)\, dx. \qquad (4.85)$$

You would naturally try to do this integral by parts. But then you will run into a problem since the CDF $Q_{\max}(x \to \infty, N) = 1$. In order that the "source term" vanishes in the integration by parts, a trick is to first rewrite

$$\frac{\mathrm{d}}{\mathrm{d}x} Q_{\max}(x, N) = -\frac{\mathrm{d}}{\mathrm{d}x}(1 - Q_{\max}(x, N)), \tag{4.86}$$

and then do the integration by parts. Then, the source terms vanish and one gets

$$\langle X_{\max}\rangle(N) = \int_0^\infty (1 - Q_{\max}(x, N))\, \mathrm{d}x. \tag{4.87}$$

Taking the generating function with respect to N on both sides and using the explicit result for $\tilde{Q}_{\max}(x, s)$ from Eq. (4.84) one gets

$$\sum_{N=1}^\infty \langle X_{\max}\rangle(N) s^N = \ell_0 \left[\frac{1}{(1-s)^{3/2}} - \frac{1}{1-s}\right]. \tag{4.88}$$

Expanding in powers of s and matching both sides, one gets, for all $N \geq 0$,

$$\langle X_{\max}\rangle(N) = \ell_0 \left(\frac{2}{\sqrt{\pi}} \frac{\Gamma(N + \frac{3}{2})}{\Gamma(N+1)} - 1\right). \tag{4.89}$$

In particular, for large N, it behaves as

$$\langle X_{\max}(N)\rangle \sim 2\ell_0 \sqrt{\frac{N}{\pi}}, \quad N \to \infty. \tag{4.90}$$

This is consistent with the Brownian limit in Eq. (4.83) for $k = 1$, once we identify $2DT$ with $N\sigma^2 = 2N\ell_0^2$.

Let us end this discussion by commenting on two interesting features of the tails of the CDF $Q_{\max}(x, N)$, respectively in the $x \to 0$ and $x \to \infty$ limits, which can be derived from Eq. (4.84) (we leave this as an exercise for the reader). First, in the case $x = 0$, the CDF $Q_{\max}(x = 0, N)$ is non-zero and is given for all N by

$$Q_{\max}(x = 0, N) = \frac{1}{2^{2N}} \binom{2N}{N}. \tag{4.91}$$

In particular, in the large-N limit, it decays as $Q_{\max}(x = 0, N) \sim 1/\sqrt{\pi N}$. Note that the event $X_{\max} = 0$ corresponds to the trajectories of the random walk, starting at the origin, that stays on the negative side up to step N. As discussed earlier, this probability in Eq. (4.91) is universal for any N, i.e., independent of the jump distribution, as long as $f(\eta)$ is symmetric and continuous—this is the Sparre Andersen theorem in Eq. (3.101). Note that, in contrast, for the continuous-time Brownian motion, $Q_{\max}(x = 0, T) = 0$ as is easily seen by putting $x = 0$ in Eq. (4.81). This is because if the Brownian motion starts exactly at the origin, it crosses and recrosses the origin infinitely often and therefore the probability weight for the path to stay below the origin for a finite time T vanishes identically.

Second, as discussed in Section 3.2, in the scaling limit where x is large and N is large but keeping $x/\sqrt{N}$ fixed, the CDF $Q_{\max}(x, N)$ approaches the Brownian scaling form, namely [see Eq. (3.76)]

$$Q_{max}(x, N) \to \text{erf}\left(\frac{x}{\sqrt{4\ell_0^2 N}}\right).$$ (4.92)

Lattice random walk. Using the results for the survival probability for the lattice random walk in Eq. (3.93), we then obtain the generating function for the CDF of the maximum,

$$\sum_{N=0}^{\infty} Q_{max}(x, N)s^N = \frac{1 - [w_-(s)]^{x+1}}{1 - s}, \quad \text{where } w_-(s) = \frac{1}{s}(1 - \sqrt{1 - s^2}).$$ (4.93)

Once again, consider first the limit $x = 0$ in this formula. Setting $x = 0$ we get

$$\sum_{N=0}^{\infty} Q_{max}(x = 0, N)s^N = \frac{\sqrt{1 + s} - \sqrt{1 - s}}{s\sqrt{1 - s}}.$$ (4.94)

Note that this is different from the Sparre Andersen result given in Eq. (3.100), which is valid for symmetric and continuous jump distributions. In fact, for large N, analyzing the behavior of the generating function in Eq. (4.94), one gets

$$Q_{max}(x = 0, N) \approx \sqrt{\frac{2}{\pi N}},$$ (4.95)

which differs by a factor of $\sqrt{2}$ from the Sparre Andersen result in Eq. (3.102). In the opposite scaling limit where $x \to \infty$ and $N \to \infty$ keeping $x/\sqrt{N}$ fixed, the CDF of the maximum converges to the Brownian result as discussed in Section 3.2.3.

Little exercise 4.14 *Compute the generating function of the moments of the maximum for a random walk starting at the origin from Eq. (4.93). If you are brave enough, try to invert this generating function for the first few moments and obtain them as a function of N.*

Arbitrary symmetric and continuous jump distribution. So far, we have discussed three specific examples of the jump distributions where we could compute the full distribution of the maximum. What can we say about the statistics of the maximum for a general continuous and symmetric jump distribution $f(\eta)$? A useful starting point is a formula derived by Spitzer (1957) for the joint distribution of the maximum X_{max} and the final position x_N of a random walk of N steps with a general continuous jump distribution $f(\eta)$. If one integrates over the final position, one can obtain the marginal distribution of X_{max}. Unfortunately, this marginal distribution is not explicit. However, one can obtain an explicit expression for the generating function of its Laplace transform (do not get scared yet!). More precisely, it reads (Spitzer, 1957)

$$\sum_{N=0}^{\infty} s^N \langle e^{-pX_{\max}} \rangle = \frac{1}{\sqrt{1-s}} \exp\left[\sum_{n=1}^{\infty} \frac{s^n}{n} \int_0^{\infty} dy\, G(y,0,n) e^{-py} \right], \tag{4.96}$$

where $G(y,0,n)$ is the propagator of the free random walk, i.e., the probability density that the walker arrives at y at step n, starting from 0. We remind the reader [see Eq. (2.40)] that it reads

$$G(y,0,n) = \int_{-\infty}^{\infty} \frac{dk}{2\pi} e^{-iky} [\hat{f}(k)]^n. \tag{4.97}$$

Note that in Eq. (4.96), the notation $\langle e^{-pX_{\max}} \rangle$ stands for

$$\langle e^{-pX_{\max}} \rangle = \int_0^{\infty} P_{\max}(x,N) e^{-px}\, dx, \tag{4.98}$$

where $P_{\max}(x,N)$ is the PDF of the maximum. We do not give the derivation of Eq. (4.96) here. Rather, we take it as the starting point for further analysis. This formula looks a bit formal and complicated. However, it is possible to extract from it at least the expected value of the maximum, as we show now. To do this, we take the derivative of Eq. (4.96) with respect to p and set $p = 0$. This gives

$$\sum_{N=0}^{\infty} s^N \langle X_{\max} \rangle(N) = \frac{1}{2(1-s)} \sum_{n=1}^{\infty} \frac{s^n}{n} \langle |x_n| \rangle, \tag{4.99}$$

where we used the fact that $\int_0^{\infty} G(y,0,n) y\, dy = \langle |x_n| \rangle / 2$, which follows from the symmetry of the jump distribution $f(\eta)$. We see that Eq. (4.99) is in the same spirit as the generalized Sparre Andersen result [see Eqs. (3.109) and (3.110)] in the sense that it relates a global quantity, here the global maximum after N steps, to a local quantity, namely the expected distance (absolute value) of the walker from the origin at a given step n. Matching powers of s, we arrive at the well-known formula of Spitzer (1957),

$$\langle X_{\max} \rangle(N) = \frac{1}{2} \sum_{n=1}^{N} \frac{\langle |x_n| \rangle}{n}, \tag{4.100}$$

where we recall that $\langle |x_n| \rangle$ can be evaluated from the free propagator in Eq. (4.97) as $\langle |x_n| \rangle = \int_{-\infty}^{\infty} dx\, |x| G(x,0,n)$. From Eq. (4.100) one can evaluate, in principle, the expected maximum for any finite N and for any symmetric jump distribution $f(\eta)$.

Unfortunately, this formula is not particularly suitable for extracting the asymptotic behavior of $\langle X_{\max} \rangle(N)$ for large N. To extract the large-N asymptotics, it is more convenient to start from a slightly different representation of the Spitzer formula in Eq. (4.96). We substitute the expression for $G(y,0,n)$ from Eq. (4.97) on the right-hand side of Eq. (4.96), carry out first the y-integral and then perform the sum over n (assuming that they commute). This gives (check this!)

$$\sum_{N=0}^{\infty} s^N \int_0^{\infty} P_{\max}(x, N) e^{-px} \, dx = \frac{1}{\sqrt{1-s}} \varphi(s, p), \qquad (4.101)$$

where

$$\varphi(s, p) = \exp\left[-\frac{p}{\pi} \int_0^{\infty} \frac{\ln(1 - s\hat{f}(k))}{p^2 + k^2} \, dk \right] \qquad (4.102)$$

with $\hat{f}(k) = \int_{-\infty}^{\infty} f(\eta) e^{ik\eta} \, d\eta$ being the Fourier transform of the jump length distribution.

This alternative representation of Spitzer's formula in Eqs. (4.101) and (4.102) can also be derived starting from the Pollaczek–Spitzer formula for the survival probability $Q(x, N)$ as described in Eq. (3.95). To see this, we recall the Pollaczek–Spitzer formula:

$$\int_0^{\infty} \left[\sum_{N=0}^{\infty} Q(x, N) \, s^N \right] e^{-px} \, dx = \frac{1}{p\sqrt{1-s}} \varphi(s, p). \qquad (4.103)$$

We recall that $Q(x, N)$ is the probability that the walker, starting at $x \geq 0$, does not cross the origin up to step N. But as explained earlier, this is also the cumulative distribution of the maximum $X_{\max}$ of the process starting at 0. In other words,

$$Q(x, N) = Q_{\max}(x, N) = \text{Prob}[X_{\max} \leq x, N \mid x_0 = 0], \quad x \geq 0. \qquad (4.104)$$

Note that $Q(x, N) = Q_{\max}(x, N) = 0$ for $x < 0$. This is because if the walker starts at the origin, the maximum is necessarily non-negative. However, $Q(0, N) = Q_{\max}(0, N)$ is finite since this measures the probability that the walker, starting at the origin, stays on the negative side of the origin up to step N. In fact, this is exactly the Sparre Andersen result,

$$Q(0, N) = Q_{\max}(0, N) = 2^{-2N} \binom{2N}{N}. \qquad (4.105)$$

To compute the PDF of the maximum $X_{\max}$, we need to take the derivative of $Q_{\max}(x, N)$ with respect to x. This is a bit subtle since $Q_{\max}(x, N)$ has a theta function $\Theta(x)$ in it implicitly. Let us evaluate the LHS of Eq. (4.103) by parts. This gives

$$\sum_{N=0}^{\infty} s^N \left[\frac{1}{p} Q_{\max}(0, N) + \frac{1}{p} \int_0^{\infty} \frac{\partial Q_{\max}(x, N)}{\partial x} e^{-px} \, dx \right] = \frac{1}{p\sqrt{1-s}} \varphi(s, p). \qquad (4.106)$$

Next, we identity the PDF of $X_{\max}$ as consisting of two parts,

$$P_{\max}(x, N) = \frac{\partial Q_{\max}(x, N)}{\partial x} \Theta(x) + Q_{\max}(0, N)\delta(x). \qquad (4.107)$$

Using this in Eq. (4.106) we recover Spitzer's alternative formula, Eq. (4.101). Finally, taking the derivative of Eq. (4.101) with respect to p and setting $p = 0$, we obtain

$$\sum_{N=0}^{\infty} s^N \langle X_{\max}\rangle(N) = -\frac{1}{\sqrt{1-s}}\frac{\partial\varphi(s,p)}{\partial p}\bigg|_{p=0}, \tag{4.108}$$

where $\varphi(s,p)$ is given in Eq. (4.102). It turns out that, to extract the large-N behavior of $\langle X_{\max}\rangle(N)$, the representation in Eq. (4.108) is more convenient.

Asymptotics of $\langle X_{\max}\rangle(N)$ for large N. Obtaining the large-N asymptotics of $\langle X_{\max}\rangle(N)$ from Eq. (4.108) is still not simple and requires some manipulations that we will not detail here—we refer the interested reader to Comtet and Majumdar (2005). Here we just highlight the main results. First, consider symmetric jump distributions $f(\eta)$ with finite variance $\sigma^2 = \int_{-\infty}^{\infty}\eta^2 f(\eta)\,\mathrm{d}\eta$. In this case, the expected maximum, for large N, behaves as

$$\langle X_{\max}\rangle(N) = \sigma\sqrt{\frac{2N}{\pi}} - c + O\left(\frac{1}{\sqrt{N}}\right). \tag{4.109}$$

The leading term reproduces the Brownian result $\langle X_{\max}\rangle(T) = \sqrt{4DT}/\pi$ [see Eq. (4.83) with $k = 1$ in the limit $N \to \infty$, $\sigma \to 0$ but keeping the product $N\sigma^2 = 2DT$ fixed]. While the leading term is relatively easy to extract from Eq. (4.108), the subleading constant term is more difficult to compute. However, it turns out to have an explicit expression (Comtet and Majumdar, 2005):

$$c = -\frac{1}{\pi}\int_0^{\infty}\frac{\mathrm{d}q}{q^2}\ln\left(\frac{1-\hat{f}(q)}{\sigma^2 q^2/2}\right). \tag{4.110}$$

One can show that the constant $c > 0$ for any $f(\eta)$ with a finite σ. For example, for the Gaussian distribution with $\hat{f}(k) = \mathrm{e}^{-\sigma^2 k^2/2}$, the constant c reads (Comtet and Majumdar, 2005)

$$c = -\frac{\zeta(1/2)}{\sqrt{2\pi}}\sigma = 0.582\,59\ldots\sigma. \tag{4.111}$$

This constant c in Eq. (4.109) for a general $f(\eta)$ has an interesting history. In some problems, such as in the Smoluchowski flux problem, this constant is related to the Milne extrapolation length, while in the radiative transfer of photons, this constant is sometimes known as the Hopf constant. It has appeared in several problems in different disguises. For example, it appeared in the rectangle-packing algorithm in computer science (Coffman et al., 1998), in the flux of independent random walkers to a spherical trap in three dimensions (Ziff, 1991; Majumdar *et al.*, 2006; Ziff *et al.*, 2007, 2009) and a point trap in one dimension (Franke and Majumdar, 2012), in the survival probability of a random walk in one dimension (Majumdar *et al.*, 2017) and in the computation of the mean perimeter of the convex hull of a two-dimensional random walk (Grebenkov et al., 2017). Interestingly, it turns out that an accurate estimate of this constant to many decimal places was needed to compute the critical mass of uranium-232 in the Manhattan project.[1] For a brief account of this history, see Majumdar *et al.* (2017).

When σ^2 is not finite, things are a bit different. For example, consider Lévy flights where the jump distribution has a fat tail $f(\eta) \propto |\eta|^{-1-\mu}$ as $\eta \to \pm\infty$ where $0 < \mu < 2$. This means that the Fourier transform of the jump distribution has the small-k behavior $\hat{f}(k) \simeq 1 - |\ell k|^\mu$ as $k \to 0$, where ℓ is the typical scale of the jump. In this case, the expected maximum turns out to be infinite for $0 < \mu \le 1$, while for $1 < \mu < 2$ it behaves, to leading order for large N, as (Comtet and Majumdar, 2005)

$$\langle X_{\text{max}} \rangle (N) = \frac{\ell\mu}{\pi}\Gamma\left(1 - \frac{1}{\mu}\right)N^{1/\mu} + o(N^{1/\mu}). \tag{4.112}$$

The subleading correction term can also be computed and has different behavior depending on the details of the small-k behavior of $\hat{f}(k)$ (Grebenkov et al., 2017).

As the reader can see, extracting the asymptotic behavior of just the expected maximum is already quite hard and technical. Naturally, extracting the asymptotics of the higher moments is even more difficult and challenging. For jump distributions with a finite variance σ^2, the leading asymptotic behavior for large N is not hard to guess, as it must converge to the Brownian answer in Eq. (4.83). In that formula, if one replaces $2DT$ by $N\sigma^2$, that would give us the leading behavior of the moments for the discrete-time random walker, i.e.

$$\langle X_{\text{max}}^k \rangle \simeq \frac{\sigma^k}{\sqrt{\pi}}\Gamma\left(\frac{k+1}{2}\right)(2N)^{k/2}. \tag{4.113}$$

However, estimating the subleading terms for higher moments is still a challenging open problem. For Lévy flights with $0 < \mu < 2$, this question does not arise since the higher moments are infinite.

Going beyond the moments, one can ask how the full cumulative distribution $Q_{\text{max}}(x, N)$ behaves as a function of x for large N. Let us recall that this is exactly the survival probability of a random walker starting at $x \ge 0$ up to N steps, i.e., the probability that the walker stays non-negative up to step N. In this latter context, as discussed in Section 3.2.5, this quantity has been studied as a function of x for fixed but large N. For instance, for $x = 0$, one obtains the universal Sparre Andersen result $Q_{\text{max}}(0, N) = 2^{-2N}\binom{2N}{N}$. On the other hand, when $x = O(N^{1/2})$ (for a jump distribution with finite σ) one would expect to recover the Brownian result in Eq. (4.81) with $2DT$ replaced by $N\sigma^2$, namely $Q_{\text{max}}(x, N) \simeq \text{erf}(x/\sqrt{2\sigma^2 N})$. How does $Q_{\text{max}}(x, N)$ interpolate between the Sparre Andersen limit (for $x \to 0$) and the Brownian limit (when $x = O(\sqrt{N})$)? This question was studied in detail in Majumdar *et al.* (2017), which demonstrated explicitly how this interpolation works. The same story also happens for Lévy flights, and we refer the interested reader to Majumdar *et al.* (2017) for details.

---

[1]We are grateful to V. V. Ivanov for sharing with us his private notes on the Hopf constant and its interesting history.

4.3.2 Statistics of the minimum

We consider again the random walk sequence $x_n = x_{n-1} + \eta_n$, starting from $x_0 = 0$, where the η_n are the IID jumps, each drawn from a symmetric distribution $f(\eta)$. Here, we are interested in the distribution of the minimum of this sequence up to N steps,

$$X_{\min} = \min\{x_0 = 0, x_1, \ldots, x_N\}. \tag{4.114}$$

This section is obviously going to be very short. This is because, for symmetric jump distributions $f(\eta)$, the statistics of $X_{\max}$ and $-X_{\min}$ are identical. All the results that were derived for $X_{\max}$ in the previous section can then be straightforwardly translated to the corresponding results for $X_{\min}$.

4.3.3 Joint statistics of the maximum and the minimum

In the previous two subsections we have discussed separately the statistics of $X_{\max}$ and $X_{\min}$. As discussed in the introduction of this section, in general these two random variables are correlated. Hence, one would be interested in computing the joint statistics of $X_{\max}$ and $X_{\min}$ for a general random walk sequence. However, even for just $X_{\max}$ (or alternatively for just $X_{\min}$), we have seen that the computation of their marginal distributions is already very hard for a general random sequence. Computing their joint distribution for a general random sequence is even harder and there are hardly any known results—except for the double-exponential jump distribution and the lattice random walk (Mori *et al.*, 2019, 2020). Instead, we focus here on a simpler case, namely the continuous-time limit of random walks (i.e., the Brownian limit) where one can calculate the joint distribution explicitly (Weiss and Rubin, 1983; Hughes, 1995; Kundu *et al.*, 2013; Biroli *et al.*, 2022) and extract from it the distribution of the span and the imbalance (as was done for the IID sequence).

The Brownian limit. We thus consider a one-dimensional Brownian motion whose position $x(t)$ evolves as

$$\frac{\mathrm{d}x(t)}{\mathrm{d}t} = \sqrt{2D}\eta(t), \tag{4.115}$$

where $\eta(t)$ is Gaussian white noise with zero mean and correlator $\langle \eta(t)\eta(t')\rangle = \delta(t-t')$. Here, D is the diffusion constant and we assume that the particle starts at the origin $x(0) = 0$. We consider the trajectory of the Brownian motion up to time T. Let $X_{\max}(T)$ and $X_{\min}(T)$ denote the global maximum and the global minimum of the trajectory up to time T (see Fig. 4.5). These are random variables, as they vary from one trajectory of the Brownian motion to another. We are interested in the joint probability distribution,

$$P_{\text{joint}}(\ell_1, \ell_2, T) = \text{Prob}[X_{\max}(T) = \ell_1, X_{\min}(T) = -\ell_2]. \tag{4.116}$$

Note that, compared to the definition in Eq. (4.76), here we set $x = \ell_1$ and $y = -\ell_2$ (where $\ell_2 > 0$) for convenience. It turns out to be more convenient to first compute the joint cumulative distribution,

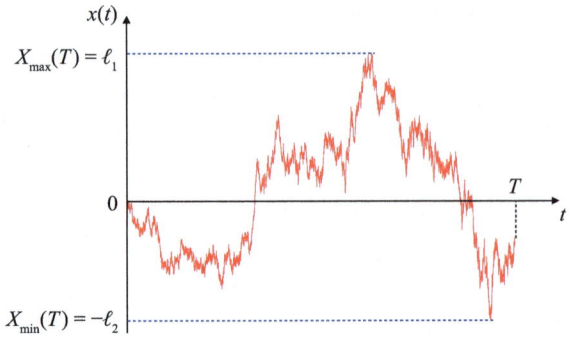

Fig. 4.5 Trajectory of a Brownian motion on the time interval $[0, T]$ for which the maximum $X_{\max}(T) = \ell_1$ and $X_{\min}(T) = -\ell_2$.

$$J(\ell_1, \ell_2, T) = \text{Prob}[X_{\max}(t) \leq \ell_1, X_{\min}(T) \geq -\ell_2]. \qquad (4.117)$$

These two quantities $P_{\text{joint}}(\ell_1, \ell_2, T)$ and $J(\ell_1, \ell_2, T)$ are simply related by the fact that

$$J(\ell_1, \ell_2, T) = \int_0^{\ell_1} d\ell_1' \int_0^{\ell_2} d\ell_2' \, P_{\text{joint}}(\ell_1', \ell_2', T). \qquad (4.118)$$

Note that the limits in these integrals follow from the fact that, since the particle starts at the origin, $X_{\max}(T) \geq 0$ and $X_{\min}(T) \leq 0$. By taking successive derivatives with respect to ℓ_1 and ℓ_2, one then gets

$$P_{\text{joint}}(\ell_1, \ell_2, T) = \frac{\partial^2 J(\ell_1, \ell_2, T)}{\partial \ell_1 \partial \ell_2}. \qquad (4.119)$$

As we said before, we will first compute $J(\ell_1, \ell_2, T)$ and then use the relation in Eq. (4.119) to compute the probability distribution $P_{\text{joint}}(\ell_1, \ell_2, T)$.

To compute $J(\ell_1, \ell_2, T)$, we note that this is just the probability that the walker, starting at the origin, does not leave the interval $[-\ell_2, \ell_1]$ up to time T (see Fig. 4.6). This is therefore just the survival probability inside this interval up to time T and it can be computed by using the backward Fokker–Planck approach introduced in Chapter 3. To proceed, we first define $Q(x_0, T \mid \ell_1, \ell_2)$ as Prob[the particle, starting at x_0, stays within the interval $[-\ell_2, \ell_1]$ up to time T]. In the backward Fokker–Planck approach, the initial position x_0 is used as a variable. After solving for $Q(x_0, T \mid \ell_1, \ell_2)$ for general x_0, we will set $x_0 = 0$ to get

$$J(\ell_1, \ell_2, T) = Q(x_0 = 0, T \mid \ell_1, \ell_2). \qquad (4.120)$$

To simplify the notation, we will omit the explicit dependence on ℓ_1 and ℓ_2 of the survival probability and we will simply denote it by $Q(x_0, T)$. The backward Fokker–Planck equation for $Q(x_0, T)$ is simply the diffusion equation,

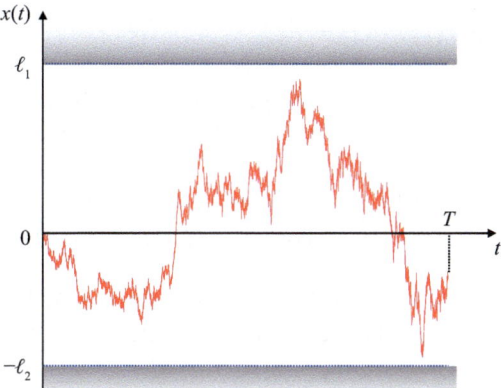

Fig. 4.6 Trajectory of a Brownian motion on the time interval $[0, T]$, starting at $x = 0$ and staying inside the box $[-\ell_2, +\ell_1]$.

$$\frac{\partial Q(x_0, T)}{\partial T} = D\frac{\partial^2 Q(x_0, T)}{\partial x_0^2} \quad \text{with } x_0 \in [-\ell_2, \ell_1]. \tag{4.121}$$

In addition, we have the absorbing boundary conditions at $x_0 = -\ell_2$ and $x_0 = \ell_1$, namely

$$Q(x_0 = -\ell_2, T) = 0, \qquad Q(x_0 = \ell_1, T) = 0, \tag{4.122}$$

together with the initial condition

$$Q(x_0, T = 0) = 1 \quad \text{for all} \ -\ell_2 < x_0 < \ell_1. \tag{4.123}$$

There are different ways to solve this partial differential equation. The simplest is to work in the Laplace space with respect to the time variable. We define the Laplace transform

$$\tilde{Q}(x_0, s) = \int_0^\infty e^{-sT} Q(x_0, T) \, dT. \tag{4.124}$$

Taking the Laplace transform of Eq. (4.121), we get an ordinary inhomogeneous differential equation,

$$-1 + s\tilde{Q}(x_0, s) = D\frac{\partial^2 \tilde{Q}(x_0, s)}{\partial x_0^2}, \tag{4.125}$$

where we have used $Q(x_0, T = 0) = 1$ for $-\ell_2 < x_0 < \ell_1$. The boundary conditions in Eq. (4.122) translate into

$$\tilde{Q}(x_0 = -\ell_2, s) = 0, \qquad \tilde{Q}(x_0 = \ell_1, s) = 0. \tag{4.126}$$

By performing the shift $\tilde{Q}(x_0, s) = 1/s + \tilde{r}(x_0, s)$, one finds that $\tilde{r}(x_0, s)$ satisfies a homogeneous differential equation which is simple to solve. Hence, the most general solution of Eq. (4.125) can be expressed as

$$\tilde{Q}(x_0, s) = \frac{1}{s} + A e^{\sqrt{s/D}x_0} + B e^{-\sqrt{s/D}x_0}, \tag{4.127}$$

where the two constants A and B are determined from the two boundary conditions in Eq. (4.126). After some straightforward algebra, one gets

$$\tilde{Q}(x_0, s) = \frac{1}{s}\left[1 - \frac{\cosh(\sqrt{s/D}(x_0 - (\ell_1 - \ell_2)/2))}{\cosh(\sqrt{s/D}((\ell_1 + \ell_2)/2))}\right]. \tag{4.128}$$

Putting $x_0 = 0$ into Eq. (4.128) and using the relation in Eq. (4.120) we get the Laplace transform of the cumulative joint distribution:

$$\tilde{J}(\ell_1, \ell_2, s) \equiv \int_0^\infty e^{-sT} J(\ell_1, \ell_2, T)\, dT = \frac{1}{s}\left[1 - \frac{\cosh(\sqrt{s/D}((\ell_1 - \ell_2)/2))}{\cosh(\sqrt{s/D}((\ell_1 + \ell_2)/2))}\right]. \tag{4.129}$$

Taking the Laplace transform with respect to T of the relation in Eq. (4.119) and taking two successive derivatives of Eq. (4.129) with respect to ℓ_1 and ℓ_2, one gets (try it!)

$$\tilde{P}_{\text{joint}}(\ell_1, \ell_2, s) \equiv \int_0^\infty e^{-sT} P_{\text{joint}}(\ell_1, \ell_2, T)\, dT$$

$$= \frac{1}{2D}\cosh\left(\sqrt{\frac{s}{4D}}(\ell_1 - \ell_2)\right)\operatorname{sech}^3\left(\sqrt{\frac{s}{4D}}(\ell_1 + \ell_2)\right). \tag{4.130}$$

By integrating Eq. (4.130) over $\ell_1 \in [0, +\infty)$ and $\ell_2 \in [0, +\infty)$ one gets $1/s$, which checks that $P_{\text{joint}}(\ell_1, \ell_2, T)$ is normalized to unity. From this joint distribution, one can easily calculate the marginal distributions $P_{\max}(\ell_1, T)$ and $P_{\min}(\ell_2, T)$. For example, by integrating Eq. (4.130) over $\ell_2 \in [0, +\infty)$ one finds

$$\int_0^\infty e^{-sT} P_{\max}(\ell_1, T)\, dT = \frac{1}{\sqrt{sD}} e^{-\sqrt{s/D}\,\ell_1}. \tag{4.131}$$

By inverting the Laplace transform, we find that

$$P_{\max}(\ell_1, T) = \frac{1}{\sqrt{\pi DT}} e^{-\ell_1^2/4DT}, \quad \ell_1 \geq 0. \tag{4.132}$$

By symmetry, $X_{\max}(T)$ and $-X_{\min}(T)$ have the same statistical laws. Indeed, by integrating Eq. (4.130) over $\ell_1 \in [0, +\infty)$, we can verify that $P_{\min}(\ell_2, T) = P_{\max}(\ell_2, T)$, with the latter given in Eq. (4.132). From Eq. (4.132), one can compute the moments of $X_{\max}(T)$, as given in Eq. (4.83). We invite the reader to verify this.

In principle, one can invert the Laplace transform in Eq. (4.130) using the Bromwich formula in the complex s-plane, but it is not very illuminating. However, one can calculate moments of $X_{\max}(T)$ and $X_{\min}(T)$ easily from Eq. (4.132). For example, the covariance between $X_{\max}(T)$ and $X_{\min}(T)$ (recall that $X_{\max}(T) \geq 0$

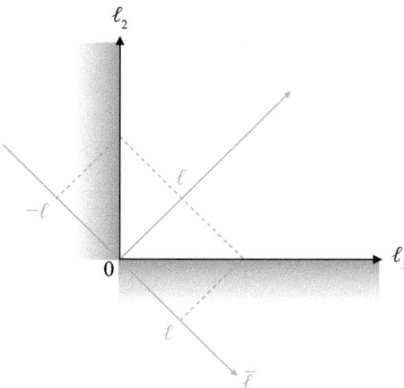

Fig. 4.7 The change of variables $(\ell_1, \ell_2) \to (\bar{\ell} = \ell_1 - \ell_2, \ell = \ell_1 + \ell_2)$ which is useful for computing the distribution of the span in Eq. (4.135). Since both ℓ_1 and ℓ_2 are constrained to be positive, $\ell_1, \ell_2 \geq 0$, this implies that $-\ell \leq \bar{\ell} \leq \ell$, while clearly $\ell \geq 0$.

and $X_{\min}(T) \leq 0$) can be computed from Eq. (4.130). We get (the reader should check this out!)

$$\langle X_{\max}(T) X_{\min}(T) \rangle = -\langle \ell_1 \ell_2 \rangle = -2D(\ln 4 - 1)T. \tag{4.133}$$

As we will see soon, the result in Eq. (4.130) will be very useful for computing the distribution of other observables, such as the span, $S(T) = X_{\max}(T) - X_{\min}(T)$, or the imbalance, $\tilde{S}(T) = X_{\max}(T) + X_{\min}(T)$, up to T.

Distribution of the span. One useful application of the joint distribution in Eq. (4.130) is in deriving the distribution of the span $S(T) = X_{\max}(T) - X_{\min}(T)$ up to the final time T. This means that we want to compute the distribution of $\ell_1 + \ell_2$ from the joint distribution in Eq. (4.130). In other words, in the Laplace space, we have

$$\tilde{P}_{\mathrm{span}}(S, s) = \int_0^\infty P_{\mathrm{span}}(S, T) e^{-sT} \, dT = \int_0^\infty d\ell_1 \int_0^\infty d\ell_2 \, \tilde{P}(\ell_1, \ell_2, s) \delta(\ell_1 + \ell_2 - S), \tag{4.134}$$

where $\tilde{P}(\ell_1, \ell_2, s)$ is given in Eq. (4.130). To carry out this integral, it is convenient to make the change of variables $\ell = \ell_1 + \ell_2$ and $\bar{\ell} = \ell_1 - \ell_2$. Note that $d\ell d\bar{\ell} = 2d\ell_1 d\ell_2$. Hence, the double integral in Eq. (4.134) becomes

$$\tilde{P}_{\mathrm{span}}(S, s) = \frac{1}{4D} \int_0^\infty d\ell \, \delta(\ell - S) \mathrm{sech}^3\left(\sqrt{\frac{s}{4D}} \ell\right) \int_{-\ell}^{\ell} \cosh\left(\sqrt{\frac{s}{4D}} \bar{\ell}\right) d\bar{\ell}. \tag{4.135}$$

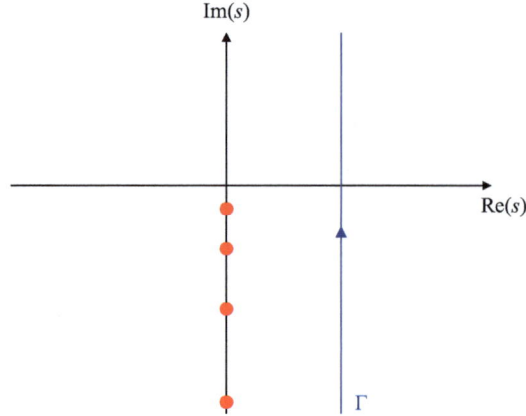

Fig. 4.8 The Bromwich contour Γ used to compute the inverse Laplace transform in Eq. (4.137). The red dots indicate the first poles s_m in Eq. (4.139) for $m = 0, 1, 2, 3$.

Note that for fixed ℓ in the (ℓ_1, ℓ_2)-plane, the variable $\bar{\ell}$ can vary only in the range $[-\ell, +\ell]$ (see Fig. 4.7). Performing this integral explicitly, we get (Biroli *et al.*, 2022)

$$\tilde{P}_{\text{span}}(S, s) = \frac{1}{\sqrt{sD}} \tanh\left(\sqrt{\frac{s}{4D}} S\right) \operatorname{sech}^2\left(\sqrt{\frac{s}{4D}} S\right). \tag{4.136}$$

Check that, once integrated over S, one gets $\int_0^\infty \tilde{P}_{\text{span}}(S, s)\, \mathrm{d}S = 1/s$, which confirms that the span distribution $P_{\text{span}}(S, T)$ is correctly normalized to unity. By formally inverting the Laplace transform, we get

$$P_{\text{span}}(S, T) = \int_\Gamma \frac{\mathrm{d}s}{2\pi \mathrm{i}} e^{sT} \frac{1}{\sqrt{sD}} \tanh\left(\sqrt{\frac{s}{4D}} S\right) \operatorname{sech}^2\left(\sqrt{\frac{s}{4D}} S\right), \tag{4.137}$$

where Γ is the vertical Bromwich contour in the complex s-plane such that all the singularities of the integrand are located on the left of Γ (see Fig. 4.8). We first note that there are no branch cuts at $s = 0$, although it may appear so naively from the appearance of $\sqrt{s}$ in the integrand in Eq. (4.137). One can easily check that the only singularities are the poles, which are located on the negative real axis in the complex s-plane. Indeed, the poles occur when $\cosh(a\sqrt{s}) = 0$, with $a = S/\sqrt{4D}$. This means that

$$e^{a\sqrt{s}} + e^{-a\sqrt{s}} = 0 \implies e^{2a\sqrt{s}} = -1 = e^{\mathrm{i}(2m+1)\pi}, \quad \text{with } m = 0, 1, 2, \ldots \tag{4.138}$$

Therefore, the poles are located at (see Fig. 4.8)

$$s_m = -\frac{\pi^2}{4a^2}(2m+1)^2, \quad \text{with } m = 0, 1, 2, \ldots \tag{4.139}$$

Evaluating the residues using Cauchy's formula, we compute the inverse Laplace transform explicitly (the residue calculation is a bit laborious!) and find that it can be expressed in a nice scaling form,

$$P_{\text{span}}(S, T) = \frac{1}{\sqrt{4DT}} f_{\text{span}}\left(\frac{S}{\sqrt{4DT}}\right), \tag{4.140}$$

where the scaling function $f_{\text{span}}(z)$ is supported over $z \geq 0$ and is given by (Kundu *et al.*, 2013; Weiss and Rubin, 1983; Hughes, 1995; Biroli *et al.*, 2022)

$$\begin{aligned} f_{\text{span}}(z) &= 2\sum_{m=0}^{\infty} \left(\frac{(2m+1)^2\pi^2}{z^5} - \frac{2}{z^3}\right) \exp\left\{-\frac{(2m+1)^2\pi^2}{4z^2}\right\} \\ &= \sum_{m=-\infty}^{\infty} \left(\frac{(2m+1)^2\pi^2}{z^5} - \frac{2}{z^3}\right) \exp\left\{-\frac{(2m+1)^2\pi^2}{4z^2}\right\}, \end{aligned} \tag{4.141}$$

where, in the second line, we have used the symmetry of the summand as a function of m. From this expression, one can easily read off the asymptotic behavior of $f_{\text{span}}(z)$ for small z. In fact, the dominant contribution comes from the $m = 0$ term and it gives $f_{\text{span}}(z) \simeq 2\pi^2 z^{-5}e^{-\pi^2/(4z^2)}$ as $z \to 0$. Thus, the scaling function vanishes extremely rapidly (with an essential singularity) as $z \to 0$. What happens for large z? To extract the large-z asymptotic behavior from the expression in Eq. (4.141) is rather tricky because the sum diverges if one takes the limit $z \to \infty$ inside the summation. One then needs a different representation of the same function $f_{\text{span}}(z)$ that is more suited to taking the large-z limit. Indeed, this is achieved by the famous Poisson summation formula which says that, for any function $g(x)$,

$$\sum_{n=-\infty}^{\infty} g(n) = \sum_{k=-\infty}^{\infty} \hat{g}(k), \quad \text{where } \hat{g}(k) = \int_{-\infty}^{\infty} g(x)e^{-\text{i}2\pi kx}\, \text{d}x. \tag{4.142}$$

Applying the Poisson summation formula to the sum in Eq. (4.141), one can show (we leave it as an exercise) that

$$f_{\text{span}}(z) = \frac{8}{\sqrt{\pi}} \sum_{m=1}^{\infty} (-1)^{m+1} m^2 e^{-m^2 z^2}. \tag{4.143}$$

Now, the asymptotic behavior for large z is easily read off from this expression since it is given by the $m = 1$ term in Eq. (4.143). This leads to a Gaussian decay, $f_{\text{span}}(z) \simeq (8/\sqrt{\pi})e^{-z^2}$, for large z. Hence, summarizing, the asymptotic behaviors of $f_{\text{span}}(z)$ in the two limits are given by

$$f_{\text{span}}(z) \simeq \begin{cases} \dfrac{2\pi^2}{z^5} e^{-\pi^2/4z^2}, & z \to 0, \\[3mm] \dfrac{8}{\sqrt{\pi}} e^{-z^2}, & z \to \infty. \end{cases} \tag{4.144}$$

For a plot of this function, see Fig. 4.9.

Distribution of the imbalance. For the Brownian motion, we start from the Laplace transform (with respect to time T) of the joint distribution of ℓ_1 and ℓ_2 in

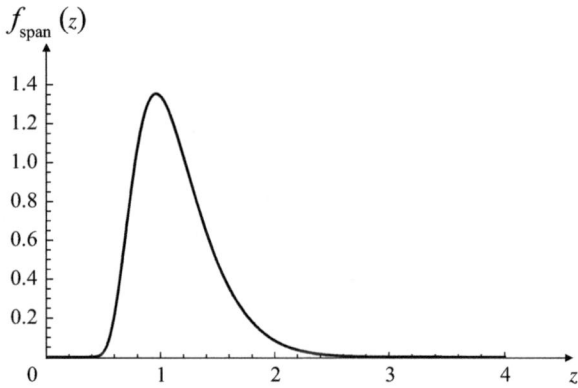

Fig. 4.9 Plot of the scaling function $f_{\text{span}}(z)$ associated with the distribution of the span of the Brownian motion and given in Eq. (4.143) vs. z.

Eq. (4.130). We recall that, while the span is $S = \ell_1 - \ell_2$, the imbalance is defined as $\tilde{S} = \ell_1 + \ell_2$. It is then convenient to write the joint distribution of S and $\tilde{S}$ in terms of the joint distribution of ℓ_1 and ℓ_2:

$$P(S, \tilde{S}, T) = \frac{1}{2} P_{\text{joint}}\left(\ell_1 = \frac{S + \tilde{S}}{2}, \ell_2 = \frac{S - \tilde{S}}{2}, T\right), \qquad (4.145)$$

where the Laplace transform of $P_{\text{joint}}(\ell_1, \ell_2, T)$ with respect to T is given in Eq. (4.130) and the factor of $1/2$ in Eq. (4.145) comes from the Jacobian of the change of variables. Therefore, the Laplace transform of $P(S, \tilde{S}, T)$, using Eq. (4.130), is given by

$$\tilde{P}_{\text{joint}}(S, \tilde{S}, s) = \int_0^\infty P(S, \tilde{S}, T) e^{-sT}\, dT$$

$$= \frac{1}{4D} \cosh\left(\sqrt{\frac{s}{4D}} \tilde{S}\right) \text{sech}^3\left(\sqrt{\frac{s}{4D}} S\right). \qquad (4.146)$$

We now integrate over S, keeping $\tilde{S}$ fixed. Note that while the span $S \geq 0$, the imbalance $\tilde{S}$ can take both positive and negative values. In fact, for a given $\tilde{S}$, it is easy to check that the inequality $-S \leq \tilde{S} \leq +S$ is satisfied. Hence, for a fixed $\tilde{S}$, we need to integrate over all values of S such that $S \geq |\tilde{S}|$. The Laplace transform of the marginal distribution of imbalance is therefore given by

$$\int_0^\infty P_{\text{imb}}(\tilde{S}, T) e^{-sT}\, dT = \frac{1}{4D} \cosh\left(\sqrt{\frac{s}{4D}} \tilde{S}\right) \int_{|\tilde{S}|}^\infty \text{sech}^3\left(\sqrt{\frac{s}{4D}} S\right) dS. \qquad (4.147)$$

The inversion of this Laplace transform is rather complicated and we refer the reader to Biroli *et al.* (2022) for the details. We just quote the main results here. One finds that the distribution of the imbalance can be written in the scaling form, for all T,

$$P_{\text{imb}}(\tilde{S}, T) = \frac{1}{\sqrt{DT}} f_{\text{imb}}\left(\frac{\tilde{S}}{\sqrt{DT}}\right),$$ (4.148)

where the scaling function $f_{\text{imb}}(z)$ is symmetric in z and is given in Biroli *et al.* (2022, Eq. (133)). Its asymptotic tails are given by

$$f_{\text{imb}}(z) \simeq \begin{cases} \dfrac{\sqrt{\pi}}{8} - \dfrac{\sqrt{\pi}}{128} z^2, & |z| \to 0, \\[2mm] \dfrac{2}{3\sqrt{\pi}} e^{-z^2/4}, & |z| \to \infty. \end{cases}$$ (4.149)

For a plot of this function, see Biroli *et al.* (2022, Fig. 6).

5

Time of the Maximum and the Minimum

So far, we have been discussing the distribution of extreme values, i.e., the distribution of the maximum $X_{\max}$ and the minimum $X_{\min}$ of the entries of a time series $\{x_1, x_2, \ldots, x_N\}$ of size N. As discussed in the introduction, equally important observables are the random variables $t_{\max}$ and $t_{\min}$ at which the maximum and the minimum occur respectively (see Fig. 1.3). We will denote the PDFs of $t_{\max}$ and $t_{\min}$ by $P(t_{\max} \mid N)$ and $P(t_{\min} \mid N)$. Note that for a time series of size N the random variables $t_{\max}$ and $t_{\min}$ are integers, taking values between 1 and N. However, for time series whose entries form a continuous-time process (such as in Brownian motion over an interval $[0, T]$), the corresponding variables $t_{\max}$ and $t_{\min}$ are continuous random variables, and their PDFs $P(t_{\max} \mid T)$ and $P(t_{\min} \mid T)$ are supported over the interval $[0, T]$. Below, we discuss the distributions of $t_{\max}$ and $t_{\min}$ for both IID and random walk sequences, and then present various examples where these random variables $t_{\max}$ and $t_{\min}$ appear naturally.

5.1 The statistics of $t_{\max}$ and $t_{\min}$ for IID random variables

Here we consider the case of N IID random variables $\{x_1, x_2, \ldots, x_N\}$ whose joint distribution factorizes into

$$P_{\text{joint}}(x_1, x_2, \ldots, x_N) = \prod_{i=1}^{N} p(x_i), \tag{5.1}$$

where $p(x)$ is an arbitrary continuous distribution. In this case, one can guess, without any computation, the distributions $P(t_{\max} \mid N)$ and $P(t_{\min} \mid N)$. For instance, consider the case of the maximum. Since any one of the N variables can be the maximum with equal probability $1/N$, it is clear that

$$P(t_{\max} \mid N) = \frac{1}{N}, \quad \text{with } 1 \le t_{\max} \le N. \tag{5.2}$$

This means that $t_{\max}$ is uniformly distributed over the interval $[1, N]$. If you do not believe this argument, let us compute it more laboriously. For this, it is convenient

Statistics of Extremes and Records in Random Sequences. Satya N. Majumdar and Grégory Schehr, Oxford University Press.
© Satya N. Majumdar and Grégory Schehr (2024). DOI: 10.1093/9780191838781.003.0005

to first compute the joint distribution of X_{max} and t_{max} and then integrate out over X_{max} to obtain the marginal distribution of t_{max}. This joint distribution can be expressed as

$$P(X_{max}, t_{max} \mid N) = p(X_{max}) \left[\int_{-\infty}^{X_{max}} p(x') \, dx' \right]^{N-1}. \tag{5.3}$$

This follows simply from the fact that, given the value of X_{max}, the values of the other $N-1$ entries must be less than X_{max}. Thus, integrating over X_{max} gives us the marginal distribution $P(t_{max} \mid N)$,

$$P(t_{max} \mid N) = \int_{-\infty}^{\infty} dX_{max} \, p(X_{max}) \left[\int_{-\infty}^{X_{max}} p(x') \, dx' \right]^{N-1} = \frac{1}{N}, \tag{5.4}$$

where we made the customary change of variable $\int_{-\infty}^{X_{max}} p(x') \, dx' = y$ in doing the integral. A similar argument shows that

$$P(t_{min} \mid N) = \frac{1}{N}, \quad \text{with } 1 \leq t_{min} \leq N. \tag{5.5}$$

Thus, both the marginal distributions $P(t_{max} \mid N)$ and $P(t_{min} \mid N)$ are completely universal, i.e., independent of $p(x)$ as long as it is continuous. The moments of t_{max} (and identically for t_{min}) can be trivially computed from Eq. (5.4).

Little exercise 5.1 *Show that the mean and the variance read*

$$\langle t_{max} \rangle = \langle t_{min} \rangle = \frac{N+1}{2}, \tag{5.6}$$

$$\mathrm{Var}(t_{max}) = \mathrm{Var}(t_{min}) = \frac{N^2 - 1}{12}. \tag{5.7}$$

Correlations between t_{max} and t_{min}. Obviously, t_{max} and t_{min} must be anticorrelated since the maximum and minimum cannot occur at exactly the same time. One can compute this correlation function easily. Indeed, the joint distribution of X_{max}, X_{min}, t_{max} and t_{min} (see Fig. 1.3) can be written, for $t_{max} \neq t_{min}$, as

$$P(X_{max}, X_{min}, t_{max}, t_{min}) = \left[\int_{X_{min}}^{X_{max}} p(x') \, dx' \right]^{N-2} p(X_{max}) p(X_{min}) \Theta(X_{max} - X_{min}). \tag{5.8}$$

This can be easily explained as follows. Given the values X_{max} and X_{min} (where X_{max} is necessarily bigger than X_{min} and this is enforced by the theta function), the remaining $N-2$ random variables must take values only in the range $[X_{min}, X_{max}]$—this explains the integral in Eq. (5.8). By integrating out first X_{min} and then over X_{max} (we leave this as a little exercise!), one obtains the joint distribution of t_{max} and t_{min} as

$$P(t_{\max}, t_{\min} \mid N) = \begin{cases} \dfrac{1}{N(N-1)}, & t_{\max} \neq t_{\min}, \\ 0, & t_{\max} = t_{\min}. \end{cases} \tag{5.9}$$

The result in the last line is due to the fact that the event when $t_{\max} = t_{\min}$ occurs with zero probability, as mentioned earlier. Clearly, from the exact formula in Eq. (5.9), we see that $P(t_{\max}, t_{\min} \mid N) \neq P(t_{\max} \mid N)P(t_{\min} \mid N)$, indicating the presence of non-zero correlations between $t_{\max}$ and $t_{\min}$. Indeed, from Eq. (5.9) one can compute the connected correlation function (check it out!):

$$\langle t_{\max} t_{\min} \rangle - \langle t_{\max} \rangle \langle t_{\min} \rangle = -\frac{N+1}{12}. \tag{5.10}$$

The negative sign in Eq. (5.10) clearly demonstrates that $t_{\max}$ and $t_{\min}$ are anticorrelated. The correlation coefficient between $t_{\max}$ and $t_{\min}$, using the result for the variance in Eq. (5.7), is given by

$$\rho_{t_{\max}, t_{\min}} = \frac{\langle t_{\max} t_{\min} \rangle - \langle t_{\max} \rangle \langle t_{\min} \rangle}{\sqrt{\mathrm{Var}(t_{\max})}\sqrt{\mathrm{Var}(t_{\min})}} = -\frac{1}{N-1}. \tag{5.11}$$

Thus, when $N \to \infty$, the correlation coefficient $\rho_{t_{\max}, t_{\min}} \to 0$. This is expected since we may recall that the source of the correlation comes from the effective repulsion between $t_{\max}$ and $t_{\min}$ as shown in Eq. (5.10). This correlation coefficient decreases with increasing N as $-1/N$ for large N.

Distribution of the interval between $t_{\max}$ and $t_{\min}$.   Another interesting observable is the time difference $\tau = t_{\min} - t_{\max}$. From the joint distribution of $t_{\min}$ and $t_{\max}$ in Eq. (5.9), we can compute the distribution of τ as follows. Since $t_{\min} = t_{\max} + \tau$, it follows that

$$P(\tau \mid N) = \sum_{t_{\max}=1}^{N-\tau} P(t_{\max}, t_{\max} + \tau \mid N) = \frac{N - |\tau|}{N(N-1)} \quad \text{for } 0 < |\tau| \leq N-1, \tag{5.12}$$

$$P(\tau = 0 \mid N) = 0. \tag{5.13}$$

Note that the allowed range of τ is $\tau \in [-(N-1), N-1]$. One can check that $P(\tau \mid N)$ is normalized, since

$$\sum_{\tau=-(N-1)}^{N-1} P(\tau \mid N) = 2 \sum_{\tau=1}^{N-1} \frac{N-\tau}{N(N-1)} = 1. \tag{5.14}$$

Hence, $P(\tau \mid N)$ has a triangular shape with a hole at the center $\tau = 0$ (see Fig. 5.1). This distribution is clearly universal, i.e., independent of $p(x)$. For later comparison to the random walk result, we note in particular that the probability distribution

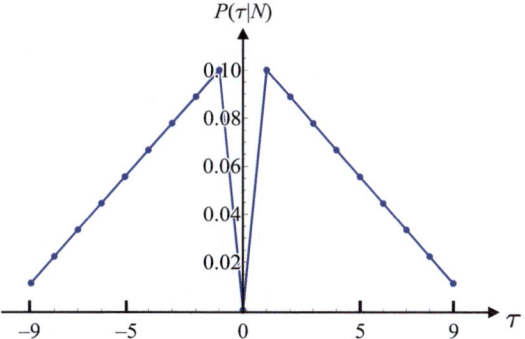

Fig. 5.1 Plot of $P(\tau \mid N)$ as given in Eqs. (5.12) and (5.13) vs. τ for $N = 10$.

$P(\tau \mid N)$ when τ takes its maximal value $N - 1$ is given by

$$P(\tau = N - 1 \mid N) = \frac{1}{N(N-1)}. \tag{5.15}$$

This probability thus decays as $P(\tau = N - 1 \mid N) \sim 1/N^2$ for large N. Later, for random walks, we will see that the same quantity is also universal, i.e., independent of the jump distribution of an N-step random walk, but is much larger than the IID case, since for random walks it decays as $1/(2N)$ for large N.

5.2 The statistics of $t_{\max}$ and $t_{\min}$ for random walks

We now consider a time series whose entries correspond to the position of a random walker on the line, starting at $x_0 = 0$ and evolving via the jump process

$$x_n = x_{n-1} + \eta_n, \tag{5.16}$$

where the η_n's are drawn independently at each step from a symmetric and continuous distribution $f(\eta)$. For this walk, we have already seen in Chapter 3 that the survival probability $q(n)$ of the walk starting at the origin is universal (i.e., independent of the jump distribution $f(\eta)$) due to the Sparre Andersen theorem in Eq. (3.101) and is given by

$$q(n) = \frac{1}{2^{2n}} \binom{2n}{n}. \tag{5.17}$$

We will now see that this $q(n)$ is a crucial ingredient for computing the distribution $P(t_{\max} \mid N)$.

Let us recall some useful observables for a discrete-time random walk as discussed in Chapter 3. We start with the restricted Green's function defined for $x \geq 0$ as

$$G_+(x, n \mid x_0, 0) = \mathrm{Prob}[x_1 \geq 0, x_2 \geq 0, \ldots, x_{n-1} \geq 0, x_n = x \mid x_0], \tag{5.18}$$

which denotes the probability density for the walker arriving at $x \geq 0$ at step n, starting from $x_0 \geq 0$, and staying non-negative in between. This is the key object. Many other observables can be expressed in terms of this restricted Green's function. For instance, let us define

$$p_n(x) = G_+(x, n \mid 0, 0) = \text{Prob}[x_1 \geq 0, x_2 \geq 0, \ldots, x_{n-1} \geq 0, x_n = x \mid x_0 = 0], \tag{5.19}$$

which denotes the probability density of arriving at x at step n, starting from the origin $x_0 = 0$ and staying non-negative in between. The survival probability $q(n)$ defined above is then simply obtained by integrating $p_n(x)$ over the final position x,

$$q(n) = \int_0^\infty p_n(x)\,\mathrm{d}x = \int_0^\infty G_+(x, n \mid 0, 0)\,\mathrm{d}x. \tag{5.20}$$

The Sparre Andersen theorem tells us that, even though $G_+(x, n \mid 0, 0)$ depends on the jump distribution $f(\eta)$ explicitly, once integrated over the final position x, the answer for $q(n)$ is completely independent of $f(\eta)$ and is given by Eq. (5.17).

After this little detour recalling previous results, let us now get back to the distribution of t_max for an N-step random walker starting at the origin. We consider the random walker in Eq. (5.16) starting at the origin. Let t_max denote the time at which the walker reaches its maximum (see Fig. 5.2). Clearly t_max is a random variable, taking integer values in the range $0 \leq t_\text{max} \leq N$. Let $P(n \mid N)$ denote the probability that t_max takes the value n with $0 \leq n \leq N$. Let us first remark that, although we assumed that the walker starts at the origin, i.e., $x_0 = 0$, it is easy to see that this probability $P(n \mid N)$ does not depend on the actual value of x_0. This simply follows from the global translational invariance of the walk in the x-direction. To compute this distribution, we first calculate the probability of the event that the trajectory arrives at x at step N and attains its maximum displacement $X_\text{max} = M$ at an intermediate step $t_\text{max} = n$ (see Fig. 5.2 for a typical trajectory),

$$P(n, M, x \mid N) = \text{Prob}[t_\text{max} = n, X_\text{max} = M, x_N = x]. \tag{5.21}$$

Note that by definition the allowed range of x is $(-\infty, M]$ since M is the largest possible position up to step N, including the last step. This probability can be related to $p_n(x)$ defined in Eq. (5.19) via

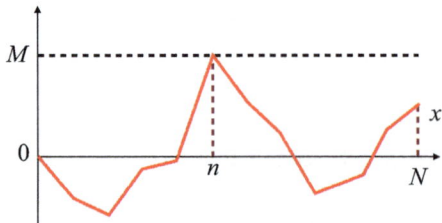

Fig. 5.2 A typical trajectory of a random walker from Eq. (5.16). starting from $x_0 = 0$, that arrives at x at step N and attains its maximum $X_\text{max} = M$ at an intermediate step $t_\text{max} = n$.

$$P(n, M, x \mid N) = p_n(M)p_{N-n}(M - x). \tag{5.22}$$

This relation looks nice and compact. Now let us try to figure out where it comes from. The best way to understand this relation is by investigating a typical path of the walker as depicted in Fig. 5.2. The first step is to split the path into two halves: one over the time interval $[0, n]$ (left half) and the other over the interval $[n, N]$ (right half). Once we specify the position of the walker at step n, namely $x_n = M$, these two halves become completely independent due to the Markov property of the random walk. We can then compute the probability of each half separately, given the value $x_n = M$. Let us start with the left half. Let us shift the origin of the walk to the level M in Fig. 5.2 and invert the time, with the origin of the time at step n. Operationally, this means that we define a new random walk of n steps with positions

$$y_k = M - x_{n-k} \quad \text{with } k = 0, 1, \ldots, n. \tag{5.23}$$

This walk starts at $y_0 = 0$, arrives at $y_n = M$ at step n and stays non-negative at all intermediate steps since $x_{n-k} \leq M$ for all $k = 0, 1, \ldots, n$. This walk evolves via

$$y_k = y_{k-1} + \eta_{n-k}, \tag{5.24}$$

which follows simply from the definition in Eq. (5.23) and the evolution equation, Eq. (5.16), of the x_k's. Therefore, the y_k's perform, statistically, the same random walk as the x_k's. Therefore, the probability of the left half of the path in Fig. 5.2, when translated in terms of the y_k's, is simply

$$\text{Prob[``left half'']} = \text{Prob}[y_1 \geq 0, y_2 \geq 0, \ldots, y_{n-1} \geq 0, y_n = M \mid y_0 = 0]$$
$$= p_n(M). \tag{5.25}$$

This explains the first factor in Eq. (5.22).

To explain the second factor, we now turn to the second half of the path in Fig. 5.2. Here we do not have to reverse time but instead slide the time, shifting the time origin to n, and also shift the origin to M as before. In other words, we define a new random walk of $N - n$ steps with positions

$$z_k = M - x_{n+k}. \tag{5.26}$$

This random walk starts at $z_0 = 0$ (since $x_n = M$), arrives at $z_{N-n} = M - x$ (since $x_N = x$) and stays positive in between. This evolves via

$$z_k = z_{k-1} - \eta_{k-1+1} \quad \text{with } k = 0, 1, \ldots, N - n. \tag{5.27}$$

Since the jump variables are symmetrically distributed, the variables η_k and $-\eta_k$ have the same distributions. Hence, it follows that the random walk z_k is statistically the same as the original walk x_k. Thus, the probability of the right half of the path in Fig. 5.2 can be expressed as

$$\text{Prob[“right half”]} = \text{Prob}[z_1 \geq 0, z_2 \geq 0, \ldots, z_{N-n-1} \geq 0, z_{N-n} = M - x \mid z_0 = 0]$$
$$= p_{N-n}(M - x). \tag{5.28}$$

Taking the product of the left and right halves in Eqs. (5.25) and (5.28) respectively then leads to the result in Eq. (5.22).

Our task is not yet complete! To compute the marginal probability of $t_{\max}$, namely

$$P(n \mid N) = \text{Prob}(t_{\max} = n \mid N), \tag{5.29}$$

we still need to integrate the joint probability $P(n, M, x \mid N)$ in Eq. (5.22) over the final position x at step N and the value of the maximum M. Therefore, integrating, we get

$$P(n \mid N) = \int_0^\infty dM \, p_n(M) \int_{-\infty}^M dx \, p_{N-n}(M - x). \tag{5.30}$$

Note that the range of integration of M is over $[0, +\infty)$. This simply follows from the fact that the walker starts at the origin. Hence the maximum value of the walk M cannot be negative. By making the change of variable $y = M - x$ we note that the second integral over x in Eq. (5.30) becomes completely independent of M. Hence, we get

$$P(n \mid N) = \int_0^\infty dM \, p_n(M) \int_0^\infty dy \, p_{N-n}(y) = q(n)q(N - n), \tag{5.31}$$

where we used the relation in Eq. (5.20). We show next how this result can be exploited to provide a simple derivation of the Sparre Andersen theorem in Eq. (5.17).

A simple derivation of the Sparre Andersen theorem.   Up to now, we have not used the explicit expression for $q(n)$, though it is known, as discussed in Section 3.2.5—see Eqs. (3.100) and (3.101). Here, we want to show that the Sparre Andersen theorem in Eq. (3.101) can be derived directly from the relation in Eq. (5.31) (Dembo *et al.*, 2013; Mounaix *et al.*, 2020). We first remark that the relation in Eq. (5.31) is completely universal and does not depend on the specific form of the jump distribution $f(\eta)$, as long as it is symmetric and continuous. Since $P(n \mid N)$ is a probability, it has to be normalized to unity, i.e., $\sum_{n=0}^N P(n \mid N) = 1$, which implies, using Eq. (5.31),

$$\sum_{n=0}^N q(n)q(N - n) = 1. \tag{5.32}$$

We now multiply both sides by s^N and sum over all $N = 0, 1, 2, \ldots$ This gives

$$\sum_{N=0}^\infty s^N \sum_{n=0}^N q(n)q(N - n) = \frac{1}{1 - s}. \tag{5.33}$$

Decomposing $s^N = s^n s^{N-n}$ on the LHS, and performing the double sum using a change of variables $(n, N) \to (n, m = N - n)$, one finds that the double sum factorizes into a product of single sums over n and m. This gives

$$\sum_{N=0}^{\infty} s^N \sum_{n=0}^{N} q(n)q(N-n) = \left[\sum_{n=0}^{\infty} s^n q(n) \right]^2 = \frac{1}{1-s}. \tag{5.34}$$

Consequently, the generating function of $q(n)$ is simply

$$\sum_{n=0}^{\infty} s^n q(n) = \frac{1}{\sqrt{1-s}}. \tag{5.35}$$

Expanding in powers of s gives the universal Sparre Andersen result

$$q(n) = \binom{2n}{n} 2^{-2n}. \tag{5.36}$$

To our knowledge, this constitutes perhaps the simplest derivation of the Sparre Andersen theorem.

Substituting the explicit expression for $q(n)$ from Eq. (5.36) into Eq. (5.31) gives

$$\text{Prob}(t_{\max} = n \mid N) = P(n \mid N) = \frac{1}{2^{2N}} \binom{2n}{n} \binom{2(N-n)}{N-n}. \tag{5.37}$$

Note that this result is again completely universal for any n and N, i.e., independent of the continuous and symmetric jump distribution $f(\eta)$ (including even Lévy flights!). The result in Eq. (5.37) was first derived by Sparre Andersen (Sparre Andersen, 1954). We can easily check that $P(n \mid N)$ in Eq. (5.37) is normalized to unity, that is, $\sum_{n=0}^{N} P(n \mid N) = 1$.

Little exercise 5.2 *Calculate the first few higher moments of $P(n \mid N)$ explicitly, and show that*

$$\langle t_{\max} \rangle = \frac{N}{2}, \tag{5.38}$$

$$\langle t_{\max}^2 \rangle = \frac{1}{8}(3N^2 + N), \tag{5.39}$$

$$\langle t_{\max}^3 \rangle = \frac{1}{16}(5N^3 + 3N^2), \tag{5.40}$$

$$\langle t_{\max}^4 \rangle = \frac{1}{128}(35N^4 + 30N^3 + N^2 - 2N), \tag{5.41}$$

$$\vdots$$

Therefore the variance is given by

$$\text{Var}(t_{\max}) = \frac{1}{8}(N^2 + N). \tag{5.42}$$

By comparing with the IID case in Eq. (5.7) we see that the variance of $t_{\max}$ is slightly enhanced in the random walk case by a multiplicative factor of $4/3$ in the large-N limit. From the expressions of the moments in the large-N limit in Eq. (5.38), we see that $t_{\max}$ scales linearly with N for large N and hence it is natural to look for a scaling form for $P(n \mid N)$ in the large-N limit with $n = O(N)$. We show below that this is indeed the case by analyzing directly the formula for $P(n \mid N)$ in Eq. (5.37) in the scaling regime.

As mentioned earlier, the universal result for $\text{Prob}(t_{\max} = n \mid N)$ in Eq. (5.37) was originally derived by Sparre Andersen (1954). In fact, the same paper showed that the same distribution also appears for two other random variables, namely: (i) the occupation time, i.e., the total number of positive (equivalently negative) points visited by N-step random walks, and (ii) the time for the last passage to the origin before step N.

Asymptotic limit: Arcsine law.   We can plot the distribution $P(n \mid N)$ as a function of n (see Fig. 5.3 for a plot for $N = 20$). For a fixed N, it typically has a "U-shaped" form symmetric around $N/2$ (strictly speaking, the integer part of it) with a minimum at the center and a maximum at the two endpoints $n = 0$ and $n = N$. It is natural to wonder whether there is a limiting scaling form of this probability in Eq. (5.37) when both n and N are large but when the ratio $\tau = n/N$ is fixed. In Eq. (5.37) we then replace n by τN (with $\tau = O(1)$) and take the large-N limit using Stirling's formula:

$$N! \simeq \sqrt{2\pi} N^{N+1/2} e^{-N}. \tag{5.43}$$

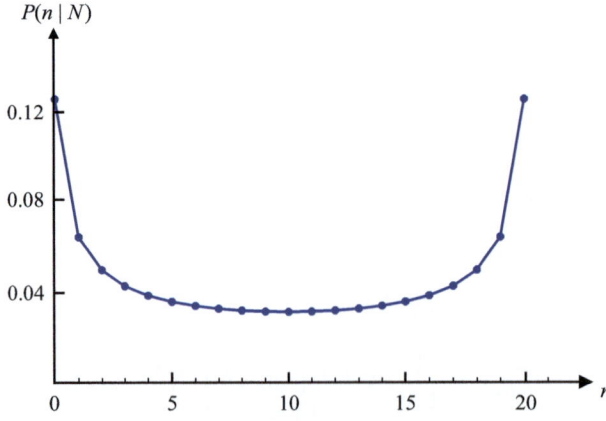

Fig. 5.3 Plot of the distribution $P(n \mid N) = \text{Prob}(t_{\max} = n \mid N)$ given in Eq. (5.37) vs. n, for $N = 20$.

One then finds (check it out!) that

$$P(n \mid N) \simeq \frac{1}{N} f_{\mathrm{AS}}\left(\frac{n}{N}\right), \qquad \text{where } f_{\mathrm{AS}}(\tau) = \frac{1}{\pi\sqrt{\tau(1-\tau)}}, \quad 0 \leq \tau \leq 1. \quad (5.44)$$

Here, the subscript AS stands for the "arcsine law" because the cumulative distribution has an arcsine form,

$$\int_0^\tau f_{\mathrm{AS}}(\tau')\, d\tau' = \frac{2}{\pi} \sin^{-1}(\sqrt{\tau}). \quad (5.45)$$

Note that the limiting arcsine law in Eq. (5.44) is also universal, i.e., it holds not just for Brownian motion but for any arbitrary random walk with symmetric and continuous jump distribution, including, for instance, Lévy flights. The U-shaped form of $P_{\mathrm{AS}}(\tau)$ is remarkable and somewhat unexpected. It implies that the PDF of $t_{\max}$ is peaked at the two edges at $\tau = 0$ and $\tau = 1$ where $\tau = t_{\max}/N$, and not at its average value $\langle \tau \rangle = N/2$ (where indeed the PDF has its minimum value!). To understand the implications of this, we note that the event $\tau = 0$ corresponds to trajectories starting at $x = 0$ that stay non-positive up to step N. Similarly, the event $\tau = 1$ corresponds to trajectories which achieve their maximum only at the last step $n = N$. Thus, one finds that such "stiff" trajectories are most likely ("typical") since the PDF is maximal for such trajectories. This is somewhat counterintuitive because naively one would have expected the typical trajectories to be the ones that cross and recross the origin many times. But this result shows that such trajectories are relatively rare, i.e., "atypical." We will see later that even Brownian motion shares this general property of stiffness of random walks.

Little exercise 5.3 *Show that the kth moment of the arcsine law is given by*

$$\langle \tau^k \rangle = \int_0^1 f_{AS}(\tau)\tau^k\, d\tau = \frac{\Gamma(k + \frac{1}{2})}{\sqrt{\pi}\,\Gamma(k+1)}, \quad k > -\frac{1}{2}. \quad (5.46)$$

Keeping only the leading-order terms for large N in the polynomial expansions of the moments in Eq. (5.38), show that they are reproduced by the exact formula in Eq. (5.46).

Probability distribution of $t_{\min}$**.**   Because of the symmetry of the random walk under the transformation $x \to -x$, the probability distribution of the time of the minimum $t_{\min}$ is actually exactly the same as the statistics of $t_{\max}$. Thus this distribution is again given by Eq. (5.37) and is universal, i.e., independent of the symmetric and continuous jump distribution $f(\eta)$. In particular, in the scaling limit $N \to \infty$, $n \to \infty$ keeping $\tau = n/N$ fixed, the distribution of $t_{\min}$ approaches the universal arcsine law,

$$\mathrm{Prob}(t_{\min} = n \mid N) \simeq \frac{1}{N} f_{\mathrm{AS}}\left(\frac{n}{N}\right), \quad (5.47)$$

where the scaling function $f_{\mathrm{AS}}(\tau)$ is given in Eq. (5.44).

Correlations between $t_{\max}$ and $t_{\min}$. As in the IID case, we expect that $t_{\max}$ and $t_{\min}$ are anticorrelated because if the global maximum happens at a given time $t_{\max}$, it is very unlikely that the global minimum will occur right before or right after this. In other words, the random variables $t_{\max}$ and $t_{\min}$ "repel" each other. To compute the connected correlation function between $t_{\max}$ and $t_{\min}$, we need to compute the joint distribution of $t_{\max}$ and $t_{\min}$, denoted by $P(t_{\max}, t_{\min} \mid N)$. The marginal distributions $P(t_{\max} \mid N)$ and $P(t_{\min} \mid N)$ can be obtained by integrating over $t_{\min}$ and $t_{\max}$ respectively. We have seen above that these marginal distributions of $t_{\max}$ and $t_{\min}$ are universal for all N, i.e., independent of the jump distribution $f(\eta)$. Unfortunately, it turns out that their joint distribution $P(t_{\max}, t_{\min} \mid N)$ does not have this universality at arbitrary N and is very hard to compute explicitly for arbitrary $f(\eta)$. An enthusiastic reader may try to derive this joint distribution $P(t_{\max}, t_{\min} \mid N)$ for the special case of a double-exponential jump distribution $f(\eta) = 1/(2\ell_0)e^{-|\eta|/\ell_0}$. As discussed several times in this book, this particular jump distribution is solvable for many observables, including this joint distribution of $t_{\max}$ and $t_{\min}$ (Mori *et al.*, 2019, 2020). In the next subsection we show that more detailed results on this joint distribution can be derived explicitly in the Brownian limit.

5.2.1 The statistics of $t_{\max}$ and $t_{\min}$ for Brownian motion

For a random walk with a symmetric jump distribution $f(\eta)$ such that the second moment $\sigma^2 = \langle \eta^2 \rangle$ is finite, let us recall that the Brownian limit emerges when $\sigma \to 0$ and the number of steps of the walk $n \to \infty$, keeping the product $n\sigma^2 = 2Dt$ fixed. In this limit, the random walk converges to a Brownian motion $x(t)$ in continuous time t with a diffusion constant D, as in Eq. (2.60). In this scaling limit, we then have a Brownian motion of duration T. In this case, the marginal distribution $P(t_{\max} \mid T)$ was computed by Lévy (1940) and was shown to have the arcsine form,

$$P(t_{\max} \mid T) = \frac{1}{T} f_{\mathrm{AS}}\left(\frac{t_{\max}}{T}\right), \qquad \text{where } f_{\mathrm{AS}}(\tau) = \frac{1}{\pi\sqrt{\tau(1-\tau)}}, \quad 0 \le \tau \le 1. \quad (5.48)$$

As expected, it coincides with the random result in the limit n large, N large with the ratio $n/N = \tau$ fixed, as discussed in detail in Eq. (5.44). This is one of the classical results in the fluctuation theory of random walks (Feller, 2008b). However, the joint distribution $P(t_{\max}, t_{\min} \mid T)$ of $t_{\max}$ and $t_{\min}$ has not been computed in the probability literature (as far as we know). Indeed, in the Brownian limit, using a path integral method the joint distribution $P(t_{\max}, t_{\min} \mid T)$ was recently computed explicitly. We will not provide the details of these calculations here; instead, we refer the interested reader to the original articles (Mori *et al.*, 2019, 2020).

Let us briefly outline the strategy underlying this computation. The idea is to first compute a more general quantity $P(X_{\max}, X_{\min}, t_{\max}, t_{\min} \mid T)$ that denotes the joint distribution of the values of the maximum, minimum and their respective times of occurrences. If we can compute this, the joint distribution $P(t_{\max}, t_{\min} \mid T)$ can be obtained by integrating over $X_{\max}$ and $X_{\min}$. This more general joint distribution $P(X_{\max}, X_{\min}, t_{\max}, t_{\min} \mid T)$ can be computed by the path decomposition method. For example, suppose that $t_{\max} < t_{\min}$ (see Fig. 5.4). We then split the interval $[0, T]$ into three segments: $[0, t_{\max}]$, $[t_{\max}, t_{\min}]$ and $[t_{\min}, T]$. In the first interval,

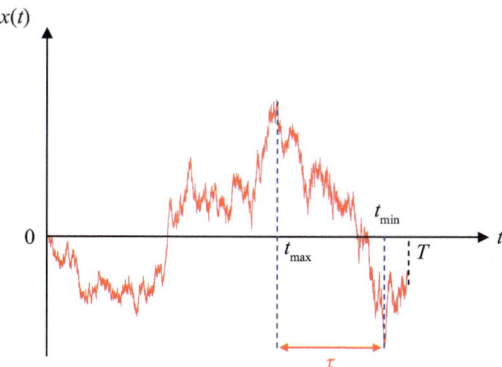

Fig. 5.4 A Brownian trajectory $x(t)$ over the time interval $[0, T]$ for which $t_{\min} > t_{\max}$. The observable of interest is $\tau = t_{\min} - t_{\max}$.

$[0, t_{\max}]$, the Brownian motion $x(t)$ starting at the origin at $t = 0$ arrives at $x = X_{\max}$ for the first time at time $t = t_{\max}$, while staying inside the box $[X_{\min}, X_{\max}]$ during the time interval $[0, t_{\max}]$. In the second interval, $[t_{\max}, t_{\min}]$, the walker starting at $X_{\max}$ at $t = t_{\max}$ arrives for the first time at $x = X_{\min}$ at time $t = t_{\min}$ while staying inside the box $[X_{\min}, X_{\max}]$ in between. Finally, in the third interval, $[t_{\min}, T]$, the walker, starting at $X_{\min}$ at $t = t_{\min}$, propagates up to $t = T$ while staying within the box $[X_{\min}, X_{\max}]$ during this time interval. Each of these three propagators can be computed separately using basic properties of Brownian motion as discussed earlier in the book. The net probability is then the product of these three, since one can use the Markov property of Brownian motion. After a few steps of algebra, one arrives at the following final result for $t_{\max} < t_{\min}$ (Mori *et al.*, 2019, 2020):

$$P_<(t_{\max}, t_{\min} \mid T) = \frac{4}{\pi^2} \sum_{n_1, n_2, n_3 = 1}^{\infty} \frac{(-1)^{n_2+1} n_2^2 [1 - (-1)^{n_1}][1 - (-1)^{n_3}]}{[n_1^2 t_{\max} + n_2^2(t_{\min} - t_{\max}) + n_3^2(T - t_{\min})]^2}, \quad (5.49)$$

where the subscript $<$ indicates that $t_{\max} < t_{\min}$. A similar expression can be derived in the case $t_{\max} > t_{\min}$ just by exchanging $t_{\max}$ and $t_{\min}$ in Eq. (5.49). By integrating the expression in Eq. (5.49) over $t_{\min}$ (or $t_{\max}$), one can recover the arcsine law of the marginal distribution $P(t_{\max} \mid T)$ (or $P(t_{\min} \mid T)$) as given in Eq. (5.44). From the joint distribution in Eq. (5.49), one can compute the connected correlation function exactly for all T. This is given by (Mori *et al.*, 2019, 2020)

$$\langle t_{\max} t_{\min} \rangle - \langle t_{\max} \rangle \langle t_{\min} \rangle = -\frac{7\zeta(3) - 6}{32} T^2 = (-0.0754\ldots) T^2, \quad (5.50)$$

where $\zeta(z)$ is the Riemann zeta function. The negative sign of the covariance clearly demonstrates that $t_{\max}$ and $t_{\min}$ are anticorrelated.

Distribution of the interval between $t_{\max}$ and $t_{\min}$. As in the IID case, the other interesting related observable is the time interval $\tau = t_{\min} - t_{\max}$. Note that $\tau \in [-T, T]$ can be positive or negative. This quantity τ has a natural application in finance. Let us consider the price of a stock over a period of time T. Then if $t_{\max} < t_{\min}$, as in Fig. 5.4, an agent would try to sell his/her shares at time $t_{\max}$ when the price is the highest and then wait up to time $t_{\min}$ to re-buy at the best price. Thus, $\tau = t_{\min} - t_{\max}$ represents the time the agent has to wait before re-buying his/her shares in order to maximize the gain. From the expression for the joint distribution in Eq. (5.49), one can compute the distribution $P(\tau \,|\, T)$ of τ as

$$P(\tau \,|\, T) = \int_0^T dt_{\max} \int_0^T dt_{\min}\, P_<(t_{\max}, t_{\min} \,|\, T)\delta(t_{\min} - t_{\max} - \tau). \qquad (5.51)$$

Substituting the expression in Eq. (5.49) into Eq. (5.51) and carrying out the integrals, one finds a compact and nice result (Mori *et al.*, 2019, 2020):

$$P(\tau \,|\, T) = \frac{1}{T} f_{\mathrm{BM}}\left(\frac{\tau}{T}\right), \qquad (5.52)$$

where the scaling function $f_{\mathrm{BM}}(y)$ is given by

$$f_{\mathrm{BM}}(y) = \frac{1}{|y|} \sum_{n=1}^{\infty} (-1)^{n+1} \tanh^2\left(\frac{n\pi}{2}\sqrt{\frac{|y|}{1-|y|}}\right), \qquad (5.53)$$

where $-1 \leq y \leq 1$. The function $f_{\mathrm{BM}}(y)$ is symmetric around $y = 0$ and is non-monotonic as a function of y (see Fig. 5.5). Moreover, $f_{\mathrm{BM}}(y)$ has the asymptotic behaviors

$$f_{\mathrm{BM}}(y) \underset{y \to 0^+}{\approx} \frac{8}{y^2} e^{-\pi/\sqrt{y}}, \qquad f_{\mathrm{BM}}(y) \underset{y \to 1}{\approx} \frac{1}{2}. \qquad (5.54)$$

One can also show that the scaling function $f_{\mathrm{BM}}(y)$ satisfies the following integral relation (Mori *et al.*, 2019, 2020):

$$\int_0^1 dy\, \frac{f_{\mathrm{BM}}(y)}{1+uy} = \int_0^{\infty} dz\, \frac{1}{\sinh(z)} \tanh^2\left(\frac{z}{2\sqrt{1+u}}\right). \qquad (5.55)$$

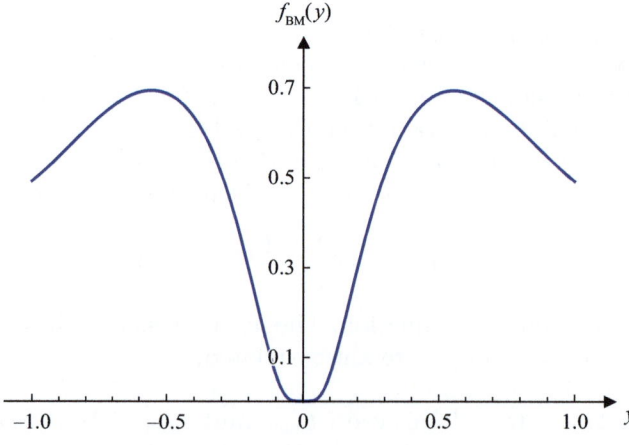

Fig. 5.5 Plot of the scaling function $f_{\mathrm{BM}}(y)$ given in Eq. (5.53) vs. y.

This integral identity turns out to be very useful for computing the moments of τ explicitly in an efficient way. For example, for BM, the first few moments of τ are given explicitly by

$$\langle |\tau| \rangle = \frac{4\ln(2) - 1}{3}T = (0.5908\ldots)T,$$

$$\langle \tau^2 \rangle = \frac{7\zeta(3) - 2}{16}T^2 = (0.4009\ldots)T^2,$$

$$\langle |\tau|^3 \rangle = \frac{147\zeta(3) - 34}{480}T^3 = (0.2972\ldots)T^3,$$

$$\langle \tau^4 \rangle = \frac{1701\zeta(3) - 930\zeta(5) - 182}{3840}T^4 = (0.2339\ldots)T^4,$$

where $\zeta(z)$ is the Riemann zeta function.

Interestingly, in the limit when $\tau \to T$, we see from Eqs. (5.51) and (5.54) that

$$P(\tau = T \mid T) = \frac{1}{2T}. \tag{5.56}$$

Since $\tau = t_{\min} - t_{\max}$, the event $\tau = T$ is achieved only when $t_{\min} = T$ and $t_{\max} = 0$. Thus, the event $\tau = T$ is achieved by trajectories whose maximum occurs in the beginning at $t = 0$, while the minimum occurs at the very end, $t = T$. The fact that this probability is simply $1/(2T)$ [as in Eq. (5.56)] does not have a trivial explanation. In fact, it turns out that this is much more general and holds even for an N-step random walk with arbitrary symmetric and continuous jump distribution $f(\eta)$, for which one can show (Mori *et al.*, 2019, 2020), using a combinatorial lemma due to Spitzer (1956)—see Mori (2022) for further details—that

$$\text{Prob}(t_{\max} = 0, t_{\min} = N \mid N) = \frac{1}{2N} \tag{5.57}$$

for all $N \geq 1$. The result in Eq. (5.56) for Brownian motion is consistent with this general result for random walks. This formula should be compared to the result obtained in the IID case in Eq. (5.15). One sees that, in the IID case, this probability decays much faster, as $1/N^2$, while in the random walk case strong correlations between $t_{\max}$ and $t_{\min}$ lead to a slower $1/(2N)$ decay for large N.

Finally, one can also compute the joint distribution $P(t_{\max}, t_{\min} \mid T)$ for a Brownian bridge, where the Brownian motion, starting at the origin at time $t = 0$, is conditioned to come back to the origin after a fixed time T. The method is more or less similar to the Brownian motion discussed above. These results for $t_{\max}$ and $t_{\min}$ for Brownian motion and Brownian bridge have interesting applications in the context of heights of fluctuating interfaces on a substrate of size L in the stationary state. We do not give the details here and the interested reader may consult Mori *et al.* (2019, 2020)for more details.

Applications of the statistics of $t_{\max}$ and $t_{\min}$. These two random variables have many applications, ranging from finance all the way to disordered systems. Here are some illustrative examples.

The distribution of $t_{\max}$ and $t_{\min}$ have also been computed exactly for Brownian motions with constraints, such as the Brownian bridge, the Brownian excursion, the Brownian meander and the reflected Brownian motion. These results show deviations from Lévy's arcsine law due to the presence of the constraints. They were computed by path integral methods (Majumdar *et al.*, 2008) as well as by the real-space renormalization group method developed in the context of disordered systems (Le Doussal and Monthus, 2003; Schehr and Le Doussal, 2010).

For a Brownian motion with a drift the distribution can be computed exactly using probabilistic methods (Shepp, 1979; Buffet, 2003), as well as path integral methods (Majumdar and Bouchaud, 2008). These results have been applied to analyze financial data.

The statistics of $t_{\max}$ has been shown to play an important role in the study of the mean area of the convex hull of two-dimensional stochastic processes, including, for instance, M independent Brownian motions (Randon-Furling *et al.*,2009; Majumdar *et al.*, 2010*a*), random acceleration processes (Reymbaut *et al.*, 2011), branching Brownian motions (Dumonteil *et al.*, 2013), continuous-time random walks (Luković *et al.*, 2013), run-and-tumble processes in two dimensions (Hartmann *et al.*, 2020; Singh *et al.*, 2022) and resetting Brownian motion in two dimensions (Majumdar *et al.*, 2021*a*).

The distribution of $t_{\max}$ has also been computed for non-intersecting Brownian motions (Rambeau and Schehr, 2011) using path integral methods, with applications to stochastic growth processes (Rambeau and Schehr, 2010; Schehr, 2012).

It has also been studied for several non-Markov processes. This includes, for example, the random acceleration process, for which the distribution of $t_{\max}$ was computed exactly (Majumdar *et al.*, 2010*b*). Another example of a non-Markovian process is fractional Brownian motion with Hurst index $0 < H < 1$, for which a systematic perturbation theory (around the Brownian value $H = 1/2$) was developed to study the distribution of $t_{\max}$ (Delorme and Wiese, 2016*b*,*a*; Delorme *et al.*, 2017; Sadhu *et al.*, 2018).

The statistics of $t_{\max}$ was recently computed for stationary processes arising both in- and out-of-equilibrium systems (Mori *et al.*, 2021*c*, 2022). For instance, for a particle confined in an external potential $V(x)$, the distribution of $t_{\max}$ in $[0, T]$ was found to have a universal edge behavior near 0 and T. A similar interesting edge behavior was also found for out-of-equilibrium systems, such as resetting Brownian motion, as well as the run-and-tumble particle in an external potential—both of these models have non-equilibrium stationary states.

In this short chapter we have provided partial glimpses of the multiple applications of $t_{\max}$ and $t_{\min}$. Indeed, this is a rapidly growing area of research and describing these results in detail is beyond the scope of this book. A short account of these works on $t_{\max}$ and $t_{\min}$ can be found in the recent pedagogical review on extreme value statistics in Majumdar *et al.* (2020). Interested readers may also consult the original publications on this subject.

6

Order Statistics

In the previous chapters we have focused only on the statistics of the *global* maximum $X_{\max}$ (or equivalently of the *global* minimum $X_{\min}$) of a set of N random variables $\{x_1, x_2, \ldots, x_N\}$. We considered two separate models: (i) where the entries x_i are independent random variables (the IID case), and (ii) where the entries represent the positions of a random walk with independent jumps drawn from a continuous and symmetric PDF $f(\eta)$ (the random walk case). In this chapter we go beyond the global maximum and global minimum and study the statistics of the second-largest maximum, third-largest maximum, etc. For this purpose, it is useful first to order the random variables $\{x_1, x_2, \ldots, x_N\}$ in a decreasing sequence $\{M_{1,N}, M_{2,N}, \ldots, M_{N,N}\}$ such that

$$X_{\max} = M_{1,N} > M_{2,N} > \cdots > M_{N,N} = X_{\min}. \tag{6.1}$$

Our main goal is then to understand how the statistics of the kth maximum depend on the order k of the sequence, for both the IID and random walk sequences. A related interesting observable is the gap between the values of the kth and the $(k+1)$th maximum,

$$g_{k,N} = M_{k,N} - M_{k+1,N}. \tag{6.2}$$

These are natural questions in the context of disordered systems, such as spin-glasses, polymers in random media or random matrix theory (Derrida, 1981; Bouchaud and Mézard, 1997; Forrester, 2010; Schehr and Majumdar, 2014; Majumdar *et al.*, 2020). To give a very simple example, as briefly discussed in the introduction, imagine that we have a quantum system defined on a graph with N sites and characterized by a Hamiltonian $\hat{H}$. Very generally, this Hamiltionian can be expressed in the site basis as

$$\hat{H} = \sum_{i,j} H_{i,j} |i\rangle\langle j|, \tag{6.3}$$

where i and j denote the site indices and we used the standard bra-ket notation to denote the site basis. In this representation, $H_{i,j}$ is an $N \times N$ Hermitian matrix. The matrix elements $H_{i,j}$ with $i \leq j$ are taken to be independent random variables to model the disordered quantum system. If, for example, the couplings involve only nearest-neighbour sites i and j, this is the famous Anderson model (Anderson,

Statistics of Extremes and Records in Random Sequences. Satya N. Majumdar and Grégory Schehr, Oxford University Press.
© Satya N. Majumdar and Grégory Schehr (2024). DOI: 10.1093/9780191838781.003.0006

1958), which has been studied extensively in the context of localization. On the other hand, if the couplings $H_{i,j}$ involve all sites i and j (the mean-field model), with the joint distribution of the matrix elements (both real and imaginary parts) given by the Gaussian $\mathrm{Prob}(\{H_{i,j}\}) \propto \exp\left[-\frac{1}{2}\sum_{i,j}|H_{i,j}|^2\right]$, this corresponds to the classical Gaussian unitary ensemble (GUE) of random matrix theory (Mehta, 2004; Forrester, 2010). When one diagonalizes this matrix $\hat{H}$ in Eq. (6.3), one obtains N real eigenvalues $\{x_1, x_2, \ldots, x_N\}$ describing the energy spectrum. The global minimum $X_{\min}$ is thus the ground-state energy E_0, a basic quantity in disordered systems, as it determines the behavior of the system at temperature $T = 0$ (Bouchaud and Mézard, 1997). As the disorder varies, the ground-state energy $X_{\min} \equiv E_0$ fluctuates from one sample of disorder to another, and determining its distribution is one of the fundamental problems in disordered systems. An equally important issue is what happens when the temperature increases from $T = 0$. As the temperature increases, more and more excited states will contribute to the behavior of the system. Hence, it is equally important to understand the statistics of the excited states of the spectrum. Indeed, if one orders these eigenvalues as in Eq. (6.1), then $X_{\min} \equiv M_{N,N}$ denotes the ground-state energy E_0, while $M_{N-1,N}$ represents the energy of the first excited state E_1, etc. More generally, $M_{N-k,N}$ denotes the energy of the kth excited state. Hence, in this context, we see that the statistics of the ordered maxima (or equivalently ordered minima) are highly relevant in disordered systems. Another important quantity is the gap in the spectrum between the ground state and the first excited state, denoted by $g_{N-1,N}$ as in Eq. (6.2). This gap plays an important role in the dynamics of a disordered system at very low temperature since it controls the timescale of relaxation toward the equilibrium. At finite temperature, however, the dynamics will be controlled by a number of excited states close to the ground state. The relevant quantity here is then the density of excited states near the ground state.

In general, studying the order statistics for a generic disordered Hamiltonian $H_{i,j}$ is very hard. There is, however, a very simple and useful model of disordered systems, known as the *random energy model* (REM) and introduced by Derrida (1981), where one assumes that the N levels of the spectrum are independent random variables. Even though this model is rather simple, it still reproduces many non-trivial features of more realistic disordered systems. Studying the spectrum of the REM is then equivalent to studying the order statistics in the IID model, which we discuss below in detail. Even though in the context of disordered systems it is natural to study ordered minima (the ground state, the first excited state, the second excited state, etc.), for the IID model one can equivalently study the ordered maxima. In fact, in this book, for convenience, we will study the ordered maxima (instead of the ordered minima). In addition to the statistics of ordered maxima, we will also study another relevant quantity, called the density of near-extreme events, discussed in the introduction of this book and defined in Eq. (1.7), that measures how the ordered maxima are organized close to the global maximum. In other words, how *crowded* the neighborhood of the global maximum is. When the maxima and minima have equivalent statistics (as in the IID model with a symmetric parent distribution $p(x)$), this density of near-extreme events is exactly equivalent to the density of states near the ground state in the language of disordered systems, where this quantity plays an important role in the dynamics at finite temperature, as discussed above.

After this IID example, we will also study the order statistics and other related quantities in a random walk sequence, in the same spirit as the rest of the book.

6.1 Order statistics of IID random variables

We start with the IID sequence $\{x_1, x_2, \ldots, x_N\}$ of N random variables, each drawn from the PDF $p(x)$. It will be useful to define the cumulative distribution of an entry as

$$P_<(x) = \text{Prob}(x_n \leq x) = \int_{-\infty}^{x} p(x') \, dx'. \tag{6.4}$$

We order the entries $\{x_1, x_2, \ldots, x_N\}$ in a decreasing sequence $\{M_{1,N}, M_{2,N}, \ldots, M_{N,N}\}$ as in Eq. (6.1). Let us denote the full joint PDF of the ordered maxima as

$$\text{Prob}(M_{1,N} = m_1, M_{2,N} = m_2, \ldots, M_{N,N} = m_N) = p_N(m_1, \ldots, m_N). \tag{6.5}$$

For IID random variables, it is straightforward to explicitly write down $p_N(m_1, \ldots, m_N)$ as

$$p_N(m_1, \ldots, m_N) = N! \prod_{i=1}^{N} p(m_i) \prod_{i=1}^{N-1} \Theta(m_i - m_{i+1}), \tag{6.6}$$

where the product of theta functions ensures the ordering of the variables (remember that $\Theta(x) = 1$ if $x > 0$ while $\Theta(x) = 0$ if $x < 0$). The global prefactor $N!$ comes from the fact that there are $N!$ ordered sectors.

Little exercise 6.1 *Show that the joint PDF in Eq. (6.6) is normalized to unity.*

6.1.1 Finite sample

We first consider a sequence of finite size N. We denote the CDF of the kth maximum as

$$Q_{k,N}(m) = \text{Prob}(M_{k,N} \leq m). \tag{6.7}$$

Once we know the CDF, one can compute the PDF by taking a derivative:

$$P_{k,N}(m) = \text{Prob}(M_{k,N} = m) = \frac{\partial}{\partial m} Q_{k,N}(m). \tag{6.8}$$

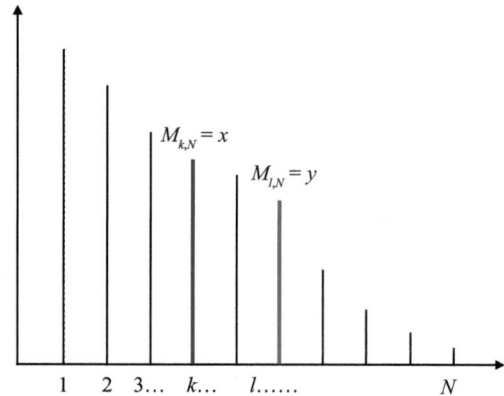

Fig. 6.1 Illustration of an ordered sequence $M_{1,N} > M_{2,N} > \cdots > M_{N,N}$ of N IID random variables.

This PDF can be obtained from the full joint PDF in Eq. (6.6) by fixing $m_k = m$ and integrating over m_i for $i \neq k$ such that $m_1 > \cdots > m_{k-1} > m > m_{k+1} > \cdots > m_N$:

$$P_{k,N}(m) = N! \, p(m) \int_m^\infty \mathrm{d}m_{k-1} \, p(m_{k-1}) \prod_{j=1}^{k-2} \int_{m_{j+1}}^\infty \mathrm{d}m_j \, p(m_j)$$

$$\times \int_{-\infty}^m \mathrm{d}m_{k+1} \prod_{j=k+2}^{N} \int_{-\infty}^{m_{j-1}} \mathrm{d}m_j \, p(m_j). \tag{6.9}$$

It is then straightforward to check (for instance by induction) that $P_{k,N}$ in Eq. (6.9) can be written as

$$P_{k,N}(m) = \frac{N!}{(k-1)!(N-k)!} p(m)[P_<(m)]^{N-k}[1 - P_<(m)]^{k-1}, \tag{6.10}$$

where $P_<(m)$ is the CDF of an entry defined in Eq. (6.4).

Note that Eq. (6.10) can also be directly obtained by noticing that the event that $m < M_{k,N} < m + \mathrm{d}m$ is the same as the event where $x_n \geq m + \mathrm{d}m$ for $(k-1)$ of the x_n, $m < x_n < m + \mathrm{d}m$ for exactly one of the x_i's and $x_n \leq m$ for the $N - k$ remaining x_n's. In particular, by setting $k = 1$ we recover the PDF of the global $X_{\max} = M_{1,N}$,

$$P_{1,N} = Np(m)[P_<(m)]^{N-1}, \tag{6.11}$$

as in Eq. (4.7).

Going beyond the one-point PDF in Eq. (6.9), we can also compute the two-point PDF, i.e., the joint PDF of $M_{k,N}$ and $M_{l,N}$, where $l \geq k+1$ (see Fig. 6.1):

$$\mathrm{Prob}(M_{k,N} = x, M_{l,N} = y) = A_{k,l}(N)[1 - P_<(x)]^{k-1}p(x)[P_<(x) - P_<(y)]^{l-k-1}$$

$$\times p(y)[P_<(y)]^{N-l}\Theta(x - y), \tag{6.12}$$

where the prefactor $A_{k,l}(N)$ is given by

$$A_{k,l}(N) = \frac{\Gamma(N+1)}{\Gamma(k)\Gamma(l-k)\Gamma(N-k+1)},\qquad(6.13)$$

with $\Gamma(z)$ being the standard gamma function. To understand Eq. (6.12) we first break the full sequence of size N into three blocks: (i) $[1, k-1]$, (ii) $[k+1, l-1]$ and (iii) $[l+1, N]$, with the kth and lth entries separating the first and the last two blocks respectively. We first choose the kth and the lth entries to have values x and y respectively. The probability of this event is proportional to $p(x)p(y)$, given that they are independent. Given these two ordered entries $M_{k,N} = x$ and $M_{l,N} = y$, with $x \geq y$, all $k-1$ entries in the first block should be bigger than x, which happens with probability $[1 - P_<(x)]^{k-1}$. Similarly, all $N - l$ entries of the last block should be less than y, which happens with probability $[P_<(y)]^{N-l}$. Finally, the $l - k - 1$ variables belonging to the middle block must lie between y and x, which happens with probability $\left[\int_y^x p(x')\,dx'\right]^{l-k-1} = [P_<(x) - P_<(y)]^{l-k-1}$. Using the independence between these three blocks of ordered entries and the kth and lth entries, one finally obtains the formula in Eq. (6.12). It then remains to fix the overall prefactor $A_{k,l}(N)$. This is done by integrating the two-point PDF in Eq. (6.12) over one of the variables, say y. This should produce the marginal one-point PDF given in Eq. (6.10). We leave it as an exercise to the reader to show that $A_{k,l}(N)$ is indeed given by Eq. (6.13) (see Exercise 6.1 at the end of this section). Moreover, the reader is expected to find a combinatorial meaning of the prefactor $A_{k,l}(N)$.

We now consider the distribution of the gap $g_{k,N} = M_{k,N} - M_{k+1,N}$ between the kth and $(k+1)$th maxima. This can be obtained from the two-point joint PDF in Eq. (6.12) by setting $l = k + 1$. The distribution of the gap is then given by

$$p_{k,N}(g) = \text{Prob}(g_{k,N} = g)$$

$$= \int_{-\infty}^{\infty}\int_{-\infty}^{\infty} dx\,dy\ \text{Prob}(M_{k,N} = x, M_{k+1,N} = y)\delta(y - x - g),\qquad(6.14)$$

where the two-point joint PDF $\text{Prob}(M_{k,N} = x, M_{l,N} = y)$ is given in Eq. (6.12). Performing the integral over y leads to

$$p_{k,N}(g) = \Theta(g)\frac{N!}{(N-k-1)!(k-1)!}\int_{-\infty}^{\infty} dx\, p(x)p(x-g)$$

$$[P_<(x-g)]^{N-k-1}[1 - P_<(x)]^{k-1}.\qquad(6.15)$$

This is an exact formula valid for arbitrary N and for all $k = 1, 2, \ldots, N-1$. Interestingly, for the positive exponentially distributed random variables $p(x) = \Theta(x)\mathrm{e}^{-x}$, the integral in Eq. (6.15) can be worked out explicitly (as a little exercise!) leading to the result

$$p_{k,N}(g) = \Theta(g)k\mathrm{e}^{-kg},\qquad(6.16)$$

which is completely independent of N.

The computation for the two-point joint PDF in Eq. (6.12) can be straightforwardly generalized to the n-point joint PDF for arbitrary $n > 2$. This can be done essentially by fixing the values of the n ordered maxima, which breaks the sequence into $(n + 1)$ independent compartments, as in the $n = 2$ case discussed in detail above. We leave it the reader as an exercise to write down this n-point joint distribution with the correct prefactor (see Exercise 6.2).

All these formulae for various order statistics of IID random variables hold for arbitrary N. However, for finite N, they depend explicitly on the parent distribution $p(x)$. As in the case of the global maximum, in the large-N limit, some universal behaviors also emerge for more general order statistics, depending on the tails of $p(x)$ at large x. We now discuss this large-N limit in detail.

6.1.2   Asymptotic results for large N

In order to analyze the large-N limit of the distribution of the kth maximum $M_{k,N}$ it is useful to recall the scaling results for $k = 1$, i.e., for the global maximum, as discussed in detail in Chapter 4. We saw there that simplifications occur in the large-N limit, leading to the emergence of three different large-N scaling behaviors of the cumulative distribution $Q_{1,N}(m)$ of the first maximum. Which of the three limiting behaviors is realized depends on the large-x tail of the parent distribution $p(x)$. More precisely, we have seen that for large N the CDF of $M_{1,N}$, properly centered and scaled, approaches the following scaling form:

$$Q_{1,N}(m) \xrightarrow[N \to \infty]{} F_\rho\left(\frac{m - a_N}{b_N}\right), \quad \rho = \mathrm{I}, \mathrm{II}, \mathrm{III}. \qquad (6.17)$$

The indices I, II and III correspond respectively to the Gumbel, Eq. (4.25), Fréchet, Eq. (4.31), and Weibull, Eq. (4.35), in Chapter 4. The scale factors a_N and b_N are non-universal and depend on the tail of $p(x)$, as discussed in Chapter 4.

We now want to do a similar analysis for other ordered maxima, going beyond $k = 1$. In other words, we want to find out the large-N scaling forms for the cumulative distribution $Q_{k,N}(m)$ for $k \geq 1$, analogous to the scaling forms in Eq. (6.17) that are valid only for $k = 1$. It turns out that a similar scaling form holds for any $k \geq 1$ with the same centering and scale factors a_N and b_N as in the $k = 1$ case, Eq. (6.17). However, the limiting scaling functions are still of three varieties but they do depend explicitly on the order k. Before stating the general results, it would perhaps be useful to the reader to work out a simple example, namely the positive exponentially distributed random variables corresponding to $p(x) = \Theta(x)e^{-x}$. In this case, the PDF $P_{k,N}$ for the kth maximum $M_{k,N}$ in Eq. (6.10) simplifies, and we get

$$P_{k,N}(m) = \frac{N!}{(k-1)!(N-k)!}e^{-km}(1 - e^{-m})^{N-k}. \qquad (6.18)$$

Now we want to take the large-N limit, keeping k fixed. In this case, in order to arrive at a non-trivial scaling form we need to explore the PDF in the region where m is also large, such that $e^{-m} = O(1/(N - k))$ (see a similar discussion around Eq. (4.10) in Chapter 4 for $k = 1$). This leads us to choose $m = \ln N + z$ (corresponding to $a_N = \ln N$ and $b_N = 1$). Substituting $m = \ln N + z$ into Eq. (6.18) and taking the large-N limit using Stirling's formula, one arrives at

$$P_{k,N}(m = \ln N + z) \xrightarrow[N \to \infty]{} \frac{1}{(k-1)!} e^{-kz} e^{-e^{-z}}, \tag{6.19}$$

which is sometimes called the *generalized Gumbel distribution*. The corresponding cumulative distribution reads

$$Q_{k,N}(m = \ln N + z) \xrightarrow[N \to \infty]{} \frac{1}{(k-1)!} \int_{-\infty}^{z} e^{-kz'} e^{-e^{-z'}} \, dz' = \frac{1}{(k-1)!} \int_{e^{-z}}^{\infty} e^{-t} t^{k-1} \, dt. \tag{6.20}$$

Using $F_I(z) = e^{-e^{-z}}$ for the Gumbel scaling function, one can express the scaling function associated with $Q_{k,N}$ in Eq. (6.20) as

$$Q_{k,N}(m = \ln N + z) \xrightarrow[N \to \infty]{} \frac{1}{(k-1)!} \int_{-\ln F_I(z)}^{\infty} e^{-t} t^{k-1} \, dt. \tag{6.21}$$

Interestingly, for all parent distributions $p(x)$ belonging to the Gumbel class, one can show that the appropriately centered and scaled CDF $Q_{k,N}(m = a_N + b_N z)$ converges, for large N, to the same scaling form as in Eq. (6.21). Here, we have demonstrated this result only for the exponential distribution, but it turns out to be much more general and holds for any $p(x)$ of this class (David and Nagaraja, 2004; Arnold *et al.*, 2008). Brave readers may venture to prove this for a general $p(x)$!

What about the other two classes, namely for $p(x)$ belonging to the Fréchet and the Weibull class? It turns out that, for arbitrary $p(x)$ belonging to these classes, the CDF $Q_{k,N}(m)$, with appropriate centering and scaling $m = a_N + b_N z$ (note that a_N and b_N do not depend on k), converges for large N to (David and Nagaraja, 2004; Arnold *et al.*, 2008)

$$\lim_{N \to \infty} Q_{k,N}(a_N + b_N z) = \frac{1}{(k-1)!} \int_{[-\ln F_\rho(z)]}^{\infty} e^{-t} t^{k-1} \, dt, \tag{6.22}$$

where $F_\rho(z)$, $\rho = \mathrm{I}, \mathrm{II}, \mathrm{III}$, correspond respectively to the Gumbel, Fréchet and the Weibull families. Note that the scaling result in Eq. (6.22) is for the CDF of $M_{k,N}$ in the large-N limit. An analogous result can be obtained for the PDF by differentiating the result in Eq. (6.22) with respect to z, leading to

$$P_{k,N}(m) = \frac{\partial}{\partial m} Q_{k,N}(m) \xrightarrow[N \to \infty]{} \frac{1}{b_N} f_{k,\rho}\left(\frac{m - a_N}{b_N}\right), \tag{6.23}$$

where the scaling function $f_{k,\rho}(z)$ reads

$$f_{k,\rho}(z) = F'_\rho(z) \frac{[-\ln F_\rho(z)]^{k-1}}{(k-1)!}, \tag{6.24}$$

with $\rho = \mathrm{I}, \mathrm{II}, \mathrm{III}$ corresponding respectively to the Gumbel, Fréchet and Weibull families. For instance, for the Gumbel case, i.e., $\rho = \mathrm{I}$, using $F_{\mathrm{I}}(z) = \mathrm{e}^{-\mathrm{e}^{-z}}$, one recovers the result in Eq. (6.19).

One can repeat this exercise to obtain the multi-point joint PDF of the ordered maxima in the scaling limit for large N, going beyond the one-point function. For example, for the two-point function, we already obtained the exact result for the joint PDF for finite N in Eq. (6.12). One can take the large-N limit of this formula as we did for the one-point case and obtain the scaling result for the two-point function. One can similarly repeat this result for an arbitrary k-point joint PDF. In particular, a simple case would be to compute the joint PDF of the first k ordered maxima $\{M_{1,N}, M_{2,N}, \ldots, M_{k,N}\}$. We do not carry out this exercise in detail here but just quote the final result (interested readers may want to carry out this exercise). In the scaling limit for large N, one can show that the ordered maxima, centered by the shift a_N and further scaled by b_N, converge jointly in law to a set of k random variables $\{W_1, W_2, \ldots, W_k\}$ (David and Nagaraja, 2004; Arnold *et al.*, 2008), i.e.,

$$\left(\frac{M_{1,N} - a_N}{b_N}, \frac{M_{2,N} - a_N}{b_N}, \ldots, \frac{M_{k,N} - a_N}{b_N} \right) \xrightarrow[N\to\infty]{} (W_1, \ldots, W_k), \qquad (6.25)$$

where the scaled variables W_i are jointly distributed via the joint PDF

$$\mathrm{Prob}(W_1 = w_1, \ldots, W_k = w_k) = F_\rho(w_k) \prod_{i=1}^{k} \frac{F_\rho'(w_i)}{F_\rho(w_i)}, \quad w_1 > \cdots > w_k, \qquad (6.26)$$

where we recall that $\rho = \mathrm{I}, \mathrm{II}, \mathrm{III}$ correspond respectively to the Gumbel, Fréchet and Weibull families. Note again that the scale factors a_N and b_N in Eq. (6.25) do not depend on k.

Little exercise 6.2: Show that, by integrating (6.26) over $w_1, w_2, \cdots, w_{k-1}$, one recovers the marginal distribution of w_k in Eq. (6.24).

Limiting distribution of the gap.   We now consider the distribution of the gap $g_{k,N} = M_{k,N} - M_{k+1,N}$ in the scaling limit for large N. In fact, for finite N, the exact formula for the gap distribution is given in Eq. (6.15). We can either take the large-N scaling limit in Eq. (6.15), but this turns out to be a bit cumbersome. It is actually easier to start already from the scaled joint distribution of $M_{k,N}$ and $M_{k+1,N}$ and obtain from that the scaled limiting distribution of the gap. We proceed as follows. We start from Eq. (6.26) with $k + 1$ variables $\{W_1, W_2, \ldots, W_k, W_{k+1}\}$ and integrate out the first $(k-1)$ variables to obtain the joint marginal distribution $p_\rho(w_k, w_{k+1})$,

$$p_\rho(w_k, w_{k+1}) = \int \mathrm{d}w_1 \cdots \mathrm{d}w_{k-1} \ \mathrm{Prob}(W_1 = w_1, \ldots, W_k = w_k, W_{k+1} = w_{k+1}).$$
$$(6.27)$$

We then substitute the expression of the joint PDF from Eq. (6.26) into the multiple integral in Eq. (6.27). This leads to an apparently complicated multiple *nested* integral. However, the good news is that this multiple integral can be explicitly performed using the "chain rule." In fact, computing such nested integrals in this way is quite common in many problems involving IID variables. Hence it is useful to illustrate this technique using this specific example.

In fact, the first step is to write down the multiple integral in Eq. (6.27) explicitly (without being afraid!). This leads to

$$p_\rho(w_k, w_{k+1}) = F_\rho'(w_{k+1}) \frac{F_\rho'(w_k)}{F_\rho(w_k)} I_{k-1}(w_k) \Theta(w_k - w_{k+1}), \qquad (6.28)$$

where

$$I_{k-1}(x) = \int_x^\infty dw_{k-1} \frac{F_\rho'(w_{k-1})}{F_\rho(w_{k-1})} \int_{w_{k-1}}^\infty dw_{k-2} \frac{F_\rho'(w_{k-2})}{F_\rho(w_{k-2})} \cdots \int_{w_2}^\infty dw_1 \frac{F_\rho'(w_1)}{F_\rho(w_1)}. \qquad (6.29)$$

To compute this nested integral, let us take a derivative with respect to x and find a nice recursion relation,

$$\frac{d}{dx} I_{k-1}(x) = -\frac{F_\rho'(x)}{F_\rho(x)} I_{k-2}(x), \quad \text{for } k \geq 2, \qquad (6.30)$$

starting from $I_0(x) = 1$. This recursion relation can be solved by the standard generating function method. However, this example is rather simple; one can obtain the solution of the recursion relation in Eq. (6.30) satisfying $I_0(x) = 1$ just by inspection:

$$I_{k-1}(x) = \frac{[-\ln F_\rho(x)]^{k-1}}{(k-1)!}, \quad k \geq 1. \qquad (6.31)$$

Substituting this expression for $I_{k-1}(x)$ into Eq. (6.28), we obtain the final expression for the scaled joint PDF of two consecutive maxima,

$$p_\rho(w_k, w_{k+1}) = F_\rho'(w_{k+1}) \frac{F_\rho'(w_k)}{F_\rho(w_k)} \frac{[-\ln F_\rho(w_k)]^{k-1}}{(k-1)!} \Theta(w_k - w_{k+1}), \qquad (6.32)$$

which is valid for all $k \geq 1$. This means that the joint PDF of $M_{k,N}$ and $M_{k+1,N}$ takes the scaling form in the large-N limit of

$$P_{k,N}^{\text{joint}}(m_k, m_{k+1}) = \text{Prob}(M_{k,N} = m_k, M_{k+1,N} = m_{k+1})$$

$$\xrightarrow[N \to \infty]{} \frac{1}{b_N^2} p_\rho \left(\frac{m_k - a_N}{b_N}, \frac{m_{k+1} - a_N}{b_N} \right), \qquad (6.33)$$

where the scaling function $p_\rho(w_k, w_{k+1})$ is given in Eq. (6.32).

Hence, the distribution of the gap $g_{k,N} = M_{k,N} - M_{k+1,N}$ is given by

$$p_{k,N}(g) = \text{Prob}(g_{k,N} = g) = \int_{-\infty}^\infty dm_k \int_{-\infty}^\infty dm_{k+1} P_{k,N}^{\text{joint}}(m_k, m_{k+1}) \delta(m_k - m_{k+1} - g), \qquad (6.34)$$

where $P_{k,N}^{\text{joint}}(m_k, m_{k+1})$ is the joint PDF of $M_{k,N}$ and $M_{k+1,N}$ that we computed above. Substituting the scaling form of the joint PDF from Eq. (6.33) into the double integral in Eq. (6.34), it is easy to check (simple exercise!) that the gap distribution also takes a scaling form,

$$p_{k,N}(g) \xrightarrow[N\to\infty]{} \frac{1}{b_N} \tilde{p}_{k,\rho}\left(\frac{g}{b_N}\right), \tag{6.35}$$

with the scaling function $\tilde{p}_{k,\rho}(z)$ given by

$$\tilde{p}_{k,\rho}(z) = \int_{-\infty}^{\infty} dw_k \int_{-\infty}^{\infty} dw_{k+1}\, p_\rho(w_k, w_{k+1})\delta(w_k - w_{k+1} - z). \tag{6.36}$$

Note that the gap distribution for large N in Eq. (6.35) does not depend on a_N, but only on b_N, as expected. Substituting the explicit expression of $p_\rho(w_k, w_{k+1})$ from Eq. (6.32) and carrying out the integration over w_k (trivially using the δ-function) and changing w_{k+1} to x, we finally get (Schehr and Majumdar, 2014)

$$\tilde{p}_{k,\rho}(z) = \frac{\Theta(z)}{(k-1)!} \int_{-\infty}^{\infty} F_\rho'(x) \frac{F_\rho'(z+x)}{F_\rho(z+x)} [-\ln F_\rho(z+x)]^{k-1}\, dx. \tag{6.37}$$

This is the final formula for the scaled gap distribution, valid for all $k \geq 1$ and for the three classes $\rho = \mathrm{I, II, III}$. We now analyze the scaling function for these three classes.

Class I (Gumbel). For the Gumbel universality class, i.e., $\rho = \mathrm{I}$, using $F_\mathrm{I}(z) = e^{-e^{-z}}$, the integral in Eq. (6.37) can be performed explicitly and we get (check it out!)

$$\tilde{p}_{k,I}(z) = \Theta(z)ke^{-kz}, \tag{6.38}$$

which is completely consistent with the exact result obtained for finite N for the exponential distribution in Eq. (6.16) upon identifying $b_N = 1$.

Class II (Fréchet). For the Fréchet universality class, i.e., $\rho = \mathrm{II}$, one finds from Eq. (6.37) using $F_\mathrm{II}(z) = \Theta(z)e^{-z^{-\alpha}}$ with $\alpha > 0$:

$$\tilde{p}_{k,II}(z) = \Theta(z)\frac{\alpha^2}{(k-1)!} \int_0^{\infty} e^{-x^{-\alpha}} x^{-\alpha-1}(x+z)^{-\alpha k-1}\, dx. \tag{6.39}$$

For generic $\alpha > 0$ this integral is difficult to perform explicitly. However, one can easily extract the large-z behavior (little exercise!):

$$\tilde{p}_{k,II}(z) \underset{z\to\infty}{\sim} \frac{\alpha}{(k-1)!} z^{-\alpha k-1}. \tag{6.40}$$

It has a power-law decay with an exponent which depends on k. For $\alpha = 1$, it turns out that the integral in Eq. (6.39) can be explicitly evaluated,

$$\tilde{p}_{k,II}(z) = \Theta(z)k(k+1)z^{-1-k}U(k+1,0,1/z), \tag{6.41}$$

where $U(a, b, x)$ is the confluent (Tricomi) hypergeometric function. Using the small-x behavior $U(k+1, 0, x = 0) = 1/(k+1)!$, one finds that the large-z behavior coincides with the result in Eq. (6.40) for $\alpha = 1$. We recall that the Fréchet universality class for the extreme statistics with parameter $\alpha = 1$ arises when the IID entries are Cauchy distributed [see Eq. (2.46)].

Class III (Weibull). Finally, for the Weibull universality class, i.e., $\rho = \mathrm{III}$, where $F_{\mathrm{III}}(z) = e^{-|z|^\alpha}$ for $z < 0$ and $F_{\mathrm{III}}(z) = 1$ for $z > 0$. This class is parameterized by $\alpha > 0$ [see Eq. (4.35)]. In this case, the integral in Eq. (6.37) reads

$$\tilde{p}_{k,\mathrm{III}}(z) = \Theta(z)\frac{\alpha^2}{(k-1)!}\int_0^\infty (x+z)^{\alpha-1}e^{-(x+z)^\alpha}x^{\alpha k-1}\,\mathrm{d}x. \tag{6.42}$$

As a little exercise, the reader should try to derive the expression in Eq. (6.42) starting from Eq. (6.37) and using the expression for $F_{\mathrm{III}}(z)$ (hint: one should notice that, given that $F'_{\mathrm{III}}(z) = 0$ for $z > 0$, the integrand in Eq. (6.37) is supported over $(-\infty, -z)$). Finally, one can make a change of variable from x to $y = -z - x$ such that the integral over y is supported over $[0, +\infty)$. For generic α it is hard to perform this integral explicitly for all z. However, one can easily extract the large-z asymptotic behavior:

$$\tilde{p}_{k,\mathrm{III}}(z) \underset{z\to\infty}{\sim} \frac{\alpha^2}{(k-1)!}\alpha^{-k\alpha}\Gamma(\alpha k)z^{(1-\alpha)(\alpha k-1)}e^{-z^\alpha}. \tag{6.43}$$

In the special case $\alpha = 1$, Eq. (6.42) simplifies and one gets a pure exponential,

$$\tilde{p}_{k,\mathrm{III}}(z) = \Theta(z)e^{-z}, \tag{6.44}$$

independent of k. Note that this case $\alpha = 1$ can be realized by uniform distribution of the IID entries in the time series.

An application. In this section we have summarized the order and gap statistics for N independent random variables. As mentioned before, this has interesting applications in disordered systems. Let us just mention briefly one recent example of these applications. Consider N IID random variables, $\{x_1, x_2, \dots, x_N\}$, and order them:

$$X_{\max} = M_{1,N} > M_{2,N} > \cdots > M_{N,N} = X_{\min}. \tag{6.45}$$

In this example, the x_i represent N energy levels of a disordered system, with $M_{N,N}$ being the lowest-energy state. We now consider K non-interacting fermions whose single-particle energy levels are represented by the $M_{j,N}$ above. In the ground state of these K fermions, we fill up the first K levels with one fermion in each level—this constraint arises from the Pauli exclusion principle. Thus, the ground-state energy of the system of K fermions is given by the partial sum

$$E_0 = \sum_{j=N-K+1}^{N} M_{j,N}. \tag{6.46}$$

This value E_0 is clearly a random variable, which fluctuates from one realization of the x_i to another. Using the results for the order statistics discussed here, the full

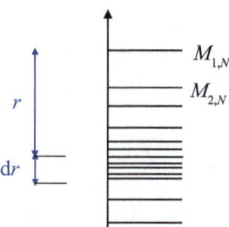

Fig. 6.2 A schematic representation of the ordered entries as levels marked by $M_{1,N} > M_{2,N} > \cdots$. The density of states $\rho(r, N)\,\mathrm{d}r$ represents the fraction of levels in the interval $[r, r + \mathrm{d}r]$ where r is measured from the maximum $X_{\max} = M_{1,N}$.

distribution of E_0 for all K and N has been computed, with interesting limiting results for large K and large N (Schawe *et al.*, 2018).

6.1.3 Statistics of near-extreme events for IID

As discussed at the beginning of this chapter, another important observable is the density of near-extremes in the IID time series $\{x_1, x_2, \ldots, x_N\}$, defined as (Sabhapandit and Majumdar, 2007)

$$\rho(r, N) = \frac{1}{N-1} \sum_{x_i \neq X_{\max}} \delta(X_{\max} - x_i - r), \qquad (6.47)$$

where $X_{\max} = M_{1,N}$ is the global maximum in the time series and the sum runs over all the entries excluding this global maximum. First, note that $\rho(r, N)$ is a random variable and thus depends on each realization of the entries. For a given realization, $\rho(r, N)$ is normalized to unity, i.e., $\int_0^\infty \rho(r, N)\,\mathrm{d}r = 1$, and $\rho(r, N)\,\mathrm{d}r$ thus measures the fraction of entries whose values lie in the interval $[r, r + \mathrm{d}r]$ where the distance r is measured from the maximum $X_{\max} = M_{1,N}$ (see Fig. 6.2). Informally speaking, $\rho(r, N)$ gives a measure of how "lonely" the maximum is at the top, i.e., it gives a measure of the "crowding" of the entries near the maximum. This justifies the name "density of near-extreme events," and its behavior for IID random variables has been studied in detail in Sabhapandit and Majumdar (2007)—here we outline the key steps of this analysis.

Averaging over the realizations of the time series, with each entry drawn independently from the PDF $p(x)$, gives the average density of states (DOS) near the maximum:

$$\overline{\rho(r, N)} = \int_{-\infty}^{\infty} \rho(r, N) \prod_{i=1}^{N} p(x_i)\,\mathrm{d}x_i, \qquad (6.48)$$

with $\rho(r, N)$ given in Eq. (6.47). This averaging can be performed in two steps: (i) we first fix the value of the global maximum $X_{\max} = x$ and calculate the conditional averaged DOS $\overline{\rho_{\mathrm{cond}}(r, N \mid x)}$ obtained by averaging over the $N-1$ entries; (ii) finally averaging $\rho_{\mathrm{cond}}(r, N \mid x)$ over the distribution of the global maximum $X_{\max} = x$.

Step (i). Once we fix the value of the global maximum to be x, the rest of the $N - 1$ entries remain IID but each of them distributed via the effective PDF

$$p_{\text{eff}}(x_i) = \frac{p(x_i)}{\int_{-\infty}^{x} p(y)\,\mathrm{d}y}, \quad x_i \in (-\infty, x]. \tag{6.49}$$

Hence, the conditional average DOS is given by

$$
\overline{\rho_{\text{cond}}(r, N \mid x)} = \frac{1}{N-1} \sum_{x_i \neq x} \int_{-\infty}^{\infty} \delta(x - x_i - r) p_{\text{eff}}(x_i)\,\mathrm{d}x_i
$$
$$
= \frac{p(x - r)}{\int_{-\infty}^{x} p(y)\,\mathrm{d}y}. \tag{6.50}
$$

Step (ii). The distribution of $X_{\max}$ for N IID variables, on the other hand, is given by [see Eq. (4.7)]

$$P_{\max}(x, N) = Np(x)\left[\int_{-\infty}^{x} p(y)\,\mathrm{d}y\right]^{N-1}. \tag{6.51}$$

Hence, averaging over this global maximum, we finally get the average DOS

$$\overline{\rho(r, N)} = \int_{-\infty}^{\infty} \overline{\rho_{\text{cond}}(r, N \mid x)} P_{\max}(x, N)\,\mathrm{d}x. \tag{6.52}$$

Using Eqs. (6.50) and (6.51) we get

$$
\overline{\rho(r, N)} = N \int_{-\infty}^{\infty} p(x - r)p(x)\left(\int_{-\infty}^{x} p(y)\,\mathrm{d}y\right)^{N-2} \mathrm{d}x
$$
$$
= \frac{N}{N-1} \int_{-\infty}^{\infty} p(x - r)P_{\max}(x, N - 1)\,\mathrm{d}x, \tag{6.53}
$$

where $P_{\max}(x, N)$ is given in Eq. (6.51). This result is valid for arbitrary N, and we now analyze its scaling behavior for large N.

We recall, from Eq. (6.17), that in the limit $N \to \infty$ the CDF of the maximum $X_{\max}$ takes the scaling form

$$\int_{-\infty}^{x} P_{\max}(x', N)\,\mathrm{d}x' \approx F_{\rho}\left(\frac{x - a_N}{b_N}\right), \tag{6.54}$$

with $\rho = \mathrm{I}, \mathrm{II}, \mathrm{III}$ for $p(x)$ belonging respectively to the Gumbel, Fréchet or Weibull universality class. Taking the derivative of Eq. (6.54) with respect to x gives the scaling form of the PDF:

$$P_{\max}(x, N) \approx \frac{1}{b_N} f_\rho\left(\frac{x - a_N}{b_N}\right), \quad f_\rho(z) = F'_\rho(z). \tag{6.55}$$

Note that the scaled PDF $f_\rho(z)$ is normalized to unity, i.e.,

$$\int_{-\infty}^{\infty} f_\rho(z)\, \mathrm{d}z = 1. \tag{6.56}$$

We now substitute the form in Eq. (6.55) into Eq. (6.53) and make the change of variable $x = a_N + b_N z$ to get, for large N,

$$\overline{\rho(r, N)} \approx \int_{-\infty}^{\infty} f_\rho(z) p(a_N + b_N z - r)\, \mathrm{d}z. \tag{6.57}$$

The asymptotic behavior of this integral depends crucially on how the scale factor b_N behaves for large N. We recall from Chapter 4 that if $p(x)$ has a tail that decays slower than e^{-x} for large x (for example as a stretched exponential $p(x) \propto \mathrm{e}^{-x^\delta}$ with $0 < \delta < 1$, or as a power law $p(x) \propto x^{-1-\alpha}$ with $\alpha > 0$) then b_N diverges as $N \to \infty$. In contrast, if $p(x)$ decays faster than exponentially (for instance as a stretched exponential $p(x) \propto \mathrm{e}^{-x^\delta}$ with $\delta > 1$, or a bounded distribution over a finite interval, such as the uniform distribution) then $b_N \to 0$ as $N \to \infty$. The case where $p(x)$ decays exponentially, i.e., $p(x) \sim \mathrm{e}^{-x}$ for large x, is a borderline "marginal" case where b_N approaches a constant as $N \to \infty$. Consequently, the asymptotic large-N behavior of $\overline{\rho(r, N)}$ in Eq. (6.57) behaves quite differently in the three cases, as discussed in detail below.

- The case where $b_N \to \infty$ (slower than exponential tail of $p(x)$): in this case we first rewrite the expression $p(a_N + b_N z - r)$ that appears inside the integrand in Eq. (6.57) as

$$p(a_N + b_N z - r) = \frac{1}{b_N} b_N p\left(b_N\left(z - \frac{r - a_N}{b_N}\right)\right). \tag{6.58}$$

When $b_N \to \infty$, if $(z - (r - a_N)/b_N) \neq 0$, the argument of the function p goes to $\pm\infty$, and consequently $p(b_N(z - (r - a_N)/b_N)) \to 0$ as $N \to \infty$. In contrast, if $(z - (r - a_N)/b_N) = 0$, then $b_N p(b_N(z - (r - a_N)/b_N)) = b_N p(0) \to \infty$ as $N \to \infty$. Note, however, that the function $b_N p(b_N(z - (r - a_N)/b_N))$ is normalized to unity when integrated over $z \in (-\infty, +\infty)$. Consequently, we conclude that

$$b_N p\left(b_N\left(z - \frac{r - a_N}{b_N}\right)\right) \xrightarrow[N \to \infty]{} \delta\left(z - \frac{r - a_N}{b_N}\right). \tag{6.59}$$

We just substitute this delta function into Eq. (6.57) to obtain the large-N scaling behavior

$$\overline{\rho(r, N)} \underset{N \to \infty}{\approx} \frac{1}{b_N} f_\rho\left(\frac{r - a_N}{b_N}\right). \tag{6.60}$$

Thus, in this case, when $b_N \to \infty$ the average DOS near the maximum is controlled by the asymptotic behavior of the maximum itself.

- The case where $b_N \to 0$ (faster than the exponential tail of $p(x)$): the distribution of the maximum becomes extremely narrow close to its peak at a_N. We can then directly put $b_N = 0$ in Eq. (6.57). Using the normalization condition of $f_\rho(z)$ in Eq. (6.56) we obtain

$$\overline{\rho(r, N)} \underset{N \to \infty}{\approx} p(a_N - r). \tag{6.61}$$

Here, the asymptotic behavior of the average DOS near the maximum is instead controlled by the fluctuations of the other entries close to the maximum and is simply given by the parent PDF of any of the variables evaluated at a distance r from the location of the maximum, which is simply a_N in this case.

- The case where $b_N \to b = const.$ (exponential tail of $p(x)$): the limiting scaled PDF of $X_{\max}$ is given by $f_I(z) = e^{-z - e^{-z}}$. Setting $b_N = b$ in Eq. (6.57), we find that the the averaged DOS near the extreme takes the limiting form

$$\overline{\rho(r, N)} \underset{N \to \infty}{\approx} g(r - a_N), \quad \text{where } g(z) = \int_{-\infty}^{\infty} e^{-y - e^{-y}} p(by - z) \, dy. \tag{6.62}$$

In this marginal case, the limiting scaling function is non-universal and depends on the full parent distribution $p(x)$. For example, for positive exponentially distributed variables with $p(x) = \Theta(x) e^{-x}$, one finds the limiting scaling function explicitly:

$$g(z) = e^z (1 - (1 + e^{-z}) e^{-e^{-z}}). \tag{6.63}$$

Example 6.1 Compute the amplitude $A_{k,l}(N)$ and show that it is indeed given by the formula in Eq. (6.13). Find also a combinatorial explanation of this amplitude $A_{k,l}(N)$.

Example 6.2 Compute the n-point joint PDF of the maxima for finite N.

6.2 Order statistics for random walks

As in the rest of the book, we now address the order statistics for a time series with strongly correlated entries, namely the positions of a random walker at discrete times on a line. Consider the time series $\{x_0 = 0, x_1, x_2, \ldots, x_N\}$ where x_n represents the position of a random walker on a line at step n, starting from the origin. We consider the walk up to N steps. The position x_n evolves via

$$x_n = x_{n-1} + \eta_n, \tag{6.64}$$

starting from $x_0 = 0$, where the noise η_n's are IID jump lengths, each drawn from a symmetric and continuous jump distribution $f(\eta)$. We first order the positions in decreasing order of magnitude (see Fig. 6.3),

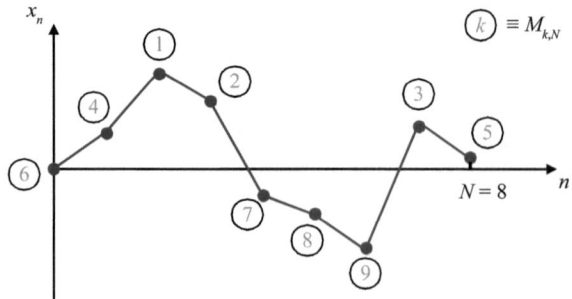

Fig. 6.3 Realization of a random walk of $N = 8$ steps. The ordered positions are marked with circles: 1 denotes the first maximum, 2 denotes the second one, etc.

$$X_{\max} = M_{1,N} > M_{2,N} > \cdots > M_{N+1,N} = X_{\min}. \qquad (6.65)$$

The distribution $X_{\max} = M_{1,N}$, i.e., the first (global) maximum, has already been studied in Chapter 4. Here we are interested in the order statistics, i.e., the statistics of the kth maximum $M_{k,N}$, for $k \geq 1$. As in the IID case in the previous section, we will also be interested in the statistics of the gap between the values of the kth and $(k+1)$th maxima, i.e.,

$$g_{k,N} = M_{k,N} - M_{k+1,N}. \qquad (6.66)$$

Below, we first discuss the statistics of the kth maximum $M_{k,N}$ and then discuss the statistics of the gap $g_{k,N}$.

In fact, the statistics of the ordered maxima $M_{k,N}$ have been studied extensively in the probability literature, under the name of "fluctuation theory," and many results are known (Wendel, 1960; Port, 1963; Dassios, 1996; Chaumont, 1999; Feller, 2008a,b), which we will recall below. However, to our knowledge the statistics of the gaps have been studied only recently and there are very few exact results known, some of which will be discussed below.

Before we start discussing the example of a random walk, let us point out in advance that the next section may appear to you a little dry and technical. The purpose of including these technical details is not just to bore you, but to illustrate the complexity that one encounters in trying to compute the order statistics of correlated systems, even for a system as simple as a random walk in one dimension. The only thing we can promise you is that at the end of all these long and boring technical details, the asymptotics of the order statistics for large N turns out to be rather simple, beautiful and universal! Unfortunately, there does not seem to be a shortcut to arrive at these final results without going through these technical ordeals. So please be patient. We hope that these technical details will be helpful to those who are interested in applying these results and techniques to other problems. Casual readers may skip the following subsection if they wish.

6.2.1 Statistics of the kth maximum

To start, our goal will be to write down an evolution equation for the cumulative distribution of the kth maximum, defined as

$$Q_{k,N}(x) = \text{Prob}(M_{k,N} \leq x). \tag{6.67}$$

Note that the event $M_{k,N} \leq x$ is equivalent to having *at most* $(k-1)$ points above the level x between step 1 and step N (see Fig. 6.4). It is natural to define the probability $U_{k,N}(x)$ as $U_{k,N}(x) = \text{Prob}[\text{there are exactly } k \text{ points above the level } x$ in an N-step trajectory starting at 0]. We have the obvious relation

$$Q_{k,N}(x) = \text{Prob}(M_{k,N} \leq x) = \sum_{m=0}^{k-1} U_{m,N}(x), \tag{6.68}$$

which can be understood very simply. If the kth maximum $M_{k,N}$ lies below x, then the number of ordered entries above x can be at most $k-1$. Note that, for $x < 0$, $U_{0,N}(x) = 0$ since, by definition, the starting point is at $x_0 = 0$. We will see that in principle it is possible to write a recursion relation for $U_{k,N}(x)$. It turns out to be a little more convenient to define the auxiliary quantity $q_{k,N}(x) = \text{Prob}[\text{there are exactly } k \text{ points below the level } 0 \text{ in an } N\text{-step trajectory starting at } x]$. For $x > 0$, it is convenient to make a shift in the original trajectory, starting at the origin, i.e., define new positions y_n such that

$$y_n = x - x_n. \tag{6.69}$$

Since $x_0 = 0$, it follows that $y_0 = x$. Also, y_k evolves by the same Markov rule, Eq. (6.64), since the jump variables are symmetric. Consequently, the event $x_n > x$ is equivalent to $y_n < 0$. Hence, for $x > 0$, we have

$$U_{k,N}(x) = q_{k,N}(x). \tag{6.70}$$

In contrast, for $x < 0$, we make the transformation

$$z_n = -x + x_n. \tag{6.71}$$

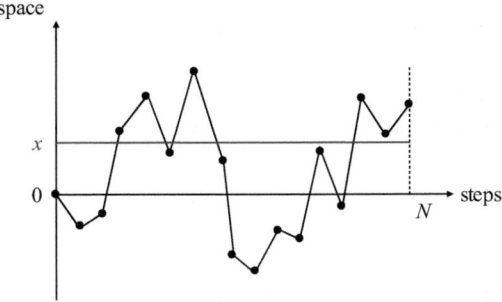

Fig. 6.4 A typical trajectory of a random walk of $N = 16$ steps with positions $\{0, x_1, x_2, \ldots, x_N\}$ marked by filled circles. In this configuration, the number of points above the level x is $k = 6$.

Therefore, the walk z_k starts at $-x > 0$ and evolves by the same Markov rule, Eq. (6.64). The probability $U_{k,N}(x)$, in terms of z_n, then corresponds to the probability that there are exactly k points above the level $z = 0$. This means that the z-trajectory, starting at $-x > 0$, has exactly $N + 1 - k$ points below $z = 0$. Hence, this gives, for $x < 0$,

$$U_{k,N}(x) = q_{N-(k-1)}(-x). \tag{6.72}$$

Therefore, using Eq. (6.68), the CDF $Q_{k,N}(x)$ can be expressed in terms of the probability $q_{k,N}(x)$ by the following relations:

$$Q_{k,N}(x) = \begin{cases} \sum_{m=0}^{k-1} q_{m,N}(x), & x > 0, \\ \sum_{m=1}^{k-1} q_{N-m+1,N}(-x), & x < 0. \end{cases} \tag{6.73}$$

These relations are valid for all $k \geq 1$. For $k = 1$, since $Q_{1,N}(x) = 0$ for $x < 0$, and the second line in Eq. (6.73) for $k = 1$ should be interpreted as zero. We now note that $q_{k,N}(x)$ is symmetrical for any x,

$$q_{k,N}(x) = q_{N+1-k,N}(-x), \tag{6.74}$$

which just follows from reversing the direction of x and the fact that there are, in total, $N + 1$ points. Using this symmetry in Eq. (6.73) we see that, for any x,

$$Q_{k,N}(x) = \sum_{m=0}^{k-1} q_{m,N}(x). \tag{6.75}$$

We now write a recursion relation for $q_{k,n}(x)$ for $x > 0$. For $x < 0$, we can use the symmetry in Eq. (6.74). By considering the stochastic jump $x \to x'$ at the first step and then subsequently using the Markov property of the evolution one gets, for $N \geq 1$ and $x > 0$ (Schehr and Majumdar, 2012; Battilana *et al.*, 2020),

$$q_{k,N}(x) = \int_0^\infty q_{k,N-1}(x')f(x' - x)\,dx' + \int_{-\infty}^0 q_{k,N-1}(x')f(x' - x)\,dx', \tag{6.76}$$

starting from $q_{0,0}(x) = 1$. The first term corresponds to a jump from $x > 0$ to $x' > 0$, see Fig. 6.5(*a*). The second term corresponds to a jump from $x > 0$ to $x' < 0$, see Fig. 6.5(*b*). For the second term, note that once the walker jumps to $x' < 0$ in the first step, for the rest of the trajectory of $N - 1$ steps it has k points below 0 (including the starting point of this trajectory starting at $x' < 0$). Changing x' to $-x'$ in the second term and using the symmetry relation in Eq. (6.74), we get

$$q_{k,N}(x) = \int_0^\infty q_{k,N-1}(x')f(x' - x)\,dx' + \int_0^\infty q_{N-k,N-1}(x')f(x' + x)\,dx', \tag{6.77}$$

where we used the symmetry of the jump distribution $f(\eta) = f(-\eta)$. To solve Eq. (6.77) it is convenient to introduce another auxiliary quantity,

$$r_{k,N}(x) \equiv q_{N-k,N}(x), \tag{6.78}$$

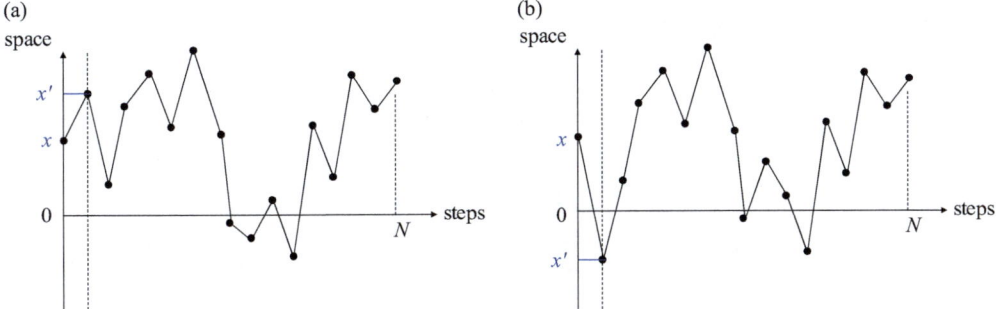

Fig. 6.5 (a) A configuration of a random walk starting at $x > 0$ and jumping at the first step to $x' > 0$. (b) A configuration of a random walk starting at $x > 0$ and jumping at the first step to $x' < 0$.

which is the probability that the random walk defined above has k points above 0 between step 1 and step N. One can then write two coupled equations for $q_{k,N}(x)$ and $r_{k,N}(x)$,

$$q_{k,N}(x) = \int_0^\infty q_{k,N-1}(x')f(x'-x)\,dx' + \int_0^\infty r_{k-1,N-1}(x')f(x'+x)dx', \qquad (6.79)$$

$$r_{k,N}(x) = \int_0^\infty r_{k-1,N-1}(x')f(x'-x)\,dx' + \int_0^\infty q_{k,N-1}(x')f(x'+x)\,dx', \qquad (6.80)$$

where the former is directly obtained from Eq. (6.77), and the latter by setting $k = N - k$ in Eq. (6.77) and using Eq. (6.78). We also have the initial conditions $q_{0,0}(x) = r_{0,0}(x) = 1$ and the convention that $q_{k,N}(x) = r_{k,N}(x) = 0$ if $k > N$.

The two coupled equations, Eqs. (6.79) and (6.80), for $q_{k,N}(x)$ and $r_{k,N}(x)$ are valid for arbitrary symmetric and continuous jump distributions $f(\eta)$. Note that these integral equations are of the Wiener–Hopf type since the limits of the integrations on the RHS range from 0 to $+\infty$ (and not from $-\infty$ to $+\infty$). Solving these coupled equations for arbitrary jump distributions $f(\eta)$ is very hard and is still an open problem. Below, we focus on a specific jump distribution (double exponential), which has served as a paradigmatic solvable example throughout this book. For this specific case, the jump distribution reads

$$f(\eta) = \frac{1}{2\ell_0}\exp\left(-|\eta|/\ell_0\right). \qquad (6.81)$$

The variance of this jump distribution is simply

$$\sigma^2 = 2\ell_0^2. \qquad (6.82)$$

We recall that the function $f(\eta)$ satisfies the identities

$$f'(\eta) = -\frac{1}{2\ell_0^2}\mathrm{sgn}(\eta)\exp\left(-|\eta|/\ell_0\right), \qquad f''(\eta) = -\frac{1}{\ell_0^2}\delta(\eta) + \frac{1}{\ell_0^2}f(\eta). \qquad (6.83)$$

Using Eq. (6.83), one can take the second derivative with respect to x of Eqs. (6.79) and (6.80) to obtain (Schehr and Majumdar, 2012)

$$\frac{\partial^2 q_{k,N}(x)}{\partial x^2} = -\frac{1}{\ell_0^2}q_{k,N-1}(x) + \frac{1}{\ell_0^2}q_{k,N}(x), \qquad (6.84)$$

$$\frac{\partial^2 r_{k,N}(x)}{\partial x^2} = -\frac{1}{\ell_0^2}r_{k-1,N-1}(x) + \frac{1}{\ell_0^2}r_{k,N}(x). \qquad (6.85)$$

One thus finds that, for the jump distribution in Eq. (6.81), the coupled integral equations yield two independent differential recurrence equations, Eq. (6.84). They can be solved by introducing the double generating functions (GFs)

$$\tilde{q}(z,s;x) = \sum_{N=0}^{\infty}\sum_{k=0}^{N} s^N z^k q_{k,N}(x), \qquad \tilde{r}(z,s;x) = \sum_{N=0}^{\infty}\sum_{k=0}^{N} s^N z^k r_{k,N}(x), \qquad (6.86)$$

with $s < 1$ and $z < 1$. Consequently, these GFs satisfy the following pair of equations in the domain $x \in [0,\infty)$ (check this as a little exercise!):

$$\ell_0^2 \frac{\partial^2 \tilde{q}(z,s;x)}{\partial x^2} = (1-s)\tilde{q}(z,s;x) - 1, \qquad (6.87)$$

$$\ell_0^2 \frac{\partial^2 \tilde{r}(z,s;x)}{\partial x^2} = (1-zs)\tilde{r}(z,s;x) - 1. \qquad (6.88)$$

One can make these inhomogeneous equations homogeneous by the shifts $\tilde{q}(z,s;x) \to \tilde{q}(z,s;x) + 1/(1-s)$ and $\tilde{r}(z,s;x) \to \tilde{r}(z,s;x) + 1/(1-zs)$. The resulting homogeneous equations have general solutions as sums of two exponentials. Retaining only the solutions that vanish when $x \to \infty$, one finds, for $x \geq 0$,

$$\tilde{q}(z,s;x) = a(z,s)e^{-\sqrt{1-s}(x/\ell_0)} + \frac{1}{1-s}, \qquad (6.89)$$

$$\tilde{r}(z,s;x) = b(z,s)e^{-\sqrt{1-zs}(x/\ell_0)} + \frac{1}{1-zs}, \qquad (6.90)$$

where, at this stage, the amplitudes $a(z,s)$ and $b(z,s)$ remain undetermined. To determine them, we first write the integral equations satisfied by the GFs $\tilde{q}(z,s;x)$ and $\tilde{r}(z,s;x)$ which are obtained from Eqs. (6.79) and (6.80). They read

$$\tilde{q}(z,s;x) = 1 + s\int_0^{\infty} f(x'-x)\tilde{q}(z,s;x')\,\mathrm{d}x' + zs\int_0^{\infty} f(x+x')\tilde{r}(z,s;x')\,\mathrm{d}x', \qquad (6.91)$$

$$\tilde{r}(z,s;x) = 1 + zs\int_0^{\infty} f(x-x')\tilde{r}(z,s;x')\,\mathrm{d}x' + s\int_0^{\infty} f(x+x')\tilde{q}(z,s;x')\,\mathrm{d}x'. \qquad (6.92)$$

If one substitutes the explicit forms of the GFs in Eqs. (6.89) and (6.90) into the integral equations in Eqs. (6.91) and (6.92) one obtains a linear system of coupled equations for $a(z,s)$ and $b(z,s)$:

$$0 = -\frac{1}{1-s} + \frac{a(z,s)}{\sqrt{1-s}-1} + \frac{z}{1-zs} + \frac{zb(z,s)}{1+\sqrt{1-zs}} \tag{6.93}$$

$$0 = -\frac{z}{1-zs} + \frac{zb(z,s)}{\sqrt{1-zs}-1} + \frac{1}{1-s} + \frac{a(z,s)}{1+\sqrt{1-s}}. \tag{6.94}$$

This system can be straightforwardly solved, yielding

$$a(z,s) = \frac{1}{\sqrt{1-s}\sqrt{1-zs}} - \frac{1}{1-s}, \qquad b(z,s) = \frac{1}{\sqrt{1-s}\sqrt{1-zs}} - \frac{1}{1-zs}. \tag{6.95}$$

One finally obtains the formula expressed in terms of $\sigma = \sqrt{2}\ell_0$ as (Schehr and Majumdar, 2012)

$$\tilde{q}(z,s,x) = \sum_{N=0}^{\infty}\sum_{k=0}^{N} s^N z^k q_{k,N}(x)$$

$$= \frac{1}{1-s} + \left(\frac{1}{\sqrt{(1-s)(1-zs)}} - \frac{1}{1-s}\right)e^{-\sqrt{2(1-s)}(x/\sigma)}. \tag{6.96}$$

Note again that this solution is valid in the range $x \in [0, \infty)$.

Now our goal is to compute the average $\langle M_{k,N}\rangle$ of the kth maximum (starting from $x_0 = 0$), using the exact generating function $\tilde{q}(z,s,x)$ in Eq. (6.96). For this, we use the fact that the CDF of the kth maximum $Q_{k,N}(x) = \text{Prob}[M_{k,N} \leq x]$ can be expressed in terms of $q_{k,N}(x)$ as in Eq. (6.75), or equivalently in Eq. (6.73). We start with the definition of $\langle M_{k,N}\rangle$,

$$\langle M_{k,N}\rangle = \int_{-\infty}^{\infty} dx\, x \frac{\partial}{\partial x} Q_{k,N}(x), \tag{6.97}$$

where we used the fact that $\partial Q_{k,N}(x)/\partial x$ is just the PDF of $M_{k,N}$. Splitting the integral into two regions, $x \in [0, \infty$ and $x \in (-\infty, 0]$, using the relation in Eq. (6.75) and, furthermore, $q_{0,N}(x < 0) = 0$ (this last relation follows from the fact that, if the walk starts on the negative side, the number of points below 0 is necessarily strictly positive), we get

$$\langle M_{k,N}\rangle = \sum_{m=0}^{k-1}\int_{0}^{\infty} x\frac{\partial}{\partial x}q_{m,N}(x)\,dx + \sum_{m=1}^{k-1}\int_{-\infty}^{0} x\frac{\partial}{\partial x}q_{m,N}(x)\,dx. \tag{6.98}$$

We now use the relation in Eq. (6.74), make a change of variable $x \to -x$ and also shift the index m to $m-1$ in the second term. This leads to an integral over the range $x \in [0, \infty)$,

$$\langle M_{k,N}\rangle = \sum_{m=0}^{k-1}\int_{0}^{\infty} x\frac{\partial}{\partial x}q_{m,N}(x)\,dx + \sum_{m=0}^{k-2}\int_{0}^{\infty} x\frac{\partial}{\partial x}q_{N-m,N}(x)\,dx. \tag{6.99}$$

Since the range of the integral is over $[0, \infty)$, we can then use the result for $q_{k,N}(x)$ in Eq. (6.96) which, we recall, is valid only for $x \geq 0$. To utilize this result, it is convenient to take the double GF of $\langle M_{k,N} \rangle$. We start by taking the double GF of the first term on the RHS of Eq. (6.99). This gives, using the result in Eq. (6.96),

$$
\sum_{N=0}^{\infty} \sum_{m=0}^{N} z^m s^N \int_0^{\infty} x \frac{\partial}{\partial x} q_{m,N}(x) \, \mathrm{d}x = \frac{\sigma}{\sqrt{2}} \left(\frac{1}{(1-s)^{3/2}} - \frac{1}{(1-s)} \frac{1}{\sqrt{1-zs}} \right)
$$

$$
= \frac{\sigma}{\sqrt{2}} \left(\frac{2}{\sqrt{\pi}} \sum_{N=0}^{\infty} \frac{\Gamma(N+3/2)}{N+1} s^N - \frac{1}{\sqrt{\pi}} \sum_{N=0}^{\infty} s^N \sum_{m=0}^{N} z^m \frac{\Gamma(m+1/2)}{\Gamma(m+1)} \right), \tag{6.100}
$$

from which one obtains, by matching powers of $z^m s^N$,

$$
\int_0^{\infty} x \frac{\partial}{\partial x} q_{m,N}(x) \, \mathrm{d}x = \begin{cases} \dfrac{\sigma}{\sqrt{2}} \left(\dfrac{2}{\sqrt{\pi}} \dfrac{\Gamma(N+3/2)}{\Gamma(N+1)} - 1 \right), & m = 0 \\[2ex] -\dfrac{\sigma}{\sqrt{2\pi}} \dfrac{\Gamma(m+1/2)}{\Gamma(m+1)}, & m > 0. \end{cases} \tag{6.101}
$$

We can now use this result to evaluate the second integral in Eq. (6.99) by replacing m by $N - m$. Finally, arranging and regrouping the terms and using the identity

$$
\sum_{k=0}^{n} \frac{\Gamma(k+1/2)}{\Gamma(k+1)} = 2 \frac{\Gamma(n+3/2)}{\Gamma(n+1)}, \tag{6.102}
$$

one arrives at a nice compact result (we leave this as a nice mathematical exercise):

$$
\langle M_{k,N} \rangle = \sigma \sqrt{\frac{2}{\pi}} \left(\frac{\Gamma(N-k+5/2)}{\Gamma(N-k+2)} - \frac{\Gamma(k+1/2)}{\Gamma(k)} \right). \tag{6.103}
$$

This result is exact for all $k = 1, 2, \ldots, N+1$ and for all N. Note that for $k = 1$, Eq. (6.103) reduces to the average global maximum computed in Eq. (4.89) for the double-exponential distribution (recalling that $\sigma = \sqrt{2}\ell_0$).

The first question one can ask is how this average kth maximum $\langle M_{k,N} \rangle$, for fixed k, behaves as $N \to \infty$, i.e., for a very long random walk. To analyze this limit from the exact result in Eq. (6.103), we need to use the leading asymptotic behavior of the gamma function

$$
\Gamma(N+a) \xrightarrow[N \to \infty]{} \sqrt{2\pi} N^{a-1/2} e^{N \ln N - N}. \tag{6.104}
$$

This gives (Schehr and Majumdar, 2012)

$$
\langle M_{k,N} \rangle \xrightarrow[N \to \infty]{} \sigma \sqrt{\frac{2N}{\pi}}, \quad \text{independent of } k. \tag{6.105}
$$

At first glance, this result, namely that the leading asymptotic behavior for large N is independent of k, is quite surprising. However, we will see shortly that the k-dependence appears only in the subleading term.

6.2.2   Statistics of the gap

We first discuss the average gap $\langle g_{k,N} \rangle = \langle M_{k,N} \rangle - \langle M_{k+1,N} \rangle$ and, for this, we can directly use the results for $\langle M_{k,N} \rangle$ from the previous subsection. Later, we will also discuss the full distribution of the gap $g_{k,N}$.

Average of the kth gap.   In fact, if we analyze the gap between the kth and the $(k+1)$th maxima, i.e., $g_{k,N} = M_{k,N} - M_{k+1,N}$, we obtain the average of the kth gap from Eq. (6.103),

$$\langle g_{k,N} \rangle = \frac{\sigma}{\sqrt{2\pi}} \left(\frac{\Gamma(k+1/2)}{\Gamma(k+1)} + \frac{\Gamma(N-k+3/2)}{\Gamma(N-k+2)} \right). \qquad (6.106)$$

Note that this result is valid for all $k = 1, 2, \ldots, N$ and all N. Furthermore, it satisfies the symmetry $\langle g_{k,N} \rangle = \langle g_{N-k+1,N} \rangle$, reflecting the up–down (max–min) symmetry of the random walk. Now, taking the $N \to \infty$ limit, we find that $\langle g_{k,N} \rangle$ approaches an N-independent (but k-dependent) value (Schehr and Majumdar, 2012)

$$\langle g_{k,N} \rangle \xrightarrow[N \to \infty]{} \frac{\sigma}{\sqrt{2\pi}} \frac{\Gamma(k+1/2)}{\Gamma(k+1)} = \overline{g_k}. \qquad (6.107)$$

Note that in the average gap $g_{k,N} = M_{k,N} - M_{k+1,N}$, the leading $\sqrt{N}$ dependence of $\langle M_{k,N} \rangle$ [see Eq. (6.105)] cancels out, and the difference of the subleading terms becomes independent of N and is given by Eq. (6.107). In other words, if one examines the walk up to step N from the global maximum, the values of the successive gaps (i.e., the "gap spectrum") becomes stationary, i.e., independent of N, for large N. In addition, from Eq. (6.107) it follows that, for large k (i.e., in the bulk of the gap spectrum), the average gap behaves as

$$\frac{\overline{g_k}}{\sigma} = \lim_{N \to \infty} \frac{\langle g_{k,N} \rangle}{\sigma} = \frac{1}{\sqrt{2\pi}} \frac{\Gamma(k+1/2)}{\Gamma(k+1)} \xrightarrow[k \to \infty]{} \frac{1}{\sqrt{2\pi k}}. \qquad (6.108)$$

The formula for the average gap in Eq. (6.106) was deduced above for the specific double-exponential jump distribution. What happens for other jump distributions $f(\eta)$? It turns out that if $f(\eta)$ has a finite variance $\sigma^2 = \int_{-\infty}^{\infty} \eta^2 f(\eta) \, d\eta$, the average gap $\langle g_{k,N} \rangle$ reaches a stationary value as $N \to \infty$ given by

$$\lim_{N \to \infty} \frac{\langle g_{k,N} \rangle}{\sigma} = \frac{\overline{g_k}}{\sigma} = \frac{1}{\sqrt{2\pi}} \frac{\Gamma(k+\frac{1}{2})}{\Gamma(k+1)} - \frac{1}{\pi k} \int_0^{\infty} \frac{dq}{q^2} \left[[\hat{f}(q)]^k - \frac{1}{(1+(\sigma^2/2)q^2)^k} \right]. \qquad (6.109)$$

Here we skip the derivation of this result but the interested reader may find the derivation in Schehr and Majumdar (2012)—see also Pitman and Tang (2023) for an alternative derivation using the tools of fluctuation theory. Interestingly, the dependence on the jump distribution $f(\eta)$ is only in the second term in Eq. (6.109), while the first term is universal, i.e., independent of $f(\eta)$. Indeed, as $k \to \infty$, the second term just drops out and we get

$$\lim_{k \to \infty} \frac{\overline{g_k}}{\sigma} \longrightarrow \frac{1}{\sqrt{2\pi k}}. \tag{6.110}$$

Thus, in an infinitely long trajectory of a random walk, the average gap in the bulk, i.e., as $k \to \infty$, becomes universal, i.e., independent of the details of $f(\eta)$, as long as the variance σ^2 is finite. Thus, as one goes away from the maximum, the gaps between successive levels become smaller and smaller. Hence, the sequence of ordered entries almost becomes continuous far away from the global maximum.

Distribution of the kth gap.   To obtain the full distribution of the kth gap, we need to know the joint distribution of the kth maximum $M_{k,N}$ and the $(k+1)$th maximum $M_{k+1,N}$,

$$P_{k,N}^{\text{joint}}(x, y) = \text{Prob}(M_{k,N} = x, M_{k+1,N} = y). \tag{6.111}$$

Knowing this joint distribution, the PDF of the kth gap $g_{k,N}$ can be obtained as

$$p_{k,N}(g) = \text{Prob}(g_{k,N} = g) = \int_{-\infty}^{\infty} dy\, P_{k,N}^{\text{joint}}(x = y + g, y). \tag{6.112}$$

For arbitrary jump distribution $f(\eta)$, it is not easy to derive the joint PDF $P_{k,N}^{\text{joint}}(x, y)$. However, for the double exponential in Eq. (6.81), it is again possible to write down an exact recursion relation for $P_{k,N}^{\text{joint}}(x, y)$, solve it by using double GF techniques (as discussed above) and finally analyze the gap distribution $p_{k,N}(g)$ in the large-N limit. We do not repeat the details here but the interested reader may consult Schehr and Majumdar (2012). For fixed k and large N, the gap distribution $p_{k,N}(g)$ becomes independent of N, i.e., $\lim_{N \to \infty} p_{k,N}(g) = p_k(g)$ and its exact GF is given by

$$\sum_{k=1}^{\infty} z^k p_k(g) = \frac{8z}{\ell_0} e^{-2x/\ell_0} \frac{u(z) - v(z)e^{-2z/\ell_0}}{[u(z) + v(z)e^{-2z/\ell_0}]^3}, \tag{6.113}$$

with $u(z) = \sqrt{1-z} + 1$ and $v(z) = \sqrt{1-z} - 1$. Extracting $p_k(g)$ for all k from Eq. (6.113) is hard. However, one can easily extract the asymptotic behavior for large k, by analyzing the $z \to 1$ limit of Eq. (6.113). This yields, for $k \gg 1$ and g fixed, $p_k(g) \sim k^{-3/2} F_{\text{gap}}(g)$, where $F_{\text{gap}}(g)$ decays exponentially for large g and represents the cut-off function. However, before the distribution gets cut off for large g, there is a scaling regime $g \sim \overline{g_k} \sim \sigma/\sqrt{2\pi k}$, with k large, where we anticipate a scaling form for the gap PDF of

$$p_k(g) \simeq (\sqrt{k}/\sigma) P_{\text{gap}}(\sqrt{k}g/\sigma), \tag{6.114}$$

and we expect that the scaling function $P_{\text{gap}}(z)$ is independent of k. Indeed, taking the $k \to \infty$ and $g \to 0$ limit in Eq. (6.113) while keeping the scaled variable

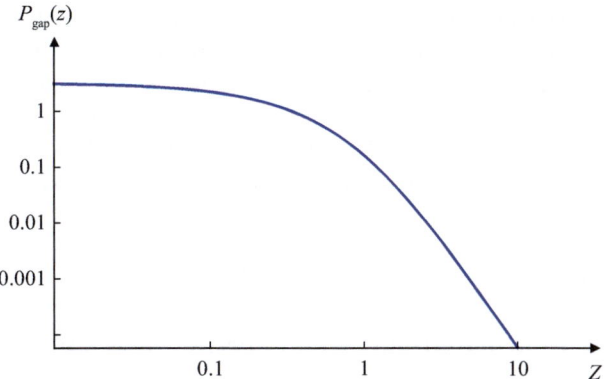

Fig. 6.6 Log–log plot of the function $P_{\text{gap}}(z)$ vs. z in Eq. (6.116). The straight line for large (log) z shows the power-law decay $P_{\text{gap}}(z) \sim z^{-4}$ in the second line of Eq. (6.117).

$\sqrt{k}g/\sigma$ fixed, one can show that the scaling function $P_{\text{gap}}(z)$ satisfies (Schehr and Majumdar, 2012)

$$\int_0^\infty e^{-z\lambda}\sqrt{z}P_{\text{gap}}(\sqrt{z})\,\mathrm{d}z = (1 + \sqrt{\lambda/2})^{-3}. \tag{6.115}$$

The Laplace transform in Eq. (6.115) can be inverted, yielding

$$P_{\text{gap}}(z) = 4\left[\sqrt{\frac{2}{\pi}}(1 + 2z^2) - e^{2z^2}z(4z^2 + 3)\mathrm{erfc}(\sqrt{2}z)\right]. \tag{6.116}$$

A plot of this function is shown in Fig. 6.6. The asymptotic behaviors of $P_{\text{gap}}(z)$ are given by (Schehr and Majumdar, 2012)

$$P_{\text{gap}}(z) \sim \begin{cases} 4\sqrt{2/\pi}, & z \to 0, \\ (3/\sqrt{8\pi})z^{-4}, & z \to \infty, \end{cases} \tag{6.117}$$

which thus exhibits a somewhat unexpected and non-trivial power-law tail $P_{\text{gap}}(z) \sim z^{-4}$ as $z \to \infty$.

We derived this non-trivial gap distribution here for large N by analyzing the explicit recursion relations satisfied by the joint distribution $P_{k,N}^{\text{joint}}(x, y)$ for the specific case of the double-exponential jump distribution. This result for the exponential distribution was recently rederived in the mathematics literature using rigorous probabilistic methods (Pitman and Tang, 2022). A natural question is: What happens for other distributions with a finite σ? Is this scaling function $P_{\text{gap}}(z)$ universal? In fact, this gap distribution was recently computed explicitly for another family of jump distributions with a finite σ, namely $f(\eta) \propto |\eta|^p e^{-|\eta|}$, parameterized by a non-negative integer p, in Battilana *et al.* (2020). And indeed, the scaled gap distribution $P_{\text{gap}}(z)$ was found to have exactly the same form as in Eq. (6.116). This provided strong support for the conjecture that, indeed, this scaled distribution is universal for any distribution with a finite σ. Numerical simulations for several jump distributions support this conjecture (Schehr and Majumdar, 2012). Some rigorous works in the probability literature have also indicated this plausible universality (Pitman and Tang, 2022, 2023). Very recently, it has been shown that this scaling function $P_{\text{gap}}(z)$ in Eq. (6.116) is indeed universal for any jump distribution $f(\eta)$ with a finite variance σ^2 (De Bruyne *et al.*, 2023). Moreover, the gap scaling function was also computed exactly for jump distributions with a power-law tail, such as in Lévy flights, where $f(\eta) \sim |\eta|^{-1-\mu}$ with $1 \leq \mu < 2$ for large $|\eta|$. This scaling function $P_{\text{gap},\mu}(z)$ is parameterized by μ and can be expressed in terms of Mittag–Leffler functions (De Bruyne *et al.*, 2023).

6.2.3 Statistics of near-extreme events for random walks and Brownian motion

In Section 6.1.3 we analyzed the DOS with respect to the maximum for a time series with IID entries. One can define this DOS near the maximum for a random walk trajectory of N steps in analogy with the IID variables in Eq. (6.47), namely

$$\rho(r, N) = \frac{1}{N} \sum_{x_i \neq X_{\text{max}}} \delta(X_{\text{max}} - x_i - r), \tag{6.118}$$

where the x_i represent the positions of an N-step random walker, evolving via $x_n = x_{n-1} + \eta_n$, starting from $x_0 = 0$ and with a symmetric and continuous jump distribution $f(\eta)$. In Eq. (6.118) X_{max} denotes the value of the global maximum $M_{1,N}$ of the walk. Note that the definition in Eq. (6.118) differs slightly from the one for IID variables in Eq. (6.47) ($N - 1$ gets replaced by N). This is simply because an N-step random walk trajectory has $N + 1$ entries as positions of the walk, $\{x_0 = 0, x_1, \ldots, x_N\}$. In principle, it may be possible to compute the average DOS $\rho(r, N)$ for a discrete-time random walker with arbitrary jump distribution $f(\eta)$, but to our knowledge this is still an open problem. Certainly, for the double-exponential jump distribution in Eq. (6.81), this should be possible using the tools discussed in this chapter because this requires the joint distribution of the ordered maximum. However, for jump distributions with a finite variance σ, we expect that in the large-N limit, where the random walk (appropriately scaled) converges to a Brownian motion of duration T, it is possible to directly compute the limiting average DOS for the Brownian motion in continuous time. This is certainly a much easier computation than for discrete-time random walkers. For Brownian motion evolving in continuous time up to a duration T, the analogue of Eq. (6.118) for the DOS near the maximum reads (Perret *et al.*, 2013, 2015)

$$\rho(r, T) = \frac{1}{T} \int_0^T \delta(X_{\text{max}} - x(t) - r) \, dt, \tag{6.119}$$

where $x(t)$ is a Brownian motion, starting at $x(0) = 0$, that evolves as

$$\frac{dx(t)}{dt} = \eta(t). \tag{6.120}$$

Here, $\eta(t)$ is Gaussian white noise of zero mean and correlator $\langle \eta(t)\eta(t') \rangle = 2D\delta(t-t')$. In Eq. (6.119), X_{max} denotes the value of the global maximum of the Brownian motion on the time interval $[0, T]$. Averaging over all trajectories of the Brownian motion gives

$$\overline{\rho(r, T)} = \frac{1}{T} \int_0^T \overline{\delta(X_{max} - x(t) - r)}\, dt. \tag{6.121}$$

The basic building block in computing the average in the integrand is actually just the Brownian propagator in Eq. (2.66), i.e.,

$$G(x, x_0, t) = \frac{1}{\sqrt{4\pi Dt}} \exp\left[-\frac{(x-x_0)^2}{4Dt}\right], \tag{6.122}$$

that denotes the probability density that the Brownian motion arrives at x at time t, starting at x_0 at time 0. The computation is somewhat involved and we refer the reader to Perret *et al.* (2013, 2015) for the details. Here, we just discuss the main result for $\overline{\rho(r, T)}$. It turns out that this average can be written in a scaling form for all T as

$$\overline{\rho(r, T)} = \frac{1}{\sqrt{2DT}} \bar{\rho}\left(\frac{r}{\sqrt{2DT}}\right), \tag{6.123}$$

where the scaling function $\bar{\rho}(z)$ is given explicitly by

$$\bar{\rho}(z) = 8[h(z) - h(2z)], \qquad h(z) = \frac{e^{-z^2/2}}{\sqrt{2\pi}} - \frac{z}{2}\mathrm{erfc}\left(\frac{z}{\sqrt{2}}\right), \tag{6.124}$$

with $\mathrm{erfc}(z) = (2/\sqrt{\pi}) \int_z^\infty e^{-y^2}\, dy$. In Fig. 6.7 we show a plot of $\bar{\rho}(z)$. Its asymptotic behaviors are given by

$$\bar{\rho}(z) \sim \begin{cases} 4z, & z \to 0, \\ z^{-2}e^{-z^2/2}, & z \to \infty. \end{cases} \tag{6.125}$$

Furthermore, it exhibits a maximum for $z_{typ} = 0.51454\ldots$, slightly smaller than the average value $z_{ave} = \sqrt{2/\pi} = 0.79788\ldots$ The fact that $\bar{\rho}(z)$ does not vanish too rapidly as $z \to 0$ indicates that, on average, there is no gap between X_{max} and the rest of the crowd: hence, "X_{max} is not lonely at the top." The average DOS for Brownian motion in Eq. (6.124) is thus quite different from the IID case discussed in Section 6.1.3. We recall that in that case, depending on whether the tail of the parent distribution of the x_i decays slower than, faster than, or as a pure exponential, the limiting mean DOS converges to three different limiting forms, which are clearly different from Eq. (6.124). Furthermore, for the Brownian motion, it is even possible to compute the full distribution of the DOS near the maximum $\rho(r, T)$ using path-integral methods, thus going beyond just its first moment $\overline{\rho(r, T)}$. Again, the details are a bit long and we refer the interested reader to Perret *et al.* (2013, 2015).

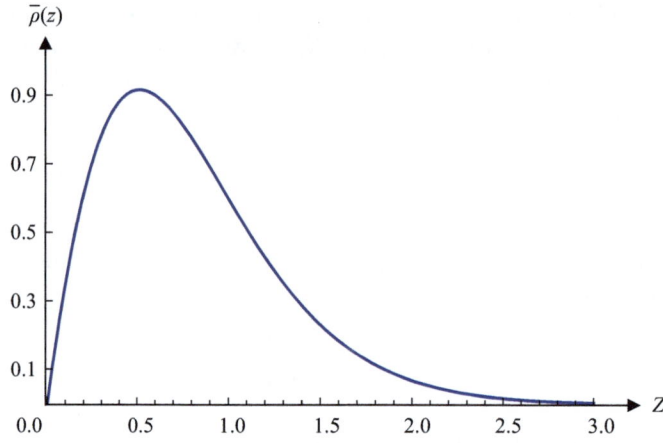

Fig. 6.7 Plot of the scaling function $\bar{\rho}(z)$ given in Eq. (6.124) vs. z.

Extensions of order statistics. In this book we have focused on the statistics of the kth maximum $M_{k,N}$ for time series with N entries in two canonical models, namely the IID model and the random walk model. In the discussions in this chapter, we have studied the asymptotic properties of $M_{k,N}$ for large N, but with a *fixed* k. There exists another interesting scaling regime where both $k \to \infty$ and $N \to \infty$, but with their ratio $k/N = \alpha$ kept fixed. In the literature, this is known as the "bulk" regime because one is probing the $k = \alpha N$-th maximum, which is very far away from the first maximum $M_{1,N}$ at the edge of the ordered entries $M_{1,N} > M_{2,N} > \cdots$. One can also consider the gap statistics in this bulk regime. For both models, the order and the gap statistics have been studied in the bulk regime. However, we do not discuss this regime in detail in this book and the interested reader may consult the relevant research articles (Dassios, 1995; Yor, 1995; Embrechts *et al.*, 1995; Lacroix-A-Chez-Toine *et al.*, 2019). Very recently, these results for the order statistics in the bulk in the IID case have been used to compute exactly the order statistics in the non-equilibrium steady state of a strongly correlated system of N Brownian particles that are reset simultaneously with rate r to the origin (Biroli *et al.*, 2023).

7
Records

Human beings are fascinated by records. If you open your newspaper or television, every day you hear of record high/low temperatures, record rainfall, records in sports, records in stock prices, ... Indeed, the Guinness books of world records happen to be some of the best-selling books in the world. A record simply means an event that has not happened before. Consider, for instance, a time series in discrete time, $\{x_1, x_2, \ldots, x_N\}$, of N steps. It can represent any time series, e.g., the price of a stock on different days, the daily temperature of a city for N successive days, the height of a river, ... An entry x_k of this time series is a record if it exceeds all the previous values: $x_k > x_1, x_k > x_2, \ldots, x_k > x_{k-1}$. This is equivalent to saying that a record occurs at step k if

$$x_k > \max\{x_1, x_2, \ldots, x_{k-1}\}. \tag{7.1}$$

How many such records occur in this time series of N steps? Let $P(M, N)$ denote the probability of having exactly M records in a series of size N. This distribution of course depends on how the entries of the time series are distributed. As discussed in the introduction, our main input is the joint distribution of the entries, $P_{\text{joint}}(x_1, x_2, \ldots, x_N)$. Given this joint distribution, our aim is to compute the record number distribution $P(M, N)$.

But this is not the only interesting observable associated with records, as we discussed in the introduction. For example, we could ask: What is the joint distribution of the record ages $\{\ell_1, \ell_2, \ldots, \ell_M\}$. Here, ℓ_k is the age of the kth record, i.e., the number of steps up to which the kth record survives before being broken by the next record (see Fig. 7.1). We can also ask about the distribution of the actual record values, denoted by $\{R_1, R_2, \ldots, R_M\}$ (see Fig. 7.1). Indeed, the value of the last record R_M is the highest possible value of the time series $\{x_1, x_2, \ldots, x_N\}$ and hence is precisely the global maximum $X_{\max}$ of the time series, i.e., $R_M = X_{\max}$. It would then be interesting to see how the actual value of the record R_k evolves with the record number k.

The study of the statistics of *records* in a discrete time series was initiated in the early fifties, mostly in the probability and statistics literature. In probability theory, the time at which a record occurs and the value of this record are usually called "ladder" epochs and "ladder" values, and this subject of research in probability theory

Statistics of Extremes and Records in Random Sequences. Satya N. Majumdar and Grégory Schehr, Oxford University Press.
© Satya N. Majumdar and Grégory Schehr (2024). DOI: 10.1093/9780191838781.003.0007

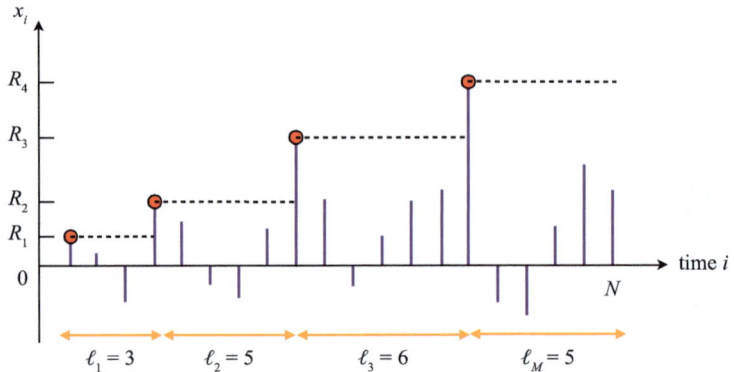

Fig. 7.1 A time series x_i vs. i with $N = 20$ entries. The records (shown by red filled circles) occur at step numbers 1, 4, 9 and 15 with record values $R_1, R_2, \ldots$ The age of the kth record is denoted by ℓ_k (for example, $\ell_1 = 3$, $\ell_2 = 5$, etc.). Note that the age of the last record before N is denoted by ℓ_M.

goes under the name of "fluctuation theory" (Feller, 2008*a*,*b*). In the statistics literature, there are quite a few books and monographs that provide detailed accounts of record statistics, but focusing mostly on the IID random sequence (Resnick, 2008; Bunge and Goldie, 2001; Nevzorov, 2001; Arnold *et al.*, 2011). It turns out that the theory of records has become fundamental and important in a wide variety of systems, including climate studies (Hoyt, 1981; Bassett Jr., 1992; Schmittmann and Zia, 1999; Benestad, 2003; Redner and Petersen, 2006; Anderson and Kostinski, 2010; Wergen and Krug, 2010; Wergen *et al.*, 2014), finance and economics (Wergen *et al.*, 2011; Sabir and Santhanam, 2014), hydrology (Matalas, 1997; Vogel *et al.*, 2001), sports (Gembris *et al.*, 2002; Ben-Naim *et al.*, 2007) and also in detecting heavy tails in statistical distributions (Franke *et al.*, 2012). Also, the study of records has found renewed interest and applications in diverse complex physical systems such as the evolution of thermoremanent magnetization in spin-glasses (Jensen, 2006; Sibani *et al.*, 2006; Sibani, 2007), the evolution of the vortex density with increasing magnetic field in type-II disordered superconductors (Oliveira *et al.*, 2005; Jensen, 2006), avalanches of elastic lines in a disordered medium (Alessandro *et al.*, 1990; Sibani and Littlewood, 1993; Fisher, 1998; Le Doussal and Wiese, 2009; Schimmenti *et al.*, 2021), the evolution of fitness in biological populations (Sibani *et al.*, 1998; Krug and Jain, 2005; Franke *et al.*, 2011; Park *et al.*, 2015), jamming in colloids (Robe *et al.*, 2016), the study of failure events in porous materials (Pál *et al.*, 2016), models of growing networks (Godrèche and Luck, 2008) and in quantum chaos (Srivastava *et al.*, 2013), among others. The purpose of this book is not to review all these applications in detail. For the literature on various interesting applications we refer the reader to two recent reviews on record statistics in the physics literature: Wergen (2013); Godrèche *et al.* (2017). The goal of this chapter, in the spirit of the rest of the book, is rather to discuss pedagogically the essential tools needed for studying record statistics, with the help of two simple canonical examples: (i) the IID case, and (ii) the random walk case. In the former the entries of the time series are completely uncorrelated, while in the latter they are strongly correlated.

7.1 Record statistics for the IID sequence

In this section we consider an IID time series whose N entries $\{x_1, x_2, \ldots, x_N\}$ are drawn from a factorized joint distribution

$$P_{\text{joint}}(x_1, x_2, \ldots, x_N) = p(x_1)p(x_2) \cdots p(x_N), \tag{7.2}$$

where the marginal distribution $p(x)$ is assumed to be continuous. We will derive the statistics of various record observables such as the number of records and their ages.

7.1.1 Statistics of the number of records

To study the number of records M in this time series, it is convenient to introduce indicator variables σ_k that take the value 0 or 1:

$$\sigma_k = \begin{cases} 1 & \text{if } x_k \text{ is a record,} \\ 0 & \text{if } x_k \text{ is not a record.} \end{cases} \tag{7.3}$$

Thus, the number of records M up to step N in any sample of the time series can be expressed in terms of these binary variables as

$$M = \sum_{k=1}^{N} \sigma_k. \tag{7.4}$$

The statistics of M clearly depends on the statistics of the σ_k. For instance, the mean number of records is given by

$$\langle M \rangle = \sum_{k=1}^{N} \langle \sigma_k \rangle. \tag{7.5}$$

Similarly, the second moment of M is given by

$$\langle M^2 \rangle = \sum_{k_1=1}^{N} \sum_{k_2=1}^{N} \langle \sigma_{k_1} \sigma_{k_2} \rangle. \tag{7.6}$$

Clearly, higher moments of M will involve higher-order correlations of the σ_k's. Thus, we need to first understand the statistics of the σ_k's in order to determine the distribution $P(M, N)$.

Statistics of σ_k's. We start with the first moment of σ_k and define

$$r_k = \langle \sigma_k \rangle, \tag{7.7}$$

where the average is taken over different realizations of the random variables $x_1, \ldots, x_N$. Hence, r_k is the rate at which a record is broken at "time" k. For IID random variables it is straightforward to compute the record rate as it is precisely the probability that the event in Eq. (7.1) happens. To compute this rate r_k, let the kth entry take the value $x_k = x$; this happens with probability $p(x)$. Given this value $x_k = x$, the kth entry will be a record if all the $(k-1)$ previous entries are less than x; this happens with probability $\left[\int_{-\infty}^{x} p(y)\, dy \right]^{k-1}$, where we used the independence of the entries. Finally, r_k is obtained by integrating over all possible values of $x_k = x$. This yields

$$r_k = \int_{-\infty}^{\infty} p(x) \left[\int_{-\infty}^{x} p(y)\, dy \right]^{k-1} dx. \tag{7.8}$$

At this stage, it looks like r_k depends explicitly on $p(x)$. However, a miracle happens by which the dependence on $p(x)$ completely disappears. To see this, we perform this integral by using the change of variable

$$u = \int_{-\infty}^{x} p(y)\, dy. \tag{7.9}$$

Clearly, the range of u is in $[0, 1]$. Thus, u is a uniform random number in $[0, 1]$. With this change of variables, the integral in Eq. (7.8) simply becomes

$$r_k = \int_0^1 u^{k-1}\, du = \frac{1}{k}. \tag{7.10}$$

Thus, the record rate r_k is completely universal for any k. From this little exercise, we locate the source of universality. It comes from the change of variables in Eq. (7.9) which maps any continuous distribution $p(x)$ to a uniform distribution over $[0, 1]$. Note that this simple result $r_k = 1/k$ can also be guessed from the fact that if the kth entry x_k is a record, then x_k must be the maximum of this set of k IID random variables. Thus, r_k is the probability that the kth member of this set is the maximum. But the maximum can be any one of these k entries with equal probability (thanks to their IID nature). Hence, clearly $r_k = 1/k$ (one out of k). However, we performed this little algebraic exercise to bring out more clearly the origin of the universality. In fact, we will see later that this change of variable will play a major role for the universality of the full distribution of M as well as that of the record ages for the IID case. Before moving on, we just make one remark on the result $r_k = 1/k$ in Eq. (7.10). As k increases, this slowly decays as a power law $1/k$, hence it becomes harder and harder to break a record as time increases.

Having obtained the mean $\langle \sigma_k \rangle$, we next compute the two-point correlation function $\langle \sigma_{k_1} \sigma_{k_2} \rangle$ needed to compute the second moment of M [see Eq. (7.6)]. We note that, since σ_k are either 0 or 1, it follows that

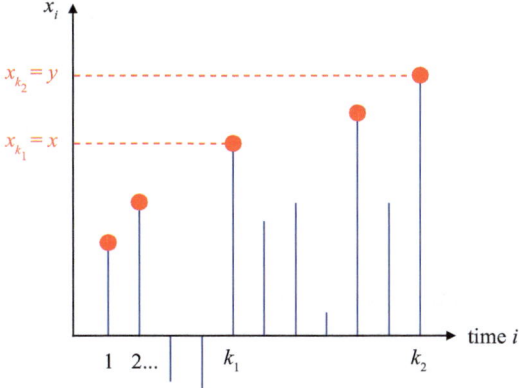

Fig. 7.2 Realization of the time series contributing to the event "a record happens at step k_1 *and* a record happens at step k_2." Note that there can be one or more records between the times k_1 and k_2.

$$\langle \sigma_{k_1} \sigma_{k_2} \rangle = \text{Prob[a record happens at step } k_1 \textbf{ and } \text{a record happens at step } k_2].$$
(7.11)

First, we note that $\langle \sigma_{k_1} \sigma_{k_2} \rangle = \langle \sigma_{k_2} \sigma_{k_1} \rangle$ and hence, without any loss of generality, we can take $k_2 \geq k_1$. The case $k_2 = k_1$ is special, since in this case $\sigma_{k_1}^2 = \sigma_{k_1}$ (this follows from the fact that $\sigma_{k_1} = 0, 1$) and hence

$$\langle \sigma_{k_1}^2 \rangle = \langle \sigma_{k_1} \rangle = r_{k_1} = \frac{1}{k_1}.$$
(7.12)

For $k_2 > k_1$, the probability of the event in Eq. (7.11) can be expressed as a double integral:

$$\langle \sigma_{k_1} \sigma_{k_2} \rangle = \int_{-\infty}^{\infty} \mathrm{d}y \, p(y) \left(\int_{-\infty}^{y} p(y') \, \mathrm{d}y' \right)^{k_2 - k_1 - 1} \int_{-\infty}^{y} \mathrm{d}x \, p(x) \left(\int_{-\infty}^{x} p(x') \, \mathrm{d}x' \right)^{k_1 - 1}.$$
(7.13)

Let us try to understand this formula. Consider the event where a record occurs at step k_1 with a record value $x_{k_1} = x$ and another record occurs at step k_2 ($k_2 > k_1$) with the record value $x_{k_2} = y$ (see Fig. 7.2). Evidently, since $k_2 > k_1$, you will necessarily have $y > x$. Given that the first record value at step k_1 is x, the $(k_1 - 1)$ previous values must be less than x; the probability for this event is $\left(\int_{-\infty}^{x} p(x') \, \mathrm{d}x' \right)^{k_1 - 1}$, which explains the last factor in Eq. (7.13). Furthermore, since a record occurs at step k_2 with a record value $x_{k_2} = y$, all the $(k_2 - k_1 - 1)$ entries sandwiched between the steps k_1 and k_2, i.e., the set $\{x_{k_1+1}, x_{k_1+2}, \ldots, x_{k_2-1}\}$, must have value less than y; the probability for this event is $\left(\int_{-\infty}^{y} p(y') \, \mathrm{d}y' \right)^{k_2 - k_1 - 1}$. We then multiply these two probabilities, since they are independent events. Furthermore, we need to multiply by the product $p(x)p(y)$ that denotes the joint probability that $x_{k_1} = x$ and $x_{k_2} = y$. Finally, you need to integrate over x and y with the constraint that $y > x$. This then produces the formula in Eq. (7.13).

Evidently, the formula in Eq. (7.13) is a bit scary but, once again, it simplifies magically if we use the same change of variables as in Eq. (7.9). More precisely, we define $u_1 = \int_{-\infty}^{x} p(x')\,dx'$ and $u_2 = \int_{-\infty}^{y} p(y')\,dy'$, in terms of which the double integral in Eq. (7.13) simplifies to

$$\langle \sigma_{k_1} \sigma_{k_2} \rangle = \int_0^1 du_2\, u_2^{k_2-k_1-1} \int_0^{u_2} du_1\, u_1^{k_1-1}. \tag{7.14}$$

The integral over u_1 trivially gives $\int_0^{u_2} du_1\, u_1^{k_1-1} = u_2^{k_1}/k_1$. Finally, performing the integral over u_2, we get the very simple result

$$\langle \sigma_{k_1} \sigma_{k_2} \rangle = \frac{1}{k_1 k_2} = \langle \sigma_{k_1} \rangle \langle \sigma_{k_2} \rangle, \quad k_1 \neq k_2. \tag{7.15}$$

Thus, very interestingly, not only is the two-point correlation of the σ_k universal, but also σ_{k_1} and σ_{k_2} are uncorrelated for $k_2 \neq k_1$.

What about the higher-order correlation functions of the σ_k? For instance, what can we say about the n-point correlation functions $\langle \sigma_{k_1} \sigma_{k_2} \cdots \sigma_{k_n} \rangle$? In fact, one can show that this universality and the factorization extends even for $n > 2$. More precisely, the following identity holds:

$$\langle \sigma_{k_1} \sigma_{k_2} \cdots \sigma_{k_n} \rangle = \langle \sigma_{k_1} \rangle \langle \sigma_{k_2} \rangle \cdots \langle \sigma_{k_n} \rangle, \tag{7.16}$$

whenever all the subscripts are different from each other, i.e., $k_1 \neq k_2 \neq \cdots \neq k_n$. We provided the proof of this identity for $n = 2$ above [see Eq. (7.15)]. The proof for $n > 2$ is more involved and the interested reader can either prove it herself/himself or consult the original article of Rényi (1962), which was recently reproduced in the review by Godrèche *et al.* (2017).

Thus, summarizing, the σ_k variables are completely uncorrelated for different k and, moreover, the statistics of the σ_k is completely universal, i.e., independent of the distribution $p(x)$ as long as it is continuous. In fact, we have implicitly assumed that the records are non-degenerate, i.e., almost surely, no two entries of the time series have the same value. This is of course guaranteed if $p(x)$ is continuous. In cases where $p(x)$ is such that two different entries may share the same value with a finite probability, the record statistics would be different. An example is $p(x) = q\delta(x) + (1-q)p_1(x)$, where $p_1(x)$ is continuous (and normalized to unity) and $0 < q < 1$. In this case, two different entries can both have values $x = 0$ with a finite probability q^2. In this case, the corresponding σ_k variables are correlated (Majumdar *et al.*, 2019*b*). Discussions of such cases go beyond the scope of this book. Here we will restrict ourselves to the case where $p(x)$ is continuous and for which we have complete knowledge of the universal statistics of the σ_k. In the following, we use this knowledge to compute the statistics of the number of records $M = \sum_{k=1}^{N} \sigma_k$.

Average of M. We start by computing the mean number of records $\langle M \rangle$ in Eq. (7.5). Using $\langle \sigma_k \rangle = r_k = 1/k$ from Eq. (7.7) we get

$$\langle M \rangle = \sum_{k=1}^{N} r_k = \sum_{k=1}^{N} \frac{1}{k} = H_N, \tag{7.17}$$

where H_N denotes the N-th harmonic number. Thus, we see that the mean number of records, for any N, is universal, i.e., independent of $p(x)$. In particular, for large N, this universal mean behaves as

$$\langle M \rangle = \ln N + \gamma_{\mathrm{E}} + \mathcal{O}(N^{-1}), \tag{7.18}$$

where $\gamma_{\mathrm{E}} = 0.577\,21\ldots$ is the Euler constant (for our enthusiastic reader, can you show this?). Thus, as the time series becomes longer and longer, the average number of records grows very slowly as $\sim \ln N$. This also reflects the fact that the breaking of records becomes rarer and rarer as N increases.

Variance of M. What about the variance $\mathrm{Var}(M) = \langle M^2 \rangle - \langle M \rangle^2$? For this, we first compute the second moment $\langle M^2 \rangle$ from Eq. (7.6), using the results for the two-point correlation of the σ_k in Eqs. (7.12) and (7.15). We first separate the $k_1 = k_2$ terms and $k_1 \neq k_2$ terms in the double sum appearing in Eq. (7.6). This gives

$$\langle M^2 \rangle = \sum_{k=1}^{N} \langle \sigma_k \rangle + \sum_{1 \leq k_1 \neq k_2 \leq N} \langle \sigma_{k_1} \sigma_{k_2} \rangle = \sum_{k=1}^{N} \frac{1}{k} + \sum_{1 \leq k_1 \neq k_2 \leq N} \frac{1}{k_1 k_2}. \tag{7.19}$$

The second term can be further simplified as

$$\sum_{1 \leq k_1 \neq k_2 \leq N} \frac{1}{k_1 k_2} = \sum_{1 \leq k_1, k_2 \leq N} \frac{1}{k_1 k_2} - \sum_{k=1}^{N} \frac{1}{k^2} = \langle M \rangle^2 - \sum_{k=1}^{N} \frac{1}{k^2}, \tag{7.20}$$

where, in the last equality, we have used the expression for $\langle M \rangle$ from Eq. (7.17). Therefore, the variance of M is given by

$$\mathrm{Var}(M) = \langle M^2 \rangle - \langle M \rangle^2 = \sum_{k=1}^{N} \left(\frac{1}{k} - \frac{1}{k^2} \right). \tag{7.21}$$

Thus, as in the case of the mean in Eq. (7.17), the variance $\mathrm{Var}(M)$ is also universal for all N. For large N, the variance behaves as (try to show this)

$$\mathrm{Var}(M) = \ln N + \gamma_{\mathrm{E}} - \frac{\pi^2}{6} + \mathcal{O}(1/N). \tag{7.22}$$

Thus, for large N, both the variance $\mathrm{Var}(M)$ in Eq. (7.22) and the mean $\langle M \rangle$ in Eq. (7.18) behave identically (both growing as $\ln N$ to leading order). This property of mean and variance being the same is a typical feature of a Poisson distribution. Indeed, we will see later that, for large N, the distribution $P(M, N)$ does converge to a Poisson distribution. However, for finite N, the mean and variance are not exactly equal, signalling a departure from the Poissonian statistics of M. This departure can be quantified more precisely by computing the Fano factor F_N defined as the

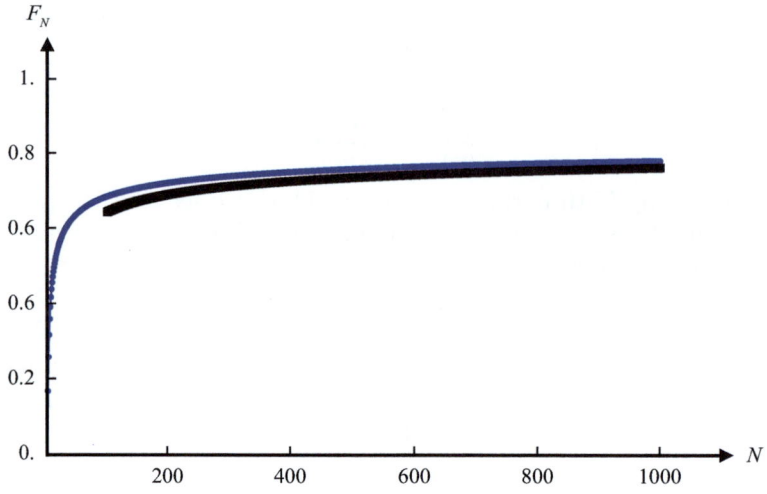

Fig. 7.3 Plot of F_N given in Eq. (7.23) vs. N (blue symbols). The black symbols indicate the asymptotic behavior in Eq. (7.24). Note that the convergence to the asymtotic value 1 is extremely slow as in Eq. (7.24).

ratio of the variance to the mean. Using the results found in Eqs. (7.17) and (7.21), we get the exact Fano factor for any N (see Fig. 7.3):

$$F_N = \frac{\mathrm{Var}(M)}{\langle M \rangle} = 1 - \frac{\sum_{k=1}^{N} 1/k^2}{\sum_{k=1}^{N} 1/k}. \tag{7.23}$$

One can show that F_N in Eq. (7.23) increases monotonically with N and approaches its Poissonian limit $F_N \to 1$ extremely slowly as

$$F_N \approx 1 - \frac{\pi^2}{6 \ln N} \tag{7.24}$$

for large N.

Full distribution of M. To go beyond the first two moments, it is useful to consider the generating function

$$\sum_{M=1}^{\infty} P(M, N) \lambda^M = \langle \lambda^M \rangle, \tag{7.25}$$

where the average, on the RHS, is over all the possible realizations of the x_i's. We first use $M = \sum_{k=1}^{N} \sigma_k$ and the property that the σ_k's are uncorrelated, Eq. (7.16). This gives

$$\sum_{M=1}^{\infty} P(M, N) \lambda^M = \langle \lambda^{\sum_{k=1}^{N} \sigma_k} \rangle = \prod_{k=1}^{N} \langle \lambda^{\sigma_k} \rangle. \tag{7.26}$$

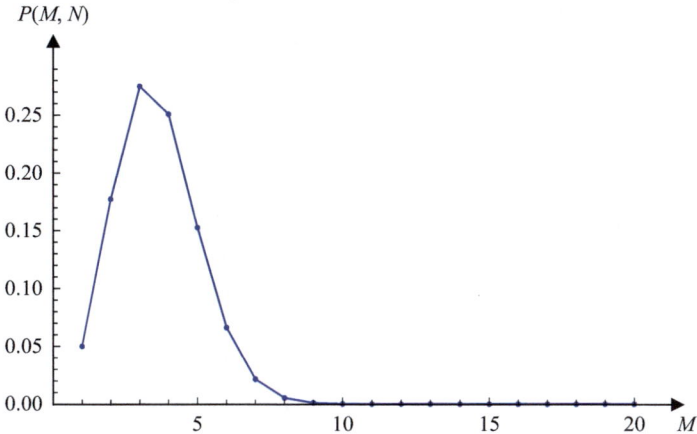

Fig. 7.4 Plot of the distribution of the number of records $P(M, N)$ in Eq. (7.30) vs. $M = 1, 2, \ldots, N$ for an IID sequence. Here, we chose $N = 20$.

Since $\sigma_k = 0, 1$, it follows that

$$\lambda^{\sigma_k} = 1 - (1 - \lambda)\sigma_k. \tag{7.27}$$

Taking the average and using $\langle \sigma_k \rangle = 1/k$ gives $\langle \lambda^{\sigma_k} \rangle = 1 + (\lambda - 1)/k$. Hence, from Eq. (7.26), we get

$$\sum_{M=1}^{\infty} P(M, N)\lambda^M = \prod_{k=1}^{N} \left(\frac{\lambda - 1}{k} + 1 \right) = \frac{\lambda(\lambda + 1) \cdots (\lambda + N - 1)}{N!}. \tag{7.28}$$

One recognizes that the rising factorial appearing in Eq. (7.28) is the generating function of the unsigned Stirling numbers of the first kind (Riordan, 2012),

$$\lambda(\lambda + 1) \cdots (\lambda + N - 1) = \sum_{M=1}^{N} \begin{bmatrix} N \\ M \end{bmatrix} \lambda^M, \tag{7.29}$$

where the unsigned Stirling numbers $\begin{bmatrix} N \\ M \end{bmatrix}$ enumerate the number of permutations of N elements with M disjoint cycles exactly. Hence, one has

$$P(M, N) = \frac{\begin{bmatrix} N \\ M \end{bmatrix}}{N!}, \quad M = 1, 2, \ldots, N, \tag{7.30}$$

which thus shows that the number of records of N IID random variables is distributed like the number of cycles in random permutations of N objects with uniform measure. We will discuss this connection with the cycles of random permutations later in more detail. A plot of $P(M, N)$ vs. M for $N = 20$ is shown in Fig. 7.4.

It turns out that there is yet another interesting representation of the generating function in Eq. (7.28) that is more amenable to asymptotic analysis for large N.

Indeed, if we consider the double generating function of $P(M, N)$ with respect to both M and N, it is easy to see from Eq. (7.28) that

$$\sum_{N=0}^{\infty} \sum_{M=0}^{\infty} P(M, N)\lambda^M z^N = (1 - z)^{-\lambda}, \qquad (7.31)$$

where we have used the convention $P(M, 0) = \delta_{M,0}$. Indeed, by expanding the RHS in power series in z, we can easily identify the coefficient of z^N to be exactly the same as the RHS of Eq. (7.28). Interestingly, writing

$$(1 - z)^{-\lambda} = e^{-\lambda \ln(1-z)} = \sum_{M=0}^{\infty} \frac{[-\ln(1 - z)]^M}{M!}\lambda^M \qquad (7.32)$$

and identifying the coefficient of λ^M on both sides of Eq. (7.31), we get yet another representation of the generating function,

$$\sum_{N=0}^{\infty} P(M, N)z^N = \frac{[-\ln(1 - z)]^M}{M!}. \qquad (7.33)$$

Let us first make an important remark. From the expression of the double generating function in Eq. (7.31), or alternatively from the two other representations respectively in Eqs. (7.28) and (7.33), we see that the full distribution $P(M, N)$ is universal for any fixed N, i.e., it is independent of $p(x)$. Even though this distribution, for any finite N, can be written formally in terms of the Stirling numbers as in Eq. (7.30), it is a bit cumbersome and not particularly illuminating. However, for large N, this universal distribution converges to a simpler form, which can be obtained by an asymptotic analysis of any of the three generating functions. It turns out, however, that the simplest is to start from the double generating function in Eq. (7.31).

Let us perform this asymptotic analysis as a nice exercise, which might be useful in preparing the reader to study other problems in the future. We start from Eq. (7.31). To proceed, it is first convenient to set $z = e^{-s}$. Now, for large N, the most important contribution comes from the vicinity of $z = 1$, i.e., $s = 0$. In this limit, the LHS of Eq. (7.31) converges to the Laplace transform with respect to N. Expanding the RHS of Eq. (7.31) for small s, one then gets, to leading order in s,

$$\int_0^{\infty} \left[\sum_{M=1}^{\infty} P(M, N)\lambda^M \right] e^{-sN} \, dN \approx \frac{1}{s^\lambda}. \qquad (7.34)$$

Inverting this Laplace transform trivially with respect to N gives, for large N,

$$\sum_{M=1}^{\infty} P(M, N)\lambda^M \approx \frac{N^{\lambda-1}}{\Gamma(\lambda)} = \frac{1}{N}\frac{\lambda N^\lambda}{\Gamma(1 + \lambda)}, \qquad (7.35)$$

where we used the identity $\Gamma(\lambda)=\Gamma(1+\lambda)/\lambda$. The next step is to expand the RHS of Eq. (7.35) in a power series in λ and identify the coefficient of λ^M. For this, we use

$$N^\lambda = e^{\lambda \ln N} = \sum_{k=0}^{\infty} \frac{(\ln N)^k}{k!} \lambda^k \qquad (7.36)$$

and the power series expansion

$$\frac{1}{\Gamma(1+\lambda)} = \sum_{m=0}^{\infty} d_m \lambda^m, \qquad (7.37)$$

where $d_0 = 1$ and the d_m for $m \geq 1$ are also known explicitly, though we will not need them here. Expanding the RHS of Eq. (7.35) using Eqs. (7.36) and (7.37) and identifying the power of λ^M gives, for fixed $M \geq 1$,

$$P(M,N) \approx \frac{1}{N} \sum_{m=0}^{M-1} \frac{(\ln N)^{M-1-m}}{(M-1-m)!} d_m. \qquad (7.38)$$

Finally, noticing that for large N the dominant contribution comes from the $m = 0$ term in the RHS of Eq. (7.38), we get, for large N and fixed $M \geq 1$,

$$P(M,N) \approx \frac{1}{N} \frac{(\ln N)^{M-1}}{(M-1)!}, \qquad (7.39)$$

which is just a Poisson distribution with parameter $\ln N$.

Finally, if one takes both M and N large, but with the ratio $x = M/\ln N$ fixed, then one finds that $P(M,N)$ admits an unusual large-deviation form,

$$P(M,N) \approx \exp\left\{-\ln N \Phi_{\text{IID}}\left(\frac{M}{\ln N}\right)\right\} = N^{-\Phi_{\text{IID}}(x)}, \qquad (7.40)$$

where the rate function $\Phi_{\text{IID}}(x)$ is given by

$$\Phi_{\text{IID}}(x) = 1 - x + x \ln x. \qquad (7.41)$$

Why do we call the large-deviation form in Eq. (7.40) unusual? Usually, in standard statistical physics problems, the probability distribution of any extensive observable M (such as energy), in a large system of volume V, admits a "canonical" large-deviation form

$$P(M,V) \approx \exp\left\{-V \Phi_0\left(\frac{M}{V}\right)\right\}, \qquad (7.42)$$

where $\Phi_0(x)$ is the associated rate function. In a usual extensive system, the volume V scales linearly with the number of degrees of freedom in the system, $V \propto N$. Hence, in terms of N, one would have expected a large-deviation form $P(M,N) \approx e^{-N\Phi_0(M/N)}$. Comparing this standard form with the result in Eq. (7.40), we see that in our case, the effective number of degrees of freedom behaves not as N but as

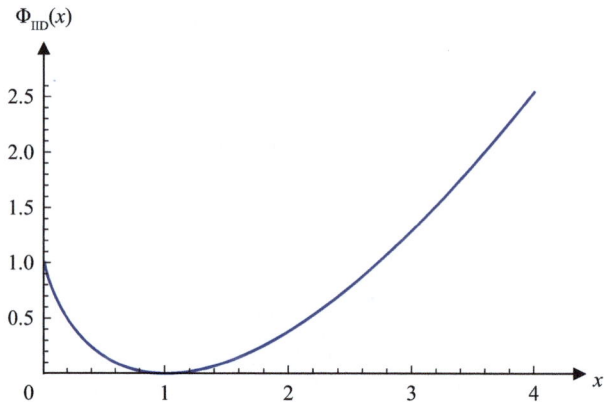

Fig. 7.5 The large-deviation function $\Phi_{\text{IID}}(x)$ in Eq. (7.41) plotted as a function of x.

$\ln N$. The reason is simply that the mean number of records scales as $\ln N$, and not as N as in a standard extensive system.

This large-deviation function $\Phi_{\text{IID}}(x)$ has a non-monotonic shape as a function of x (for a plot, see Fig. 7.5), achieving a minimum at $x = 1$, around which it behaves quadratically, i.e.,

$$\Phi_{\text{IID}}(x) \approx \frac{1}{2}(x-1)^2. \tag{7.43}$$

Substituting this quadratic form intp Eq. (7.40), we see that, near its mean, the distribution $P(M, N)$ has a Gaussian form that describes the typical fluctuations of M (see Fig. 7.6):

$$P(M, N) \sim \frac{1}{\sqrt{2\pi \ln N}} \exp\left(-\frac{(M - \ln N)^2}{2 \ln N}\right). \tag{7.44}$$

In contrast, for atypically large fluctuations, when $M \gg \ln N$ or $M \ll \ln N$, the shape $P(M, N)$ of the distribution is characterized by the more general large-deviation form in Eq. (7.40) that includes the Gaussian behavior around $M = \ln N$.

Thus, to summarize, for large N the typical fluctuations of the random variable M around its mean value $\langle M \rangle \sim \ln N$ scale as $\sqrt{\ln N}$ and, for $(M - \ln N) \sim \sqrt{\ln N}$ the PDF $P(M, N)$ has a Gaussian form as in Eq. (7.44). However, for atypically large fluctuations, when $(M - \ln N) \sim \ln N$, the PDF $P(M, N)$ has a large-deviation form as in Eq. (7.40).

7.1.2 Joint statistics of the number and ages of records

In the previous section we computed exactly the statistics of the total number of records $M = \sum_{k=1}^{N} \sigma_k$ up to step N. In computing the distribution $P(M, N)$, we used the fact that the variables σ_k's are uncorrelated. In many cases, this independence does not always hold, an example being the random walk sequence that will be discussed in Section 7.2. Hence, it is useful to have an alternative derivation that

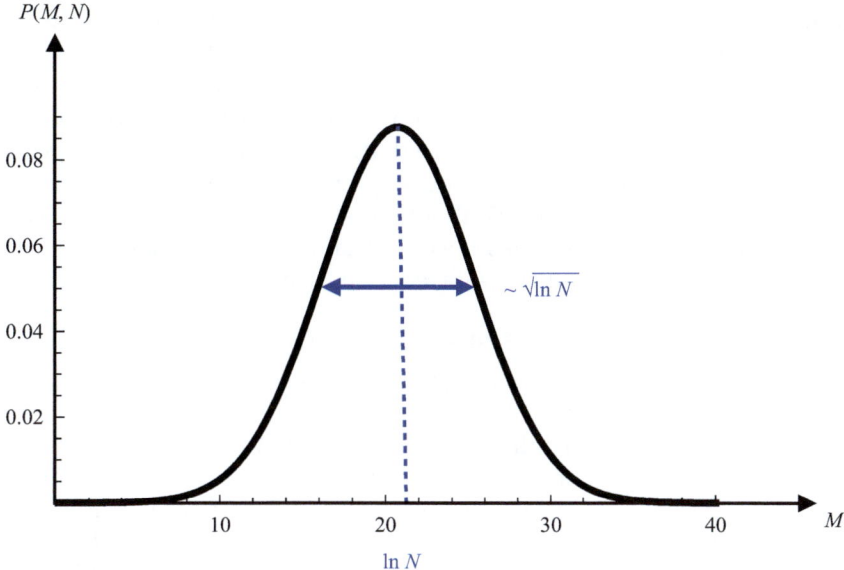

Fig. 7.6 A plot of $P(M, N)$ in Eq. (7.44) describing the typical fluctuations of M around its mean $\langle M \rangle \sim \ln N$ with a width of order $\sim \sqrt{\ln N}$.

does not directly use the σ_k variables. In addition, we would like to have an approach that not only gives access to the number of records but also to the statistics of the ages of the records.

To proceed, let us recall once more the definition of the ages of the records. Consider a particular realization of the sequence of N random variables $\{x_1, x_2, \ldots, x_N\}$ with M records as in Fig. 7.1. The records are shown by red dots. We treat the ages of the first $M - 1$ records and the last record slightly differently for a technical reason that will become clear shortly. For the first $M - 1$ records, we denote by $\ell_1, \ell_2, \ldots, \ell_{M-1}$ the time intervals between successive records—these are the ages of the broken records. The record age ℓ_k (for $k < N$) can then take only positive integer values $\ell_k = 1, 2, 3, \ldots$ Note, however, that the story of the last record (the Mth) is slightly different. This is because it still remains a record until the last step N of the sequence. The age of this last record can, for instance, even be 0 (this corresponds to the case when the Nth entry is a record). In order to put the ages of all the records, including this last one, on the same footing, it is convenient to define the age of the last record as $\ell_M - 1$ where $\ell_M = 1, 2, 3, \ldots$ We then have the sum rule

$$\ell_1 + \ell_2 + \cdots + \ell_{M-1} + \ell_M = N. \tag{7.45}$$

It is convenient to denote the collection of the ages of the records as a single vector,

$$\vec{\ell} = \{\ell_1, \ell_2, \ldots, \ell_{M-1}, \ell_M\}. \tag{7.46}$$

In a given realization, the ages $\vec{\ell}$ and the number of records M are both random variables. We denote their joint probability distribution by $P(\vec{\ell}, M \mid N)$, where $\vec{\ell}$ and M are the random variables while the length N of the sequence is fixed. If we can compute this joint distribution then summing over the ℓ_i's gives us the marginal distribution $P(M, N)$ of the number of records. Likewise, by summing over M we can obtain the joint distribution of the ages $P(\vec{\ell}, N)$. From this latter object, by successive summations, we can further compute the marginal distributions of ages $\ell_1, \ell_2, \ldots, \ell_M$. In particular, we will see that the joint distribution of ages $P(\vec{\ell}, N)$ does not factorize, indicating that the ages ℓ_i's form a correlated sequence. In the spirit of this book, it is natural to ask about the statistics of extremes of a correlated sequence. In particular, in this case, we will show how to compute exactly the statistics of the largest age and the shortest one, defined respectively as

$$\ell_{\max} = \max\{\ell_1, \ell_2, \ldots, \ell_M\}, \tag{7.47}$$

$$\ell_{\min} = \min\{\ell_1, \ell_2, \ldots, \ell_M\}. \tag{7.48}$$

To compute the joint probability distribution $P(\vec{\ell}, M \mid N)$ of the ages $\vec{\ell}$ and the number M of records, given the length N of the sequence, we need to first specify the values of the records: we denote them by $\{R_1, R_2, \ldots, R_M\}$ (see Fig. 7.1). Note that the first entry of the time series is indeed a record, i.e., $x_1 = R_1$. We clearly have the constraint $R_1 < R_2 < \cdots < R_M$. Once we have imposed this constraint, we have to make sure that all the $\ell_j - 1$ entries between R_j and R_{j+1} stay below R_j. Finally, we need to integrate over R_j for all $j = 1, 2, \ldots, M$. This yields

$$P(\vec{\ell}, M \mid N) = \int_{-\infty}^{\infty} dR_1 \cdots \int_{-\infty}^{\infty} dR_M \left[\prod_{i=1}^{M-1} \Theta(R_{i+1} - R_i) \right]$$
$$\times \prod_{j=1}^{M} p(R_j) \left[\int_{-\infty}^{R_j} p(x)\, dx \right]^{\ell_j - 1} \delta\left(\sum_{m=1}^{M} \ell_m, N \right), \tag{7.49}$$

where the Kronecker delta, $\delta(i, j) = 1$ if $i = j$ and 0 otherwise, ensures that the size of the sample is N. If one performs the change of variables $u_j = \int_{-\infty}^{R_j} p(x)\, dx$, the distribution $P(\vec{\ell}, M \mid N)$ in Eq. (7.49) can be written as

$$P(\vec{\ell}, M \mid N) = \int_0^1 du_M\, u_M^{\ell_M - 1} \prod_{j=1}^{M-1} \int_0^{u_{j+1}} du_j\, u_j^{\ell_j - 1} \delta\left(\sum_{m=1}^{M} \ell_m, N \right). \tag{7.50}$$

This multiple integral can be performed straightforwardly term by term to obtain (please check this!)

$$P(\vec{\ell}, M \mid N) = \frac{1}{\ell_1 (\ell_1 + \ell_2) \cdots (\ell_1 + \ell_2 + \cdots + \ell_M)} \delta\left(\sum_{m=1}^{M} \ell_m, N \right). \tag{7.51}$$

It is important to stress that this joint distribution is completely universal, i.e., independent of the parent distribution $p(x)$. This means that any observable depending only on the ages of the records is also universal. Quite interestingly, although the variables x_i are independent, we see in Eq. (7.51) that the ages ℓ_j are correlated, which yields non-trivial statistics for the ages in this IID case.

Recovering the distribution of the number of records. Let us first check how to recover the distribution $P(M, N)$ for the number of records from the joint distribution $P(\vec{\ell}, M \mid N)$ in Eq. (7.50). Below, we show how to recover the generating function for $P(M, N)$ in Eq. (7.33) by marginalizing the age variables $\vec{\ell}$ in the joint distribution $P(\vec{\ell}, M \mid N)$. We start from Eq. (7.50), multiply both sides by z^N and sum over $\vec{\ell}$ and N. This gives

$$\sum_{N \geq 0} P(M, N) z^N = \sum_{\vec{\ell}, N \geq 0} P(\vec{\ell}, M \mid N) z^N$$

$$= \sum_{\vec{\ell}} \int_0^1 \mathrm{d}u_M \, u_M^{\ell_M - 1} \prod_{j=1}^{M-1} \int_0^{u_{j+1}} \mathrm{d}u_j \, u_j^{\ell_j - 1} z^{\ell_1 + \ell_2 + \cdots + \ell_M}, \qquad (7.52)$$

where the sum over $\vec{\ell} = \{\ell_1, \ell_2, \ldots, \ell_M\}$ indicates that each ℓ_j runs over $\ell_j = 1, 2, \ldots$ We can carry out the sum over each of the ℓ_j since these are simple geometric series. This gives a nested integral of the form

$$\sum_{N \geq 0} P(M, N) z^N = \int_0^1 \frac{z \, \mathrm{d}u_M}{1 - z u_M} \int_0^{u_M} \frac{z \, \mathrm{d}u_{M-1}}{1 - z u_{M-1}} \cdots \int_0^{u_2} \frac{z \, \mathrm{d}u_1}{1 - z u_1}. \qquad (7.53)$$

One can now perform the integrals serially from right to left (it's a funny exercise, try it!) and we get

$$\sum_{N \geq 0} P(M, N) z^N = \frac{[-\ln(1 - z)]^M}{M!}, \qquad (7.54)$$

thus recovering exactly the formula in Eq. (7.33).

The statistics of the ages of the records. In this section, starting from the full joint PDF $P(\vec{\ell}, M \mid N)$ in Eq. (7.51), we want to derive three observables of interest: (i) the marginal distribution of the age of the kth record, (ii) the distribution of the age of the longest-lasting record and (iii) the distribution of the age of the shortest-lasting record.

Marginal probability distribution of the age of the kth record. The marginal probability distribution of ℓ_k and M can be obtained by summing the full joint distribution $P(\vec{\ell}, M \mid N)$ in Fig. (7.51) over all the ages ℓ_j with $j \neq k$:

$$P(\ell_k, M \mid N) = \sum_{\ell_1 \geq 1} \cdots \sum_{\ell_{k-1} \geq 1} \sum_{\ell_{k+1} \geq 1} \cdots \sum_{\ell_M \geq 1} P(\vec{\ell}, M \mid N). \qquad (7.55)$$

If we further sum over $M \geq k$, we can get just the marginal distribution of ℓ_k,

$$P(\ell_k, N) = \sum_{M \geq k} P(\ell_k, M \mid N) = \sum_{M \geq k} \sum_{\ell_1 \geq 1} \cdots \sum_{\ell_{k-1} \geq 1} \sum_{\ell_{k+1} \geq 1} \cdots \sum_{\ell_M \geq 1} P(\vec{\ell}, M \mid N). \qquad (7.56)$$

The full distribution $P(\vec{\ell}, M|N)$ given in Eq. (7.51) is obviously not invariant under the permutation of the ages ℓ_j and therefore the distribution of ℓ_k depends explicitly on k. Also, for ℓ_k to be defined, the number of records needs to be at least k, which implies that $M \geq k$, and naturally $N \geq k$. Note that ℓ_k, for fixed k and N, cannot take arbitrary values. This is because, for ℓ_k to exist, we must have $k - 1$ records before it, of ages $\{\ell_1, \ell_2, \ldots, \ell_{k-1}\}$, and each of them must be at least of length 1. Since the total length $\ell_1 + \ell_2 + \cdots + \ell_{k-1} + \ell_k \leq N$, it follows that $\ell_k \leq N - k + 1$. Since $\ell_k \geq 1$, it follows that

$$1 \leq \ell_k \leq N - k + 1. \tag{7.57}$$

Now remains the hard task of performing the multiple sum in Eq. (7.56). It may look very complicated but amazingly it can be done in a closed form. It just requires a few straightforward steps, as follows.

- First, in Eq. (7.56) we need to use the result for $P(\vec{\ell}, M|N)$. Instead of using Eq. (7.51), it turns out to be more convenient to use the preceding integral representation in Eq. (7.50).
- Second, upon substituting Eq. (7.50) into Eq. (7.56), we first note the presence of the delta function $\delta\left(\sum_{m=1}^{M} \ell_m, N\right)$. Whenever one encounters a delta function in a sum, it immediately suggests that we should use a generating function, i.e., multiply by z^N and sum over all N—this will replace z^N by $z^{\sum_{m=1}^{M} \ell_m}$ and, hence, it decouples the sums over different ℓ_m.
- The third step is to perform the sums over all the ℓ_j except ℓ_k. These sums are just geometric and can be performed easily. Each of them gives a factor $z/(1 - zu_j)$.
- The next step is to carry out the integrals over the u_j. For this we break the system into two halves, $\{u_1, u_2, \ldots, u_{k-1}\}$ and $\{u_{k+1}, u_{k+2}, \ldots, u_M\}$, and then carry out the nested integrals on the left side (with $u_1 \leq u_2 \leq \cdots \leq u_{k-1}$) and similarly on the right side (with $u_{k+1} \leq u_{k+2} \leq \cdots \leq u_M$). These nested integrals are actually quite simple to perform explicitly (please try it out for $M = 2, 3, 4, \ldots$ and you will immediately see the pattern). This brings us to the following formula for the generating function of $P(\ell_k, M \mid N)$ in Eq. (7.55):

$$\sum_{N=1}^{\infty} z^N P(\ell_k, M \mid N)$$

$$= \int_0^1 du_k \left[\frac{(-1)^{k-1}}{(k-1)!} \ln^{k-1}(1 - zu_k) \right] u_k^{\ell_k - 1} z^{\ell_k} \left[(-1)^{M-k} \frac{\ln^{M-k}(1-z)}{(M-k)!} \right]. \tag{7.58}$$

The first square bracket comes from the integration over the left block $\{u_1, u_2, \ldots, u_{k-1}\}$, while the second square bracket results from the integration over the right block $\{u_{k+1}, u_{k+2}, \ldots, u_M\}$.

- We now sum over all $M \geq k$, which just gives $e^{-\ln(1-z)} = 1/(1-z)$. Finally, making a change of variable $zu_k = x$ we arrive at

$$\sum_{N \geq k} P(\ell_k, N)z^N = \frac{1}{1-z} \int_0^z dx\,(1-x)\frac{[-\ln(1-x)]^{k-1}}{(k-1)!}x^{\ell_k - 1}. \qquad (7.59)$$

From this exact generating function we would next like to extract the asymptotic behavior of the marginal distribution $P(\ell_k, N)$ for large N. To extract this large-N asymptotic behavior, we need to investigate the $z \to 1$ limit of the generating function in Eq. (7.59). In the limit $z \to 1$, we see that the RHS of Eq. (7.59) behaves like $\propto (1-z)^{-1}$ when $z \to 1$, because the integral over x is convergent at $z = 1$ (check this!). What does this tell us? Suppose that $P(\ell_k, N)$ in the LHS of Eq. (7.59) becomes independent of N for large N and converges to $P(\ell_k)$. Then we can take this factor $P(\ell_k)$ out of the summation on the LHS. Performing the sum over N then predicts that the LHS should behave as $P(\ell_k)/(1-z)$. Comparing to the RHS, we see that this behavior as $z \to 1$ is indeed consistent. Moreover, it tells us that the limiting distribution is given by (Neuts, 1967; Shorrock, 1972; Diaconis and Pitman, 1986)

$$P(\ell_k) = \lim_{N \to \infty} P(\ell_k, N) = \int_0^1 dx\,(1-x)\frac{[-\ln(1-x)]^{k-1}}{(k-1)!}x^{\ell_k - 1} \qquad (7.60)$$

$$= \sum_{m=0}^{\ell_k - 1}(-1)^m \binom{\ell_k - 1}{m}\frac{1}{(2+m)^k}.$$

In order to arrive at the second line, we first performed the change of variable $u = -\ln(1-x)$ in the integral in Eq. (7.60) and then used the binomial formula to expand $x^{\ell_k - 1} = (1 - e^{-u})^{\ell_k - 1}$ in order to perform the integral over u term by term. The probability $P(\ell_k)$ is normalized to unity and decreases monotonically with increasing ℓ_k, starting from $P(\ell_k = 1) = 2^{-k}$. For large ℓ_k, its asymptotic behavior is more conveniently obtained from the integral representation in Eq. (7.60) which can be analyzed in the large-ℓ_k limit by performing the change of variable $v = (1-x)\ell_k$, yielding

$$P(\ell_k) \sim \frac{1}{(k-1)!}\frac{(\ln \ell_k)^{k-1}}{\ell_k^2}, \quad \ell_k \to \infty. \qquad (7.61)$$

A plot of this distribution is shown in Fig. 7.7 for two different values of k. Thus we see that for large ℓ_k, the stationary distribution $P(\ell_k)$ decays algebraically as ℓ_k^{-2}, but modulated by a logarithmic factor $(\ln \ell_k)^{k-1}$. This indicates that all moments $\langle (\ell_k)^p \rangle$, with $p \geq 1$, diverge in the large-N limit. Thus, interestingly, this constitutes a nice example where even though the full distribution reaches a well-defined limit as $N \to \infty$, the moments of this limiting distribution are all divergent. Similar algebraically decaying stationary distributions with diverging moments have already been discussed in this book, such as the position distribution of Lévy flights in

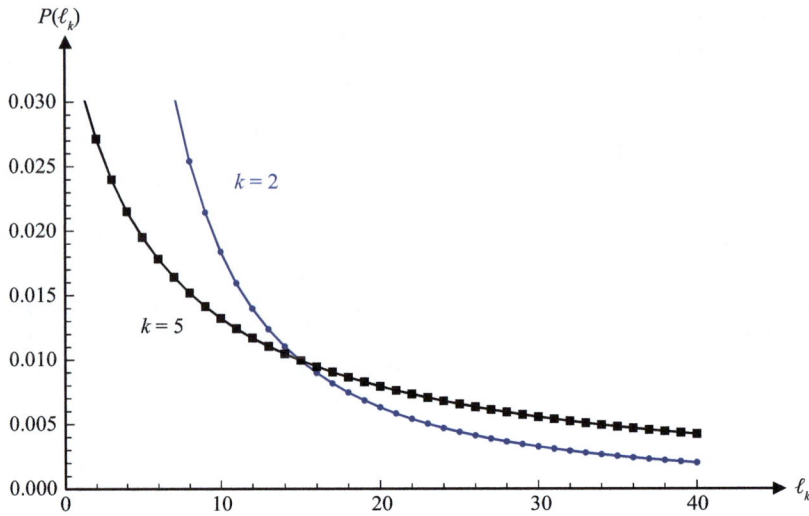

Fig. 7.7 Plot of the marginal distribution of the age of the kth record $P(\ell_k)$ given in Eq. (7.60) vs. ℓ_k for two different values of k, namely $k = 2$ (blue dots) and $k = 5$ (black squares).

Eq. (2.53) as well as the distribution of the stationary gaps for long random walks in Eq. (6.117). In fact, one can compute all the moments of ℓ_k for finite but large N from Eq. (7.59), with the result (for $p \geq 1$)

$$\langle \ell_k^p \rangle \sim \frac{[\ln N]^k}{k!} N^{p-1} \quad \text{as } N \to \infty. \tag{7.62}$$

We leave the derivation of this result as an exercise. Since $\ell_k \leq N$, one can write $P(\ell_k, N)$ as

$$P(\ell_k, N) \sim P(\ell_k)\theta(N - \ell_k), \tag{7.63}$$

where $P(\ell_k)$ is independent of N and given in Eq. (7.60). If one computes the pth moment of ℓ_k using the asymptotic form in Eq. (7.63), one indeed recovers Eq. (7.62) for large N (check!).

Distribution of the age of the longest-lasting record.   In the previous section, we characterized the distribution of the age ℓ_k of the kth record. The distribution depends strongly on k, i.e., different records have different age statistics. A natural question is then: Which one is the longest-lasting record, i.e., has the largest age? For this, we need to investigate the statistics of the following extreme observable:

$$\ell_{\max,N} = \max\{\ell_1, \ell_2, \dots, \ell_M\}. \tag{7.64}$$

Note that the variables $\ell_1, \ell_2, \dots, \ell_M$ are strongly correlated random variables, with the joint distribution given in Eq. (7.51). There are three points to note about this joint distribution: (i) the total number of relevant variables M in Eq. (7.64) is itself a random variable for fixed N, (ii) the variables ℓ_i are not independent and (iii) the marginal distribution of ℓ_k depends strongly on the label k. Hence, the extreme statistics for this example will to be very different from the three limiting universal distributions for IID variables (Gumbel, Fréchet and Weibull) discussed earlier. Nevertheless, we will see that we can compute the distribution of $\ell_{\mathrm{max},N}$ exactly in the limit of large N. This then provides an example of an exactly solvable model of EVS for strongly correlated random variables.

To proceed, as discussed in the introduction, the most convenient starting point is to investigate the cumulative distribution $\mathcal{Q}_{\mathrm{max}}(w, N) = \mathrm{Prob}(\ell_{\mathrm{max},N} \leq w)$ for $w \geq 1$. This can be obtained from the full joint distribution in Eq. (7.51) by summing over M and $\ell_1, \ldots, \ell_M$ with the constraint that each of them is less than or equal to w. This reads, for $N \geq 1$,

$$\mathcal{Q}_{\mathrm{max}}(w, N) = \sum_{M \geq 1} \sum_{\ell_1=1}^{w} \cdots \sum_{\ell_M=1}^{w} P(\vec{\ell}, M \mid N), \tag{7.65}$$

while $\mathcal{Q}_{\mathrm{max}}(w, N = 0) = 1$. This multiple sum looks very similar to Eq. (7.56) except that the sum over each ℓ_j here goes up to w. As before, the presence of the delta function in $P(\vec{\ell}, M \mid N)$ [see Eq. (7.50)] immediately suggests we compute the generating function of $\mathcal{Q}_{\mathrm{max}}(w, N)$ with respect to N. We will not repeat the algebra here but, following exactly the same steps as outlined before for the marginal distribution of ℓ_k [see the discussion after Eq. (7.57)], one can show that

$$\sum_{N \geq 0} z^N \mathcal{Q}_{\mathrm{max}}(w, N) = \exp\left(\sum_{k=1}^{w} \frac{z^k}{k}\right). \tag{7.66}$$

For later purposes, it is useful to rewrite the RHS of Eq. (7.66) as

$$\exp\left(\sum_{k=1}^{w} \frac{z^k}{k}\right) = \exp\left(\sum_{k=1}^{\infty} \frac{z^k}{k} - \sum_{k=w+1}^{\infty} \frac{z^k}{k}\right) = \frac{1}{1-z} \exp\left(-\sum_{k=w+1}^{\infty} \frac{z^k}{k}\right), \tag{7.67}$$

where we used the identity $\sum_{k=1}^{\infty} z^k/k = -\ln(1-z)$. Hence, Eq. (7.66) then reads

$$\sum_{N \geq 0} z^N \mathcal{Q}_{\mathrm{max}}(w, N) = \frac{1}{1-z} \exp\left(-\sum_{k=w+1}^{\infty} \frac{z^k}{k}\right). \tag{7.68}$$

This representation will be useful for analyzing the large-N behavior of the full CDF $\mathcal{Q}_{\mathrm{max}}(w, N)$. But before doing so, it is useful to first analyze how the average $\langle \ell_{\mathrm{max},N} \rangle$ behaves for large N. This will give us an idea of how $\ell_{\mathrm{max},N}$ scales with N for large N.

To compute $\ell_{\mathrm{max},N}$, we use the well-known identity (try to prove it!)

$$\langle \ell_{\mathrm{max},N} \rangle = \sum_{w \geq 1}(1 - \mathcal{Q}_{\mathrm{max}}(w, N)). \tag{7.69}$$

Taking the generating function on both sides of Eq. (7.69) and using Eq. (7.68), we get

$$\sum_{N \geq 0} \langle \ell_{\max,N} \rangle z^N = \frac{1}{1-z} \sum_{w \geq 1} \left[1 - \exp \left(- \sum_{k \geq w+1} \frac{z^k}{k} \right) \right]. \qquad (7.70)$$

Now, we want to obtain the large-N behavior of $\langle \ell_{\max,N} \rangle$. To extract this, we need to analyze the limit $z \to 1$ of Eq. (7.70), as seen before in the derivation of the marginal distribution of ℓ_k [see the discussion below Eq. (7.59)]. It is useful to first set $z = e^{-s}$ and take the $s \to 0$ limit. We first analyze the RHS of Eq. (7.70) in the small-s limit. Let us start with the sum inside the exponential. In the small-s limit, one can replace the sum by an integral and write

$$\sum_{k \geq w+1} \frac{e^{-sk}}{k} \approx \int_{sw}^{\infty} \frac{du}{u} e^{-u} = \mathrm{E}(sw), \qquad (7.71)$$

where we made the change of variable $u = sk$ and defined the exponential integral function $\mathrm{E}(y) = \int_y^{\infty} du \, e^{-u}/u$. This behavior in Eq. (7.71) makes sense in the limit $s \to 0$, $w \to \infty$ but with the product sw fixed. Next, we also replace the sum over w by an integral and make the change of variable $sw = y$. The RHS then becomes, to leading order for small s,

$$\sum_{N \geq 0} \langle \ell_{\max,N} \rangle e^{-sN} \approx \frac{C_{\mathrm{GD}}}{s^2}, \quad \text{where } C_{\mathrm{GD}} = \int_0^{\infty} dy \, (1 - e^{-\mathrm{E}(y)}). \qquad (7.72)$$

For small s, we can also replace the sum in the LHS of Eq. (7.72) by an integral over N, which transforms the generating function to a Laplace transform with respect to N. Inverting this Laplace transform, we get

$$\langle \ell_{\max,N} \rangle = C_{\mathrm{GD}} N + \mathcal{O}(1), \qquad (7.73)$$

where the constant C_{GD} is given in Eq. (7.72). By a simple integration by parts, this constant can also be expressed as

$$C_{\mathrm{GD}} = \int_0^{\infty} dy \, (1 - e^{-\mathrm{E}(y)}) = \int_0^{\infty} dy \, e^{-y - \mathrm{E}(y)} = 0.624\,33\ldots \qquad (7.74)$$

The constant C_{GD} is known as the Golomb–Dickman or Goncharov constant (Arratia *et al.*, 2003; Finch, 2003; Pitman, 2006). It also describes the linear growth of the longest cycle of a random permutation (Finch, 2003). It also appeared in a model of a growing network (Godrèche and Luck, 2008) and in a one-dimensional ballistic aggregation model (Majumdar *et al.*, 2009). The complete asymptotic expansion of $\langle \ell_{\max,N} \rangle$, beyond the leading order, was established in Gourdon (1997). We refer the reader to Pitman (2006) and Arratia *et al.* (2003) for more discussions of $\ell_{\max,N}$ in this IID case.

What is the physical meaning of this constant C_{GD}? First, let us note an interesting fact. We recall from the previous section that the average age of the kth record scales as $\langle \ell_k \rangle \sim (\ln N)^k / k!$ for large N [see Eq. (7.62) with $p = 1$]. The result in Eq. (7.73) shows, in contrast, that the age of the longest-lasting record scales much faster with growing N (linearly as $C_{\text{GD}} N$). The largest record age is thus anomalously large ($\sim N$) compared to the age of a typical record. This already indicates that the label of the record k has to be very large if it is to be a candidate for the longest-lasting record. In fact, most likely, the last record will be the longest one. Thus, one can ask: What is the probability that the last record (the Mth, given that M is the number of records) is the longest-lasting one in a sequence of N IID entries? Indeed, one can define this probability:

$$Q_{\text{last}}(N) = \text{Prob}(\ell_M > \max(\ell_1, \ldots, \ell_{M-1})) = \text{Prob}(\ell_{\max,N} = \ell_M). \qquad (7.75)$$

In fact, this probability $Q_{\text{last}}(N)$ turns out to be the rate of growth of $\langle \ell_{\max,N} \rangle$ with N for large N. To see this, we consider increasing the size of the sequence from N to $N + 1$ by adding one more independent entry into the sequence. Then we ask whether $\ell_{\max,N}$ changes due to the addition of this new entry. Indeed, if the last interval is the longest then adding a new entry will increase the length of the longest interval by 1. In contrast, if the last interval is not the longest one then $\ell_{\max,N+1}$ remains unchanged, i.e., $\ell_{\max,N+1} = \ell_{\max,N}$. Thus, one can write the evolution of $\ell_{\max,N}$ as

$$\ell_{\max,N+1} = \ell_{\max,N} + \xi_N, \qquad (7.76)$$

where ξ_N is a binary random variable with distribution

$$\xi_N = \begin{cases} 1 & \text{with probability } Q_{\text{last}}(N), \\ 0 & \text{with probability } 1 - Q_{\text{last}}(N). \end{cases} \qquad (7.77)$$

Taking the average on both sides of Eq. (7.76), one gets an exact equation for all N:

$$\langle \ell_{\max,N+1} \rangle - \langle \ell_{\max,N} \rangle = Q_{\text{last}}(N). \qquad (7.78)$$

For large N, substituting the asymptotic result for $\langle \ell_{\max,N} \rangle$ into Eq. (7.73), we get

$$\lim_{N \to \infty} Q_{\text{last}}(N) = C_{\text{GD}} = 0.624\,33\ldots, \qquad (7.79)$$

where C_{GD} is the Golomb–Dickman constant given in Eq. (7.74).

So far we have just computed the average value of $\ell_{\max,N}$. What about the full distribution of this random variable? Indeed, given that $\langle \ell_{\max,N} \rangle \approx C_{\text{GD}} N$ is the only length scale in the problem for large N, it is natural to anticipate that the full probability distribution should have a scaling form [see, for instance, the review in Godrèche *et al.* (2017)]

$$\text{Prob}(\ell_{\max,N} = \ell) \xrightarrow[N \to \infty]{} \frac{1}{N} f_{\max}\left(\frac{\ell}{N}\right), \qquad (7.80)$$

where $f_{\max}(x)$ is a scaling function supported over the interval $[0, 1]$, since $\ell_{\max,N} \leq N$. To check the validity of this scaling form we have numerically extracted

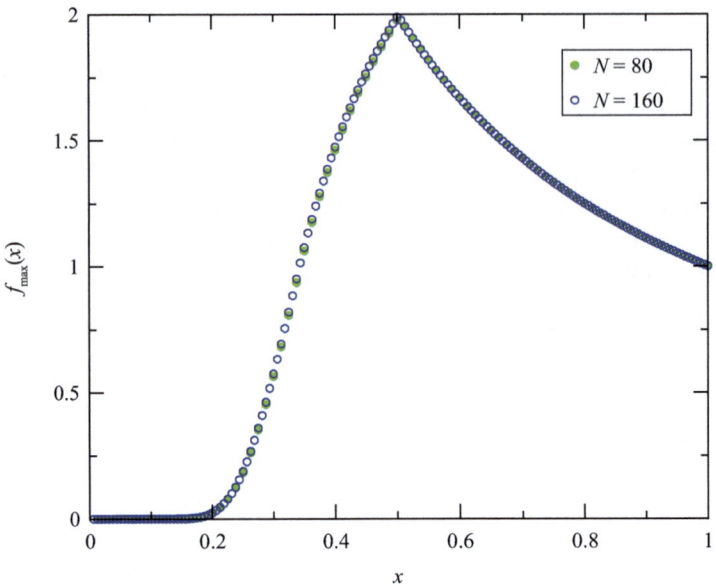

Fig. 7.8 The scaled distribution $N \, \text{Prob}(\ell_{\max,N} = \ell)$ as a function of the scaled variable $x = \ell/N$ for two different values of N ($N = 80$ and $N = 160$). The two curves collapse onto each other, thus verifying the scaling form postulated in Eq. (7.80). This figure, with slightly different notation, appeared first in Godrèche *et al.* (2016).

$\text{Prob}(\ell_{\max,N} = \ell)$ from the generating function of the CDF given in Eq. (7.66) using Mathematica for $N = 80$ (green full circles) and $N = 160$ (blue empty circles) in Fig. 7.8. The data collapse for two different values of N indeed confirms the scaling behavior in Eq. (7.80). The scaling function $f_{\max}(x)$ can in principle be extracted by analyzing Eq. (7.66) in the limit $z \to 1$. The details are a bit cumbersome and here we summarize the main properties of this scaling function. For example, $f_{\max}(x)$ has singularities at the points $x = 1/k$ where k is an integer $k = 2, 3, \ldots$ At $x = 1/k$ the $(k-1)$th derivative of $f_{\max}(x)$ is discontinuous. The singularities become weaker and weaker as k increases. For instance, one clearly sees the singularity at $x = 1/2$, i.e., $k = 2$, but the higher ones are almost invisible in the plot. Moreover, the asymptotic behaviors of the function $f_R(x)$ can also be obtained explicitly:

$$f_{\max}(x) \sim \begin{cases} \exp\left(\dfrac{1}{x}\ln x\right), & x \to 0, \\ 1, & x \to 1. \end{cases} \tag{7.81}$$

We refer the reader to Godrèche and Luck (2008) for further details on this limiting distribution.

Distribution of the age of the shortest record. We now focus on the age of the shortest record, denoted by $\ell_{\min,N}$, which is defined as

$$\ell_{\min,N} = \min\{\ell_1, \ell_2, \dots, \ell_M\}. \tag{7.82}$$

We define $Q_{\min}(\ell, N) = \mathrm{Prob}(\ell_{\min,N} \geq \ell)$, $\ell \geq 1$, and $Q_{\min}(\ell, N = 0) = 0$. Using the same lines of reasoning that led to Eq. (7.66) for $\ell_{\max,N}$, we find that the generating function of $Q_{\min}(\ell, N) = \mathrm{Prob}(\ell_{\min,N} \geq \ell)$ with respect to N satisfies

$$\sum_{N \geq 0} Q_{\min}(\ell, N) z^N = \exp\left(\sum_{k \geq \ell} \frac{z^k}{k}\right) - 1. \tag{7.83}$$

The generating function of the average value $\langle \ell_{\min,N} \rangle = \sum_{\ell \geq 1} Q_{\min}(\ell, N)$ can then be obtained from Eq. (7.83) using the same methods as for $\langle \ell_{\max,N} \rangle$. Indeed, one finds that the generating function of $\langle \ell_{\min,N} \rangle$ is given by

$$\sum_{N \geq 0} \langle \ell_{\min,N} \rangle z^N = \sum_{\ell \geq 1} \left[\exp\left(\sum_{k \geq \ell} \frac{z^k}{k}\right) - 1\right]. \tag{7.84}$$

Little exercise 7.1 *Please verify the relation in Eq. (7.84)! In addition, following the same lines that led to the asymptotic behavior for $\langle \ell_{\max,N} \rangle$ in Eq. (7.73), show that, for large N, Eq. (7.84) yields the following asymptotic behavior:*

$$\langle \ell_{\min,N} \rangle = e^{-\gamma_E} \ln N + o(\ln N), \tag{7.85}$$

with the numerical value $e^{-\gamma_E} = 0.561\,45\dots$, where $\gamma_E = 0.577\,21\dots$ is the Euler constant. The asymptotic behavior in Eq. (7.85) was derived in Shepp and Lloyd (1966).

As in the case of $\ell_{\max}$, one may wonder what the full distribution of $\ell_{\min,N}$ is. It turns out that, as $N \to \infty$, unlike the distribution of $\ell_{\max,N}$ which has a scaling form as in Eq. (7.80), the distribution of $\ell_{\min,N}$ approaches a "stationary," i.e., N-independent, form. This limiting form can be extracted from the analysis of the generating function in Eq. (7.83) as follows. Assuming that $Q_{\min}(\ell, N) \to Q_{\min}(\ell)$ for large N, the LHS of Eq. (7.83) behaves, as $z \to 1$, as

$$\sum_{N \geq 0} Q_{\min}(\ell, N) z^N \approx \frac{Q_{\min}(\ell)}{1 - z}, \quad z \to 1. \tag{7.86}$$

We then write

$$\sum_{k=\ell}^{\infty} \frac{z^k}{k} = -\ln(1 - z) - \sum_{k=1}^{\ell-1} \frac{z^k}{k}. \tag{7.87}$$

Inserting this on the RHS of Eq. (7.83) and comparing with the LHS for the leading behavior as $z \to 1$ gives the limiting cumulative distribution

$$Q_{\min}(\ell) = \exp\left(-\sum_{k=1}^{\ell-1} \frac{1}{k}\right). \tag{7.88}$$

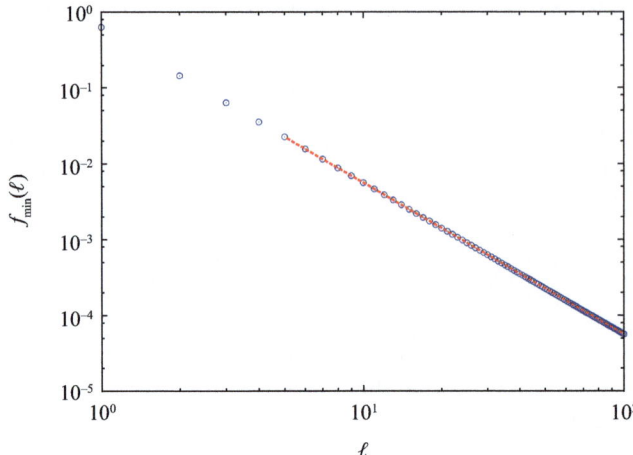

Fig. 7.9 Plot of the limiting distribution $f_{min}(\ell)$ given in Eq. (7.89) (blue circles). The dotted red line corresponds to the large-ℓ asymptotic behavior given in Eq. (7.90).

The corresponding probability distribution $\text{Prob}(\ell_{min,N} = \ell) = \mathcal{Q}_{min}(\ell) - \mathcal{Q}_{min}(\ell + 1)$ is then given by

$$\text{Prob}(\ell_{min,N} = \ell) = f_{min}(\ell) = \exp\left[-\sum_{k=1}^{\ell-1} \frac{1}{k}\right](1 - e^{-1/\ell}), \quad \ell \geq 2, \qquad (7.89)$$

while $f_{min}(\ell = 1) = 1 - e^{-1}$. The limiting distribution $f_{min}(\ell)$ is a monotonically decreasing function of ℓ and its asymptotic behaviors are given by

$$f_{min}(\ell) \approx \begin{cases} 1 - e^{-1}, & \ell \to 1, \\ \dfrac{e^{-\gamma_E}}{\ell^2}, & \ell \to \infty, \end{cases} \qquad (7.90)$$

Interestingly, this limiting distribution in the $N \to \infty$ limit has an algebraic tail for large ℓ. This $e^{-\gamma_E}\ell^{-2}$ decay for large ℓ is also consistent with the asymptotic behavior of $\langle \ell_{min,N} \rangle$ for large N in Eq. (7.85). Indeed, assuming this tail behavior in an infinite sequence, one can estimate the average $\langle \ell_{min,N} \rangle$ for a finite sequence of length $N \gg 1$ as

$$\langle \ell_{min,N} \rangle \approx \sum_{\ell=\ell_0}^{N} \frac{e^{-\gamma_E}}{\ell^2} \times \ell, \qquad (7.91)$$

where we have used a lower cut-off ℓ_0 (since we are using the large-ℓ form in estimating the average). The upper cut-off is obviously set to N since $\ell_{min,N}$ cannot be greater than N. This sum clearly yields, for large N, $\langle \ell_{min,N} \rangle \sim e^{-\gamma_E} \ln N$, consistent with Eq. (7.85). In Fig. 7.9 we show a plot of this limiting distribution $f_{min}(\ell)$, where we see in particular that the asymptotic large-ℓ behavior $\sim e^{-\gamma_E}/\ell^2$ in Eq. (7.90) already gives a quite accurate description of the exact distribution $f_{min}(\ell)$ for $\ell \gtrsim 10$.

7.2 Record statistics for random walks

Here we consider a time series of $N + 1$ entries $\{x_0 = 0, x_1, x_2, \ldots, x_k, \ldots, x_N\}$ where the x_n are the positions of a random walker on a line, starting at the initial position $x_0 = 0$. The position of the walker evolves via the Markov jump process

$$x_n = n_{k-1} + \eta_n, \tag{7.92}$$

where the η_n are IID jump variables, each drawn from a continuous and symmetric distribution $f(\eta)$ as discussed in Chapter 2 (see also Fig. 2.1). We are interested in deriving the statistics of the number of records and their ages for this sequence. As discussed in the introduction, even though the jumps η_n are uncorrelated, the entries x_n themselves are strongly correlated. A natural question is then how these strong correlations between the entries will affect the statistics of records for IID variables in the previous section.

We follow the method developed in Majumdar and Ziff (2008) for random walks with symmetric and continuous jump distributions $f(\eta)$, that also include Lévy flights. This method has been generalized to the case of discrete jump distributions (Mounaix *et al.*, 2020). Furthermore, generalizations of this basic method were also developed to study the record statistics for other types of random walks such as those with asymmetric jump distributions $f(\eta)$, in particular in the presence of a constant drift (Le Doussal and Wiese, 2009; Majumdar *et al.*, 2012), continuous-time random walks (Sabhapandit, 2011), and multiple independent random walks (Wergen *et al.*, 2012).

7.2.1 Average number of records

In this sequence, as usual, a record happens at step k if x_k exceeds all previous values, i.e., $x_k > \max\{x_0, x_1, \ldots, x_{k-1}\}$ [see Eq. (7.1)]. By convention, we choose the first entry $x_0 = 0$ as a record. The first natural question is: How many records occur in an N-step random walk? To address this question, it is useful, as in the IID case, to first introduce the indicator variables

$$\sigma_k = \begin{cases} 1 & \text{if } x_k \text{ is a record,} \\ 0 & \text{if } x_k \text{ is not a record.} \end{cases} \tag{7.93}$$

In terms of the σ_k, the number of records M in a walk of N steps reads

$$M = \sum_{k=0}^{N} \sigma_k. \tag{7.94}$$

Note that this relation holds for any configuration of the random walk. Thus, M fluctuates from one realization of the walk to another with its statistics governed by those of the σ_k.

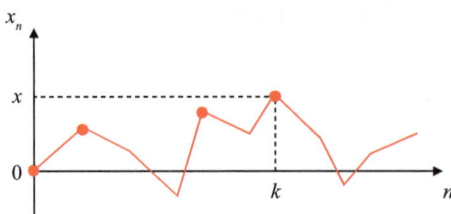

Fig. 7.10 Realization of a discrete-time random walk on a line starting at $x_0 = 0$ and jumping at each step by a random increment drawn from the PDF $f(\eta)$. The red filled circles denote the records. The value of the record at the kth step lies in $[x, x + \mathrm{d}x]$.

Let us start with the mean number of records. Taking the average on both sides of Eq. (7.94) gives

$$\langle M \rangle = \sum_{k=0}^{N} \langle \sigma_k \rangle. \tag{7.95}$$

Since σ_k is a binary variable as in Eq. (7.93), its expectation value $r_k = \langle \sigma_k \rangle$ just denotes the probability that a record occurs at step k. To compute $\langle \sigma_k \rangle$, we proceed as follows. Consider a realization of the walk starting at the origin that has a record event at step k where the record value lies in the interval $[x, x + \mathrm{d}x]$ (see Fig. 7.10). Since this is a record, this means that at all previous time steps before k, the walk must have stayed below the level x, reaching x at the kth step for the first time. To compute the probability of this event, let us undertake a little "cut and stitch" exercise with the walk configuration in Fig. 7.10.

This exercise consists of the following three steps: (i) shift the origin to the level x, (ii) change the direction of time, i.e., go backwards in time, starting from k, and (iii) reverse the vertical axis (i.e., look "down" at the walk from the original level x, i.e., the new origin)—see Fig. 7.11. Using the fact that the noise distribution is symmetric, we immediately see from the figure that this event (of a record at step k with record value lying in $[x, x + \mathrm{d}x]$) is exactly equivalent, in Fig. 7.11, to the event that the walk starting at the origin reaches x at step k without crossing the origin. But the probability of this latter event is precisely $G_+(x, 0, k)\,\mathrm{d}x$, where we recall, from the definition of the restricted Green's function in Eq. (3.12), that $G_+(x, x_0, k)$ is the probability density to arrive at x, starting from x_0, at step k while staying non-negative. Finally, integrating over the record value x, we obtain

$$r_k = \langle \sigma_k \rangle = \int_0^{\infty} G_+(x, 0, k)\,\mathrm{d}x. \tag{7.96}$$

But this integral is exactly the survival probability $Q(0, k)$ of the walk up to step k, starting at the origin [see Eq. (3.13)]. As discussed in Section 3.2.5, for a continuous and symmetric jump distribution $f(\eta)$, this survival probability is universal, i.e., independent of $f(\eta)$, and is given by the Sparre Andersen result in Eq. (3.101).

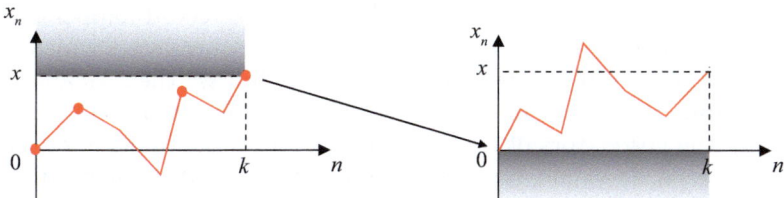

Fig. 7.11 The figure from the left panel can be transformed to the figure on the right by the following moves: (i) shift the origin in the left panel to the level x, (ii) change the direction of time starting from k backwards and (iii) reverse the vertical axis. The record at step k in the left panel gets mapped onto the origin 0 in the right panel (as shown by the arrow) as a result of these three transformations. Similarly, the shaded region above the level x in the left panel gets transformed into the shaded region below 0 in the right panel. Note that the walk in the left panel stays below the shaded region up to step k, since a record occurs at step k with record value x. Consequently, in the right panel, the walk stays above the origin up to step k.

Using this result, we then get

$$r_k = Q(0,k) = \binom{2k}{k} 2^{-2k}. \tag{7.97}$$

Therefore, the average number of records up to step N is given by

$$\langle M \rangle = \sum_{k=0}^{N} \binom{2k}{k} 2^{-2k} = (2N+1) \binom{2N}{N} 2^{-2N}. \tag{7.98}$$

This result is universal and exact for all N. In particular, for large N, it grows as

$$\langle M \rangle \approx \frac{2}{\sqrt{\pi}} \sqrt{N}, \quad N \to \infty. \tag{7.99}$$

Thus, we see that the strong correlations present in the random walk sequence enhance considerably the mean number of records compared to the IID case, where it grows slowly as $\langle M \rangle \approx \ln N$ for large N [see Eq. (7.18)].

The next natural question is: Can one compute the variance, or even the full distribution, of M? In principle, one can follow the same route as in the IID case, using the indicator variables σ_k's. For instance, the second moment of M can be obtained by squaring the relation in Eq. (7.94) and taking the average. This gives

$$\langle M^2 \rangle = \sum_{k_1=1}^{N} \sum_{k_2=1}^{N} \langle \sigma_{k_1} \sigma_{k_2} \rangle, \tag{7.100}$$

where $\langle \sigma_{k_1} \sigma_{k_2} \rangle$ is just the joint probability of having a record at step k_1 and another record at step k_2, for $k_1 \neq k_2$. For $k_1 = k_2$, $\langle \sigma_{k_1}^2 \rangle = \langle \sigma_{k_1} \rangle$ since $\sigma_k = 0, 1$. Hence, in this diagonal case, this is just the probability that a record happens at step k_1. However, unlike the IID case where the σ_k's were uncorrelated which made the computation of $\langle M^2 \rangle$ simple, here in the random walk case the σ_k's are correlated. While, in principle, the second or higher moments can be computed by this method (by computing the correlators of the σ_k's), the computation quickly becomes quite cumbersome. Hence, it would be nice to have an alternative method. Indeed, it turns out that the alternative method used in the IID case (see Section 7.1.2) for computing the joint distribution $P(\vec{\ell}, M \mid N)$ of the records' ages $\vec{\ell} = \{\ell_1, \ell_2, \dots, \ell_M\}$ and the number of records M can be adapted more easily to the random walk case, exploiting the renewal property of the random walk, as explained later. This gives access not only to the full distribution $P(M, N)$ of M, but also to the statistics of the ages of records. In the next subsection, we show in detail how this method works.

7.2.2 Joint statistics of the number and ages of records

As in the IID case, instead of trying to compute $P(M, N)$ directly, it is convenient to first consider a bigger collection of random variables in a given sequence, namely the number of records M as well as their ages denoted by the vector $\vec{\ell} = \{\ell_1, \ell_2, \dots, \ell_M\}$. The joint distribution of these random variables will be denoted by $P(\vec{\ell}, M \mid N)$ as in the IID case. The main point is that this apparently more complicated joint distribution actually has a rather simple structure, due to the renewal property (as explained below). Consequently, by integrating out the age variables $\vec{\ell}$ from the joint distribution, one can exactly obtain the marginal distribution $P(M, N)$ of the record number only.

To compute this joint distribution $P(\vec{\ell}, M \mid N)$, we will need two crucial quantities as building blocks (Majumdar and Ziff, 2008). The first quantity we need is the probability that a random walk starting at the initial position x_0 stays below x_0 up to step ℓ, i.e.,

$$S(x_0, \ell) = \text{Prob}(x_1 < x_0, \ x_2 < x_0, \ x_3 < x_0, \ \dots, \ x_\ell < x_0 \mid x_0). \tag{7.101}$$

Shifting $x_i \to x_i - x_0$ and using the translation invariance of the process, this probability is identical to

$$S(x_0, \ell) = \text{Prob}(x_1 < 0, \ x_2 < 0, \ x_3 < 0, \ \dots, \ x_\ell < 0 \mid 0). \tag{7.102}$$

First, we note that $S(x_0, \ell)$ evidently does not depend on x_0. Moreover, using the $x \to -x$ symmetry of the walk, we find

$$S(x_0, \ell) = \text{Prob}(x_1 > 0, \ x_2 > 0, \ x_3 > 0, \ \dots, \ x_\ell > 0 \mid 0) = Q(0, \ell) = q(\ell), \tag{7.103}$$

where $Q(0, \ell)$ is the survival probability of the random walk up to step ℓ, starting at the origin, as defined in Eq. (3.101). We also recall that $q(\ell)$ is given by the universal Sparre Andersen formula

$$q(\ell) = \binom{2\ell}{\ell} 2^{-2\ell}, \tag{7.104}$$

valid for all $\ell \geq 0$, independently of the jump distribution $f(\eta)$. For later purposes, let us also define its generating function,

$$\tilde{q}(z) = \sum_{\ell \geq 0} q(\ell) z^\ell = \frac{1}{\sqrt{1-z}}. \tag{7.105}$$

From Eq. (7.104), it follows that, for large ℓ, $q(\ell)$ decays in a universal algebraic fashion:

$$q(\ell) \sim \frac{1}{\sqrt{\pi \ell}}. \tag{7.106}$$

The second ingredient is the related first-passage probability $F(\ell)$ (starting at $x_0 = 0$) and defined (see also Chapter 3) as

$$F(\ell) = \text{Prob}(x_1 < 0, \, x_2 < 0, \, \ldots, \, x_{\ell-1} < 0, \, x_\ell > 0 \mid x_0 = 0). \tag{7.107}$$

As shown in Chapter 3, $F(\ell)$ is related to $q(\ell)$ simply via

$$F(\ell) = q(\ell - 1) - q(\ell) = \frac{\Gamma(\ell - 1/2)}{2\sqrt{\pi}\Gamma(\ell + 1)}, \tag{7.108}$$

where we have used the expression for $q(\ell)$ given in Eq. (7.104). Consequently, the generating function of $F(\ell)$ is related simply to that of $q(\ell)$ as

$$\tilde{F}(z) = \sum_{\ell \geq 1} F(\ell) z^\ell = 1 - (1 - z)\tilde{q}(z) = 1 - \sqrt{1 - z}. \tag{7.109}$$

Clearly, the result in Eq. (7.108) is also universal, i.e., independent of the jump distribution $f(\eta)$. Finally, for large ℓ, $F(\ell)$ has a universal power-law tail,

$$F(\ell) \sim \frac{1}{2\sqrt{\pi}\ell^{3/2}}. \tag{7.110}$$

Armed with these two probabilities $q(\ell)$ and $F(\ell)$, we can write down the joint probability distribution $P(\vec{\ell}, M \mid N)$ as (see Fig. 7.12)

$$P(\vec{\ell}, M \mid N) = F(\ell_1) F(\ell_2) \cdots F(\ell_{M-1}) q(\ell_M) \delta\left(\sum_{k=1}^{M} \ell_k, N\right). \tag{7.111}$$

Let us now explain this formula. Starting at the origin, the walk stays below the origin up to step $\ell_1 - 1$ and crosses its initial value 0 for the first time, from the negative side, thus forming the next record, at step ℓ_1. The probability of this event is clearly $F(\ell_1)$, as defined in Eq. (7.107). Now we consider the interval between the second and third records. Here, the walk starts from x_{ℓ_1}, stays below x_{ℓ_1} up to a time $\ell_2 - 1$ and crosses x_{ℓ_1} from below after exactly ℓ_2 steps. As argued before, due to the translational invariance of the walk with respect to its starting point, this probability is independent of x_{ℓ_1} and is precisely given by $F(\ell_2)$ in Eq. (7.107). This argument can be repeated for the subsequent intervals, all the way up to the Mth

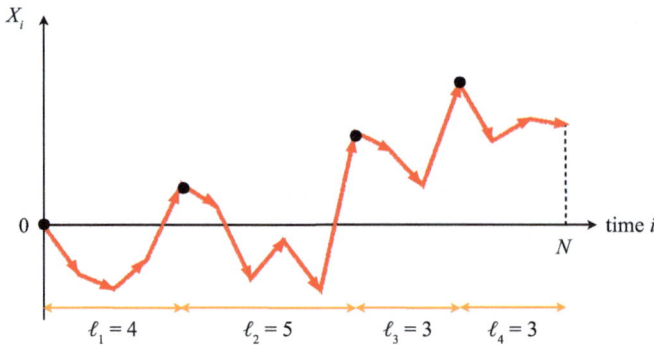

Fig. 7.12 A typical realization of the random walk sequence $\{X_0 = 0, X_1, X_2, \ldots, X_N\}$ of $N = 15$ steps with $M = 4$ records. Each record is represented by a filled circle. The set $\{\ell_1, \ell_2, \ell_3\}$ represents the time intervals between the successive records and $\ell_M = \ell_4$ is the age of the last record, which is still a record at time N.

record, which is the last one in this sequence. For example, $F(\ell_{M-1})$ corresponds to the interval between the $(M-1)$th and Mth records. However, after the last record occurs at step $\ell_1 + \ell_2 + \cdots + \ell_{M-1}$, the walk stays below the final record value (i.e., the Mth record) until step N, because we are imposing that there are exactly M records. Thus, the last interval of length ℓ_M after the Mth record is on a different footing than the preceding $M-1$ intervals. The probability of this last interval is precisely $q(\ell_M)$ as defined in Eq. (7.103). Finally, we note that the random walk defined in Eq. (7.92) is a Markov process. As a consequence, once we specify the position of the walk at any fixed time step, say k, the history of the walk following step k becomes completely decoupled from the history of the walk before step k. This means that the walk configuration after step k is statistically independent of the configuration before step k. This is referred to as the *renewal property* of a random walk (Feller, 2008b). As a result, once we specify the record events in the walk configuration, the intervals $\{\ell_1, \ell_2, \ldots, \ell_M\}$ between the record events become statistically independent, except for the overall global constraint that $\ell_1 + \ell_2 + \cdots + \ell_M = N$. In Eq. (7.111), the independence between the intervals explains the factorization, while the overall length constraint is ensured by the Kronecker delta. As we will see later, $P(\vec{\ell}, M \mid N)$ is normalized to unity when summed over $\vec{\ell}$ and M, although this is quite a non-trivial check.

The distribution of the number of records. The record number distribution is just the marginal of the joint distribution when one sums over the interval lengths,

$$P(M, N) = \sum_{\vec{\ell}} P(\vec{\ell}, M \mid N), \qquad (7.112)$$

where $P(\vec{\ell}, M \mid N)$ is given in Eq. (7.111). Due to the global constraint, this sum is most easily carried out by considering the generating function with respect to N.

Multiplying Eq. (7.111) by z^N and summing over $\vec{\ell}$ and N decouples the ℓ_i variables and we arrive at the following simple form:

$$\sum_{N \geq 0} P(M,N)z^N = \left[\sum_{\ell_1=1}^{\infty} F(\ell_1)z^{\ell_1}\right]\left[\sum_{\ell_2=1}^{\infty} F(\ell_2)z^{\ell_2}\right]\cdots\left[\sum_{\ell_M=0}^{\infty} q(\ell_M)z^{\ell_M}\right] \tag{7.113}$$

$$= [\tilde{F}(z)]^{M-1}\tilde{q}(z) \tag{7.114}$$

$$= [1 - \sqrt{1-z}]^{M-1}\frac{1}{\sqrt{1-z}}, \tag{7.113}$$

where, in going from the second to the third line, we used the relations in Eqs. (7.105) and (7.109). Inverting this generating function would give us access to the moments of M, and even the full distribution $P(M,N)$.

For instance, multiplying Eq. (7.114) by M and summing over all M, one obtains the exact generating function of the average number of records $\langle M \rangle$ in N steps,

$$\sum_{N \geq 0} \langle M \rangle z^N = \frac{1}{(1-z)^{3/2}}. \tag{7.115}$$

Expanding in powers of z gives, for arbitrary $N \geq 0$,

$$\langle M \rangle = \sum_{k=0}^{N} \binom{2k}{k}2^{-2k} = (2N+1)\binom{2N}{N}2^{-2N}, \tag{7.116}$$

thus recovering the result obtained previously in Eq. (7.98) by a different method. For large N, as given in Eq. (7.99), the mean number of records grows as

$$\langle M \rangle \approx \frac{2}{\sqrt{\pi}}\sqrt{N}. \tag{7.117}$$

Let us now go to the second moment $\langle M^2 \rangle$. Multiplying Eq. (7.113) by M^2 and summing over M, we obtain

$$\sum_{N \geq 0} \langle M^2 \rangle z^N = \frac{2}{(1-z)^2} - \frac{1}{(1-z)^{3/2}}. \tag{7.118}$$

Expanding in powers of z and using Eq. (7.116), we find the formula for general $N \geq 0$:

$$\langle M^2 \rangle = 2(N+1) - (2N+1)\binom{2N}{N}2^{-2N}. \tag{7.119}$$

We note from the discussion of Eq. (7.100) and below it that computing $\langle M^2 \rangle$ explicitly for all N was much harder using the σ_k representation. But here, we see that the renewal method allows us to extract the second moment (and even the higher ones) much more easily. The second term in Eq. (7.119) is just the mean

$\langle M \rangle$, which grows as $\sqrt{N}$ for large N. Hence, the first term dominates in the limit of large N and one has

$$\langle M^2 \rangle \approx 2N. \qquad (7.120)$$

From the second moment, we can easily compute the variance for all N. Using the asymptotic behaviors in Eqs. (7.117) and (7.120),

$$\mathrm{Var}(M) = \langle M^2 \rangle - \langle M \rangle^2 \approx 2\left(1 - \frac{2}{\pi}\right)N. \qquad (7.121)$$

From these results, we see that the variance scales as N while the mean grows as $\sqrt{N}$ for large N. This is quite different from the IID case where we recall that both the mean and the variance grow as $\ln N$ for large N. This already hints that the full distribution of M in this random walk rcase is going to be quite different from that of the IID case discussed in the previous section.

Let us then go one step further beyond the second moment and extract the full distribution $P(M, N)$ from the generation function in Eq. (7.114). This can be obtained formally by inverting Eq. (7.114) using Cauchy's formula,

$$P(M, N) = \int_{C_0} \frac{\mathrm{d}z}{2\pi i} \frac{1}{z^{N+1}} \frac{(1 - \sqrt{1 - z})^{M-1}}{\sqrt{1 - z}}, \qquad (7.122)$$

where C_0 is a contour around the origin in the complex z-plane. Now, make the change of variable $1 - z = w^2$, which reduces the complex integral (over the appropriate contour C_1, the details of which are not important) to

$$P(M, N) = 2\int_{C_1} \frac{\mathrm{d}w}{2\pi i} \frac{(1 - w)^{-(N-M+2)}}{(1 + w)^{N+1}}. \qquad (7.123)$$

Now make a further change of variable $1 + w = u$ to get

$$P(M, N) = 2^{M-N-1} \int_{C_2} \frac{\mathrm{d}u}{2\pi i u^{N+1}} \left[1 - \frac{u}{2}\right]^{-(N-M+2)}. \qquad (7.124)$$

But this contour integral, using Cauchy's theorem, is simply the coefficient of u^N in the expansion of $(1 - u/2)^{-(N-M+2)}$. This is easy to find, as one can write

$$\left(1 - \frac{u}{2}\right)^{-(N-M+2)} = \sum_{k=0}^{\infty} \frac{\Gamma(N - M + 2 + k)}{\Gamma(k+1)\Gamma(N - M + 2)} \left(\frac{u}{2}\right)^k. \qquad (7.125)$$

Reading off the coefficient of u^N and substituting in Eq. (7.124) gives the explicit result (without actually the need to perform the contour integration!)

$$P(M, N) = \binom{2N - M + 1}{N} 2^{-2N+M-1}, \qquad (7.126)$$

valid for all $1 \leq M \leq N + 1$ and $N \geq 0$. For a plot of this PDF, see the left panel of Fig. 7.13. It is important to note that this result for $P(M, N)$ does not

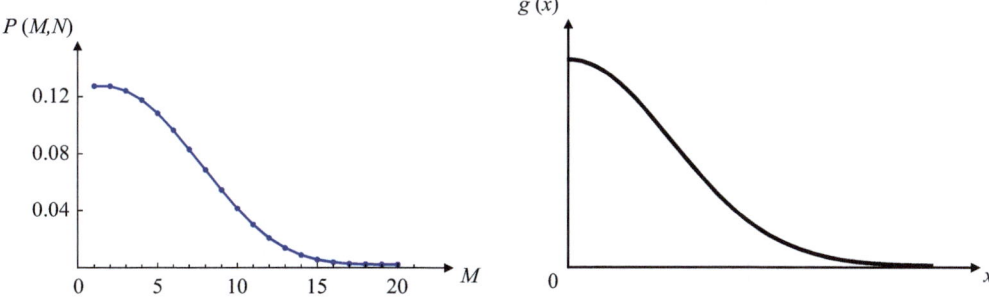

Fig. 7.13 The left panel shows a plot of $P(M,N)$ in Eq. (7.126) vs. M for $N = 20$. For large N, the distribution approaches a scaling form $P(M,N) \sim 1/\sqrt{N}g(M/\sqrt{N})$, where the scaling function $g(x) = e^{-x^2/4}/\sqrt{\pi}$ is plotted in the right panel.

depend on the jump distribution $f(\eta)$ as long as $f(\eta)$ is symmetric and continuous. Moreover, this universality holds for all N (and not just for large N). It is clear from the derivation above that the origin of this universality can be traced back to the universality of the survival probability $q(\ell)$ in Eq. (7.104), due to the Sparre Andersen theorem.

Let us now analyze how the full distribution $P(M,N)$ looks in the limit of large N. It turns out that if we want to describe the form of the distribution $P(M,N)$ for $M = \mathcal{O}(\sqrt{N})$, i.e., of the order of the typical fluctuations of M, we set $M = \sqrt{N}x$ and take the limit $M \to \infty$ and $N \to \infty$ but keeping x fixed. This limit can be taken most conveniently from the generating function in Eq. (7.114). In this equation, we set $z = e^{-s}$. The scaling limit described above is then equivalent to taking $s \to 0$, $M \to \infty$ but keeping the product $M\sqrt{s}$ fixed. The LHS of this equation becomes a Laplace transform in the limit $s \to 0$ (where we can replace the sum over N by an integral over N) and this gives

$$\int_0^\infty P(M,N)e^{-sN}\,\mathrm{d}N \approx \frac{e^{-\sqrt{s}M}}{\sqrt{s}}. \tag{7.127}$$

Inverting this Laplace transform (try it with Mathematica!) we get the following scaling form:

$$P(M,N) \approx \frac{1}{\sqrt{N}}g\left(\frac{M}{\sqrt{N}}\right), \quad \text{with } g(x) = \frac{1}{\sqrt{\pi}}e^{-x^2/4}\Theta(x), \tag{7.128}$$

where $\Theta(x)$ is the Heaviside step function (i.e., $\Theta(x) = 1$ if $x \geq 0$ and $\Theta(x) = 0$ if $x < 0$). The scaling function $g(x)$ is plotted in the right panel of Fig. 7.13. This scaling behavior of $P(M,N)$ for the random walk case is markedly different from the IID case in Eq. (7.44), where $P(M,N)$ approaches a Gaussian distribution $P(M,N) \sim \exp[-(M-\ln N)^2/2\ln N]$, with mean $\langle M \rangle = \ln N$ and standard deviation $\sigma = \sqrt{\ln N}$, as shown in Fig. 7.6.

The result in Eq. (7.128) is of course valid for $M = \mathcal{O}(\sqrt{N})$, i.e., it describes the form of the probability distribution for typical fluctuations. However, atypically

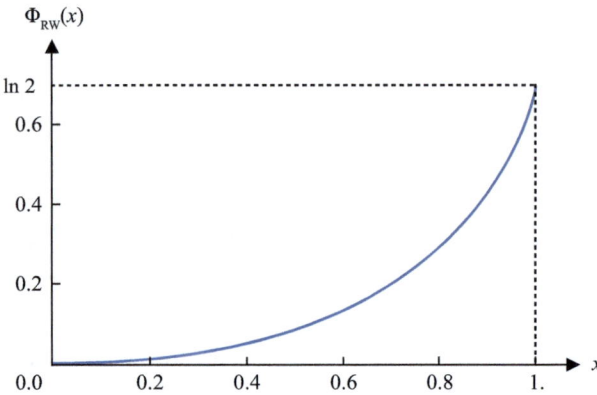

Fig. 7.14 Plot of the function $\Phi_{\mathrm{RW}}(x)$ in Eq. (7.130) vs. x. The dotted horizontal line indicates the leading asymptotic behavior $\Phi_{\mathrm{RM}}(x \to 1) = \ln 2$ as given in the second line of Eq. (7.132).

large fluctuations, say when $M = \mathcal{O}(N)$, are not described by this scaling form—this encodes only the extreme right tail of $P(M, N)$. To extract this large-deviation tail, one could again use the generating function representation in Eq. (7.122), set $M = Nx$ and perform a saddle-point analysis of this complex integral for large N. An easier route is to start from the exact formula in Eq. (7.126), set $M = xN$ and then take the large-N limit using Stirling's formula. This gives

$$P(M = xN, N) \approx \frac{1}{\sqrt{N}} A(x) e^{-N\Phi_{\mathrm{RW}}(x)} \quad \text{with } 0 \le x < 1, \tag{7.129}$$

where the rate function $\Phi_{\mathrm{RW}}(x)$ and the amplitude $A(x)$ are given by

$$\Phi_{\mathrm{RW}}(x) = 2x \operatorname{arctanh}\left(\frac{x}{4 - 3x}\right) + \ln(4(1 - x)) - 2\ln(2 - x), \tag{7.130}$$

$$A(x) = \frac{1}{\sqrt{8\pi}} \left(\frac{2 - x}{1 - x}\right)^{3/2}. \tag{7.131}$$

We leave the derivation of this result as an exercise (hint: set $M = xN$ in Eq. (7.126) and use Stirling's formula to expand it for large N, with $0 < x < 1$ fixed). A plot of the rate function $\Phi_{\mathrm{RW}}(x)$ is shown in Fig. 7.14, where we see that $\Phi_{\mathrm{RW}}(x)$ vanishes as $x \to 0$ and approaches a constant as $x \to 1$. The more precise asymptotic behaviors are given (check this!) by

$$\Phi_{\mathrm{RW}}(x) \approx \begin{cases} x^2/4, & x \to 0, \\ \ln 2 + (1 - x)\ln(1 - x), & x \to 1. \end{cases} \tag{7.132}$$

For $M \ll N$, i.e., when $x \to 0$, we obtain the behavior as stated in the first line of Eq. (7.132). Substituting this into the large-deviation form in Eq. (7.129), we get

$P(M, N) \approx e^{-M^2/(4N)}$ for $M \ll N$, which exactly matches the right tail behavior of the typical regime, i.e., when $M \gg \sqrt{N}$ in Eq. (7.128). This matching of two scaling regimes is quite common in many examples in statistical physics. A given distribution may have very different forms depending on which scale one looks at (for example, here $M = \mathcal{O}(\sqrt{N})$ or $M = \mathcal{O}(N)$), but then these different scaling forms should smoothly match each other (provided there is no phase transition) at the junction of the two scales.

What about the other limit when $x \to 1$, i.e., $M \to N$? Indeed, by putting $M = N$ in the exact formula in Eq. (7.126), we get $P(M = N, N) = (N+1)2^{-1-N} \approx e^{-N \ln 2}$ to leading order for large N. By substituting the leading term $\ln 2$ in the second line of Eq. (7.132) into the large-deviation form in Eq. (7.129), we get $P(M, N) \approx e^{-N \ln 2}$ to leading order, which matches with the leading asymptotics of the exact formula for $M = N$ and $N \to \infty$.

The statistics of the ages of the records. Apart from the number of records, another interesting observable is the ages of the records of a random walk sequence. As defined at the beginning of this section, the age ℓ_k of the kth record is the number of steps between the kth and $(k + 1)$th records, i.e., the time up to which the kth record survives (see Fig. 7.12). Note that the last record is still a record at step N and hence the last age ℓ_M is not on the same footing as the other ones. Here, as in the case of IID random variables, starting from the joint PDF $P(\vec{\ell}, M \mid N)$ in Eq. (7.111), we also compute three interesting observables: (i) the marginal distribution of the age of the kth record, (ii) the distribution of the age of the longest-lasting record and (iii) the distribution of the age of the shortest-lasting record.

Marginal probability distribution of the age of the kth record. Our starting point is the renewal equation for the joint PDF $P(\vec{\ell}, M \mid N)$ in Eq. (7.111). A first rough and naive inspection of the joint distribution of the ages in Eq. (7.111) suggests that these ages are essentially independent (assuming for the moment that the global constraint can be ignored) and also identical (except for the last interval ℓ_M which is different). Therefore, if one is interested in the distribution $P(\ell_k, N)$ of the typical age of a record, i.e., of ℓ_k with $k < M$, one naturally expects that

$$P(\ell_k) = \lim_{N \to \infty} P(\ell_k, N) = F(\ell_k), \tag{7.133}$$

where $F(\ell_k)$ is the first-passage probability given in Eq. (7.108). Since $F(\ell_k)$ is universal, i.e., independent of the jump distribution, the marginal age distribution $P(\ell_k)$ is also universal for all k.

Note that this result holds for all k (with $k < M$), which is quite different from the limiting distribution of the age of the kth record for an IID sequence as in Eq. (7.60) that depends explicitly on k. Furthermore, using $F(\ell) \propto 1/\ell^{3/2}$ for large ℓ from Eq. (7.110), one obtains that the typical age ℓ_{typ} behaves as $\ell_{\text{typ}} = \langle \ell_k \rangle = \sum_{\ell=1}^{N} \ell f(\ell) \propto \sqrt{N}$. This behavior can also be obtained by the following simple heuristic argument: given that the average number of records is $\langle M \rangle$, the typical age which is the typical time interval between two successive records is $\ell_{\text{typ}} \sim N/\langle M \rangle \propto \sqrt{N}$, where we have used that $\langle M \rangle \propto \sqrt{N}$.

Distribution of the age of the longest-lasting record. As in the case of the IID sequence, we consider the age of the longest-lasting record,

$$\ell_{\text{max},N} = \max\{\ell_1, \ell_2, \ldots, \ell_M\}, \tag{7.134}$$

and denote its cumulative distribution as $\mathcal{Q}_{\text{max}}(w, N) = \text{Prob}(\ell_{\text{max},N} \leq w)$. As in the IID case, this can be written as

$$\mathcal{Q}_{\text{max}}(w, N) = \sum_{M \geq 1} \sum_{\ell_1=1}^{w} \cdots \sum_{\ell_M=1}^{w} P(\vec{\ell}, M \mid N), \tag{7.135}$$

where $P(\vec{\ell}, M \mid N)$ is the joint distribution of the ages and the number of records given in Eq. (7.111). As usual, we take the generating function with respect to N to get rid of the delta function in Eq. (7.111). Expanding the resulting geometric series appearing in the sum in Eq. (7.135), one can check that

$$\sum_{N \geq 0} \mathcal{Q}_{\text{max}}(w, N) z^N = \frac{\sum_{m=1}^{w} q(m) z^m}{1 - \sum_{m=1}^{w} F(m) z^m}, \tag{7.136}$$

where $q(m)$ and $F(m)$ are defined respectively in Eqs. (7.104) and (7.108). To compute the average $\langle \ell_{\text{max},N} \rangle$ from the CDF $\mathcal{Q}_{\text{max}}(w, N)$ we use the relation

$$\langle \ell_{\text{max},N} \rangle = \sum_{w \geq 1} [1 - \mathcal{Q}_{\text{max}}(w, N)]. \tag{7.137}$$

Multiplying by z^N on both sides of Eq. (7.137) gives

$$\sum_{N \geq 0} z^N \langle \ell_{\text{max},N} \rangle = \sum_{\ell \geq 0} \left[\frac{1}{1 - z} - \frac{\sum_{m=1}^{\ell} q(m) z^m}{1 - \sum_{m=1}^{\ell} F(m) z^m} \right]. \tag{7.138}$$

Inserting the explicit expressions of $q(m)$ and $F(m)$ given respectively in Eqs. (7.104) and (7.108) into Eq. (7.138), one can in principle obtain the exact value of $\langle \ell_{\text{max},N} \rangle$ for arbitrary N. For instance, we get $\langle \ell_{\text{max},N} \rangle = 0, 1, 3/2, 17/8, 11/4$ respectively for $N = 0, 1, 2, 3, 4$ (Godrèche *et al.*, 2014).

The large-N behavior of $\langle \ell_{\text{max},N} \rangle$ is tricky to extract from Eq. (7.138). We provide these details here so that the reader may find it useful in other contexts. We start from Eq. (7.138) and express the RHS completely in terms of $q(m)$ only, using the relation between $q(m)$ and $F(m)$, namely $F(m) = q(m-1) - q(m)$. After some straightforward algebra, this gives

$$\sum_{N \geq 0} z^N \langle \ell_{\text{max},N} \rangle = \frac{1}{(1-z)} \sum_{\ell \geq 0} \frac{1 - z + q(\ell) z^{\ell+1}}{q(\ell) z^{\ell+1} + (1-z) \sum_{m=0}^{\ell} q(m) z^m}. \tag{7.139}$$

To extract the large-N behavior of $\langle \ell_{\text{max},N} \rangle$, we need to analyze the behavior of its generating function near $z = 1$. In fact, it is useful to write $z = e^{-s}$ and examine the behavior near $s = 0$. In this limit, the sum over ℓ on the RHS in Eq. (7.139) is dominated by the large-ℓ behavior of the summand. In the large-ℓ limit, we can replace $q(\ell)$ by its asymptotic behavior $q(\ell) \approx 1/\sqrt{\pi \ell}$. Then, in the limit $s \to 0$, $\ell \to \infty$ but with the product $s\ell = y$ fixed, the numerator of the summand takes the scaling form

$$\mathcal{N}(s, \ell) = 1 - e^{-s} + e^{-s(\ell+1)}q(\ell) \approx \sqrt{s} f_{\mathcal{N}}(s\ell), \quad \text{where } f_{\mathcal{N}}(y) = \frac{e^{-y}}{\sqrt{\pi y}}. \tag{7.140}$$

In the denominator, we can replace the sum over m by an integral in the scaling limit and this gives

$$\mathcal{D}(s, \ell) = e^{-s(\ell+1)}q(\ell) + (1 - e^{-s}) \sum_{m=0}^{\ell} q(m)e^{-sm}$$

$$\approx \sqrt{s} f_{\mathcal{D}}(s\ell), \quad \text{where } f_{\mathcal{D}}(y) = \frac{e^{-y}}{\sqrt{\pi y}}[1 + \sqrt{\pi y}\, e^{y} \operatorname{erf}(\sqrt{y})], \tag{7.141}$$

where we have used the identity

$$\frac{1}{\sqrt{\pi}} \int_0^y \frac{e^{-z}}{\sqrt{z}} \, dz = \operatorname{erf}(\sqrt{y}). \tag{7.142}$$

Therefore, taking the ratio of these two expressions in Eqs. (7.140) and (7.141) and replacing the sum over ℓ by an integral, we get

$$\sum_{N \geq 0} e^{-sN} \langle \ell_{\max,N} \rangle \approx \frac{1}{s^2} \int_0^{\infty} \frac{1}{1 + \sqrt{\pi y}\, e^{y} \operatorname{erf}(\sqrt{y})} \, dy. \tag{7.143}$$

In the small-s limit, we can replace the sum over N in the LHS by an integral and it then becomes a Laplace transform,

$$\int_0^{\infty} \langle \ell_{\max,N} \rangle e^{-sN} \, dN \approx \frac{C}{s^2}, \tag{7.144}$$

where the constant C is given by

$$C = \int_0^{\infty} \frac{1}{1 + \sqrt{\pi y}\, e^{y} \operatorname{erf}(\sqrt{y})} \, dy = 0.626\,508\ldots \tag{7.145}$$

The Laplace transform in Eq. (7.144) can be easily inverted using $\mathcal{L}_{s \to N}^{-1}(1/s^2) = N$. Hence, for large N, we obtain the leading asymptotic behavior (Majumdar and Ziff, 2008)

$$\langle \ell_{\max,N} \rangle \approx CN. \tag{7.146}$$

Hence, the longest age is much larger than the typical record age, which is of order $\mathcal{O}(\sqrt{N})$. Note that the constant C also appears in the study of the longest excursion of Brownian motion (Pitman and Yor, 1997; Godrèche *et al.*, 2009).

As in the IID case, this constant C has a physical interpretation. Indeed, let us define $Q_{\text{last}}(N)$ as the probability that the last interval is the longest record. Note that the evolution equation for $\ell_{\max,N}$ in Eq. (7.76) as well as the relation in Eq. (7.78) are generally valid, including in the random walk case. Therefore, substituting the asymptotic result from Eq. (7.146), it is clear that (Godrèche *et al.*, 2009; Godrèche *et al.*, 2014; Godrèche *et al.*, 2015)

$$\lim_{N \to \infty} Q_{\text{last}}(N) \to C = 0.626\,508\ldots \tag{7.147}$$

Interestingly, comparing the IID case in Eq. (7.79) and the random walk result in Eq. (7.147), we see that the asymptotic values of $Q_{\text{last}}(N)$ are almost the same, but not quite. They actually differ in the third decimal place. Thus, in both cases, the last interval is the longest record in about 62% of cases.

We now turn to the full distribution of $\ell_{\max,N}$, which has a rich structure. The PDF of $\ell_{\max,N}$ can be expressed in terms of the CDF $Q_{\max}(w, N)$ via the relation

$$\text{Prob}(\ell_{\max,N} = w) = Q_{\max}(w, N) - Q_{\max}(w+1, N), \tag{7.148}$$

where the generating function of $Q_{\max}(w, N)$ is given in Eq. (7.136). Note that on the RHS of Eq. (7.136), the two quantities $q(m)$ and $F(m)$ are given exactly in Eqs. (7.104) and (7.108) respectively. By substituting these exact expressions on the RHS of Eq. (7.136) and expanding in powers of z, one can obtain $Q_{\max}(w, N)$ for any finite N using Mathematica. Consequently, the PDF $\text{Prob}(\ell_{\max,N} = w)$ in Eq. (7.148) can also be obtained for any finite N. This procedure was used to exactly enumerate the PDF in Fig. 7.15.

It turns out that, for large N, the PDF $\text{Prob}(\ell_{\max,N} = w)$ has a scaling form (Lamperti, 1961; Godrèche *et al.*, 2015)

$$\text{Prob}(\ell_{\max,N} = w) \to \frac{1}{N} f_{\max}^{\text{RW}}\left(\frac{w}{N}\right), \tag{7.149}$$

where the function $f_{\max}^{\text{RW}}(x)$ (where the superscript RW refers to the random walk case) is piecewise continuous on the interval $[0, 1]$. It is continuous on each interval of the form $[1/2, 1]$, $[1/3, 1/2]$, and so on, and exhibits singularities at the points $x_k = 1/k$ with $k = 2, 3, \ldots$ (Lamperti, 1961). This is thus qualitatively very similar to the IID case. Moreover, its asymptotic behaviors can also be obtained explicitly (Lamperti, 1961; Pitman and Yor, 1997; Godrèche *et al.*, 2015):

$$f_{\max}^{\text{RW}}(x) \approx \begin{cases} 2\alpha_0 x^{-2} \exp\left(-\dfrac{\alpha_0}{x}\right), & x \to 0, \\[2mm] \dfrac{1}{\pi}(1-x)^{-1/2}, & x \to 1, \end{cases} \tag{7.150}$$

where $\alpha_0 = 0.854\,032\ldots$ is the only zero of the hypergeometric function $_1F_1(1, 1/2, -x)$ on the real axis. In Fig. 7.15, we show a plot of the scaling function $f_{\max}^{\text{RW}}(x)$.

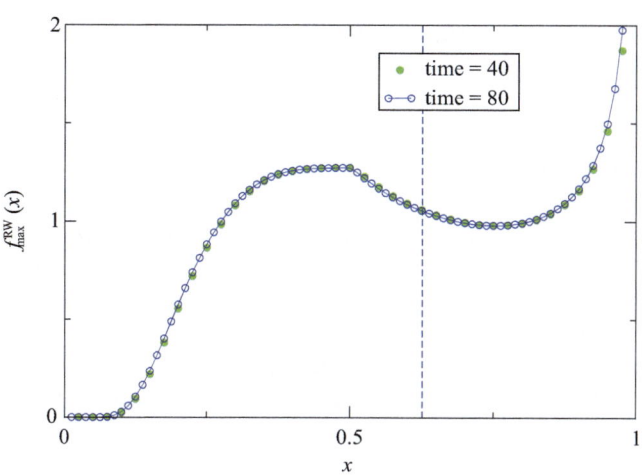

Fig. 7.15 Limiting distribution of the scaled random variable $\ell_{\max,N}/N$, see Eq. (7.149). It was obtained using an exact enumeration method discussed after Eq. (7.148) for $N = 40$ steps (green full circles) and $N = 80$ steps (blue empty circles). The continuous line is a guide for the eyes and connects the blue circles. The good collapse of the data for two different values of N confirms the scaling form in Eq. (7.149). This figure, with slightly different notation, appeared first in Godrèche *et al.* (2014).

Distribution of the age of the shortest-lasting record. We now focus on the age of the shortest record, denoted by $\ell_{\min,N}$, which is defined as

$$\ell_{\min,N} = \min\{\ell_1, \ell_2, \ldots, \ell_M\}. \tag{7.151}$$

We define $\mathcal{Q}_{\min}(\ell, N) = \text{Prob}(\ell_{\min,N} \geq \ell)$, $\ell \geq 1$, and $\mathcal{Q}_{\min}(\ell, N = 0) = 0$. To compute the generating function of $\mathcal{Q}_{\min}(\ell, N)$ with respect to N, we start from the joint distribution in Eq. (7.111). Taking the generating function of Eq. (7.111) and summing over each ℓ_i from ℓ to ∞, we obtain

$$\sum_{N \geq 0} \mathcal{Q}_{\min}(\ell, N) z^N = \frac{\sum_{m=\ell}^{\infty} q(m) z^m}{1 - \sum_{m=\ell}^{\infty} F(m) z^m}. \tag{7.152}$$

This is the analogue of Eq. (7.136) for $\ell_{\max,N}$. By analyzing this equation near $z = 1$, the large-N behavior of the statistics of $\ell_{\min,N}$ can be obtained explicitly. We do not provide the details of the derivation here, as it is very similar to that of $\ell_{\max,N}$. However, the results for large N are quite different from those of $\ell_{\max,N}$. For example, the average $\langle \ell_{\min,N} \rangle$ grows for large N as (Majumdar and Ziff, 2008)

$$\langle \ell_{\min,N} \rangle \approx \frac{1}{\sqrt{\pi}} \sqrt{N}, \tag{7.153}$$

which is thus of the same order as the *typical* record age ℓ_{typ}, i.e., $\mathcal{O}(\sqrt{N})$, but much smaller than the $\mathcal{O}(N)$ growth of $\langle \ell_{\max,N} \rangle$ in Eq. (7.146).

The distribution of $\ell_{\min,N}$ turns out to be quite simple for large N and given at leading order by

$$\text{Prob}(\ell_{\min,N} = \ell) = \delta_{\ell,1} + \mathcal{O}(N^{-1/2}). \tag{7.154}$$

This shows that the average value of $\langle \ell_{\min,N} \rangle$ in Eq. (7.153) is controlled by rare events. In fact, the main contribution to $\langle \ell_{\min,N} \rangle$ comes from the paths with a single record, $M = 1$, occurring at $X_0 = 0$ (Godrèche *et al.*, 2014). Indeed, the result in Eq. (7.153) can simply be recovered by noting that a path with $M = 1$ is such that it stays negative up to step N. Such paths occur with a probability $q(N) \approx 1/\sqrt{\pi N}$ and they contribute to a value of $\ell_{\min,N} = N$, implying precisely the result in Eq. (7.153). This shows explicitly that $\langle \ell_{\min,N} \rangle$ is dominated by rare events, such that the random walk never crosses the origin up to step N.

8
Extremes in Other Correlated Systems

In the previous chapters, we have discussed the statistics of many extreme observables in detail for two distinct time series: (i) the case of an IID random sequence and (ii) the case of a random walk sequence. The first case gives an example of the EVS of uncorrelated variables, while the second provides an example of a solvable EVS in the presence of strong correlations. There can be other situations, such as weakly correlated random variables, that have also been discussed briefly. In addition, there are many other systems where correlations are strong but not of the random walk types. In this chapter, we discuss miscellaneous other examples of correlated systems, going beyond the simple random walk example, for which the EVS can be computed exactly. Unlike the two models (IID and random walk) that have been discussed in great detail, in this chapter we do not provide much detail but just show the reader that the subject of EVS of correlated variables has far-ranging applications in various systems. In fact, the material presented in this chapter is essentially reproduced from a recent review on EVS co-authored by us (Majumdar *et al.*, 2020), in particular in Sections 8.3–8.7. However, the discussion below also contains some new examples.

8.1 Ornstein–Uhlenbeck processes: Weakly correlated time series

Here, we recall the definition of a weakly correlated time series described in Section 4.2. We consider a time series $\{x_1, x_2, \ldots, x_N\}$ and assume that the two-point correlation function $\langle x_i x_j \rangle$ (in the limit of large N) decays as $\langle x_i x_j \rangle \sim e^{-|i-j|/\xi}$ for large $|i-j|$, where ξ denotes the correlation length. An example of such a process is the so-called AR(1) process that appears in many applications (Barndorff-Nielsen and Shephard, 2001; Tankov, 2003; Majumdar and Kearney, 2007; Wergen, 2014). It interpolates between an IID sequence and a random walk sequence (which is strongly correlated). Here, the x_n evolve via the Markov rule, starting from $x_0 = 0$,

$$x_n = (1 - \alpha)x_{n-1} + \eta_n, \quad 0 \leq \alpha \leq 1, \tag{8.1}$$

Statistics of Extremes and Records in Random Sequences. Satya N. Majumdar and Grégory Schehr, Oxford University Press. © Satya N. Majumdar and Grégory Schehr (2024). DOI: 10.1093/9780191838781.003.0008

where the η_n's are IID jump variables, each drawn from a symmetric and continuous PDF $f(\eta)$, with a finite jump variance $\sigma^2 = \int_{-\infty}^{\infty} \eta^2 f(\eta)\, d\eta$. Note that the parameter α allows us to interpolate between the IID sequence (for $\alpha = 1$) and the random walk sequence (for $\alpha = 0$). To see that the process in Eq. (8.1) indeed corresponds to a weakly correlated time series, we need to compute the correlation function $\langle x_i x_j \rangle$. For this, it is first convenient to express x_i as the following sum:

$$x_i = \sum_{m=1}^{i} (1-\alpha)^{i-m} \eta_m. \tag{8.2}$$

This can be easily proved by iterating the relation in Eq. (8.1). We then take the product of x_i and x_j using the representation in Eq. (8.2). Next, we average over the noise using the property that $\langle \eta_m \eta_{m'} \rangle = \sigma^2 \delta_{m,m'}$. This gives

$$\langle x_i x_j \rangle = \sigma^2 \sum_{m=1}^{\min(i,j)} (1-\alpha)^{i+j-2m} = \frac{\sigma^2}{\alpha(2-\alpha)}\left((1-\alpha)^{|i-j|} - (1-\alpha)^{i+j}\right). \tag{8.3}$$

One can easily check that, in the limit $\alpha \to 0$, one recovers the random walk correlation, namely, $\langle x_i x_j \rangle = \sigma^2 \min(i,j)$ [see Eq. (2.34)]. In the opposite limit $\alpha \to 1$, it is easy to see that $\langle x_i x_j \rangle \to \sigma^2 \delta_{i,j}$, recovering the IID limit. For general $0 < \alpha < 1$, the correlation function (in the limit when both i and j are large, keeping the difference $|i-j|$ fixed) behaves as

$$\langle x_i x_j \rangle \approx \frac{\sigma^2}{\alpha(2-\alpha)}(1-\alpha)^{|i-j|} \sim \frac{\sigma^2}{\alpha(2-\alpha)} e^{-|i-j|/\xi}, \tag{8.4}$$

where the correlation length ξ is given by

$$\xi = -\frac{1}{\ln(1-\alpha)}. \tag{8.5}$$

Thus, for $0 < \alpha < 1$, the sequence in Eq. (8.1) is weakly correlated, with a finite correlation length ξ. In the IID limit $\alpha \to 1$, the correlation length $\xi \to 0$, as expected. In contrast, when $\alpha \to 0$ (the random walk limit), the correlation length ξ diverges, indicating a strongly correlated time series.

For the weakly correlated time series in Eq. (8.1) with $0 < \alpha < 1$, we would like to compute the distribution of the global maximum up to N steps, $X_{\max} = \max\{x_0 = 0, x_1, \ldots, x_N\}$. In Chapter 4, we argued that for such a weakly correlated sequence where $\xi \ll N$, the limiting distribution of the global maximum, appropriately centered and scaled, should be given by one of the three EVS distributions for IID variables (Gumbel, Fréchet or Weibull). It would be great to find an example of a weakly correlated sequence where one can demonstrate this explicitly by computing the distribution of the maximum exactly. Unfortunately, for the AR(1) process described in Eq. (8.1), the analytical computation of the distribution of $X_{\max}$ for a general $0 < \alpha < 1$ is hard and only numerical results are available (Zarfaty *et al.*, 2022). Instead, below we show that there is a limit when

$\alpha \to 0$, $\sigma \to 0$ with σ^2/α fixed where the AR(1) process converges to the Ornstein–Uhlenbeck (OU) process in continuous time. In this limit, one can compute the distribution of the maximum of the OU process over a fixed time $[0, t]$. This exact computation in fact confirms the results obtained from the heuristic renormalization group argument in Chapter 4.

To see the mapping to the OU process, we first rewrite the sequence in Eq. (8.1) as

$$x_n - x_{n-1} = -\alpha x_{n-1} + \eta_n. \tag{8.6}$$

Next, we set $\alpha = \mu \Delta t$ and simultaneously $\sigma = \sqrt{2D\Delta t}$, and then take the limit $\Delta t \to 0$. This gives

$$\frac{dx}{dt} = -\mu x + \eta(t), \tag{8.7}$$

where we set $x_n \to x(t)$ and define $\eta(t) = \eta_n/\Delta t$. Using the correlation function $\langle \eta_m \eta_n \rangle = \sigma^2 \delta_{m,n}$ and $\sigma^2 = 2D\Delta t$, we get the correlation function of the noise $\eta(t)$ in continuous time (we identify t with m and t' with n):

$$\langle \eta(t)\eta(t') \rangle = \frac{\sigma^2}{(\Delta t)^2} \delta_{m,n} = \frac{2D}{\Delta t} \delta_{m,n} \to 2D\delta(t - t'), \tag{8.8}$$

where the Kronecker delta function $\delta_{m,n}$, divided by Δt, converges to the Dirac delta function $\delta(t-t')$. Moreover, one can show that, in this continuous-time limit $\Delta t \to 0$, all higher cumulants of $\eta(t)$ vanish, rendering $\eta(t)$ a Gaussian white noise with zero mean and correlator $\langle \eta(t)\eta(t') \rangle = 2D\delta(t - t')$. Note that the continuous-time Langevin equation in Eq. (8.7) is the famous OU process, corresponding to the noisy, overdamped equation of a particle in a harmonic potential $V(x) = (\mu/2)x^2$. Since Eq. (8.7) for $x(t)$ is linear, and $\eta(t)$ is a Gaussian white noise, it follows that $x(t)$ is a Gaussian process with zero mean and its two-time correlation function $C(t_1, t_2)$ (which completely characterizes a Gaussian process) can be easily computed by taking the continuous-time limit in Eq. (8.3); this gives (we leave this as a little exercise)

$$C(t_1, t_2) = \langle x(t_1)x(t_2) \rangle = \frac{D}{\mu}[e^{-\mu|t_1 - t_2|} - e^{-\mu(t_1 + t_2)}]. \tag{8.9}$$

In particular, setting $t_1 = t_2 = t$, this gives the time-dependent variance

$$\sigma^2(t) = \frac{D}{\mu}(1 - e^{-2\mu t}). \tag{8.10}$$

Thus, in the long-time limit, the variance becomes independent of time, i.e., $\sigma^2(t \to \infty) \to D/\mu$. Consequently, at late times, the position distribution $P(x, t)$ of this Gaussian process approaches the stationary form

$$\lim_{t \to \infty} P(x, t) = P_{\text{stat}}(x) = \sqrt{\frac{\mu}{2\pi D}} e^{-\mu x^2 / 2D}. \tag{8.11}$$

Coming back to the two-time correlation function $C(t_1, t_2)$ in Eq. (8.9), we see that it reduces to the Brownian limit when $\mu \to 0$, since in this limit $C(t_1, t_2) \to 2D\min(t_1, t_2)$ and the system becomes strongly correlated. In contrast, for non-zero $\mu > 0$, the correlation function, at large times $t_1, t_2 \gg 1/\mu$, decays exponentially with the time difference $C(t_1, t_2) \approx (1/2\mu)e^{-\mu|t_1 - t_2|}$ with a correlation length $\xi = 1/\mu$. From our arguments about weakly correlated random variables (see Section 4.2), we would then expect to get the limiting Gumbel distribution for $\mu > 0$ (since $x(t)$ is a Gaussian random variable of variance $\sqrt{D/\mu}$ for large t). We demonstrate below how this Gumbel distribution emerges by solving exactly the EVS for the OU process.

Consider the OU process $x(t)$ evolving via Eq. (8.7) on the interval $[0, T]$, starting from $x(0) = 0$. Let $X_{\max}$ denote the maximum of the process over $[0, T]$, i.e.,

$$X_{\max} = \max_{0 \leq t \leq T} x(t). \tag{8.12}$$

Let $Q_{\max}(z, T)$ denote the cumulative distribution of $X_{\max}$ over the interval $[0, T]$, i.e.,

$$Q_{\max}(z, T) = \text{Prob}(X_{\max} \leq z, T). \tag{8.13}$$

This is equivalent to considering the process starting at 0 and evolving via Eq. (8.7) but with an absorbing barrier at z, that selects only those trajectories that stay below the level z up to time T. To compute $Q_{\max}(z, T)$, we proceed as follows. We consider the OU process evolving in continuous time t via Eq. (8.7) and define the restricted propagator of the OU process in continuous time t as follows: let $G_+(x, t \mid z)$ denote the probability density for the particle, starting at $x(0) = 0$, to arrive at $x \leq z$ at time t, while staying below the level z. This restricted propagator satisfies the forward Fokker–Planck equation in the domain $x \in (-\infty, z]$,

$$\frac{\partial G_+}{\partial t} = D \frac{\partial^2 G_+}{\partial x^2} + \mu \frac{\partial}{\partial x}(x G_+), \tag{8.14}$$

with the initial condition $G_+(x, 0 \mid z) = \delta(x)$ and the boundary conditions $G_+(x, t \mid z) = 0$ as $x \to -\infty$ together with the absorbing condition at level z, i.e., $G_+(x = z, t \mid z) = 0$ for all t. Once we know the solution for the restricted propagator, the cumulative distribution (i.e., the survival probability in the presence of an absorbing barrier at z) is obtained by integrating the final position of the particle at time T over $x \in (-\infty, z]$, i.e.,

$$Q_{\max}(z, T) = \int_{-\infty}^{z} dx \, G_+(x, T \mid z). \tag{8.15}$$

For simplicity, we set $D = 1/2$. We note that, unlike in the Brownian case ($\mu = 0$), for $\mu > 0$ we can no longer use the method of images (as discussed in Chapter 3) due to the presence of the harmonic force $-\mu x$ in Eq. (8.7). To solve the linear Fokker–Planck equation in Eq. (8.14), one can use an eigenfunction expansion (this is quite standard in solving the time-dependent Schrödinger equation in quantum mechanics),

$$G_+(x, t \mid z) = \sum_\lambda a_\lambda \psi_\lambda(x) e^{-\lambda t}, \tag{8.16}$$

where the coefficients a_λ can be fixed from the initial condition $G_+(x, 0 \mid z) = \delta(x)$ and using the orthonormality properties of the eigenfunctions over the space $x \in (-\infty, z]$. Substituting this form into Eq. (8.14) and setting $D = 1/2$, one finds that the function $\psi_\lambda(x)$ satisfies an eigenvalue equation,

$$\frac{1}{2}\psi_\lambda''(x) + \mu x \psi_\lambda'(x) + (\mu + \lambda)\psi_\lambda(x) = 0. \tag{8.17}$$

This differential equation is valid in the domain $x \in (-\infty, z]$ with the boundary conditions $\psi_\lambda(x = z) = 0$ and $\psi_\lambda(x) \sim e^{-\mu x^2}$ as $x \to -\infty$. The former condition comes from the fact that there is an absorbing boundary at $x = z$. The latter condition can be understood as follows. When $x \to -\infty$, the particle does not feel the absorbing boundary at $x = z$ and therefore we expect $\psi_\lambda(x) \sim e^{-\mu x^2}$. This is like the unconstrained OU process with $D = 1/2$.

Equation (8.17) can now be reduced to a known form by substituting $\psi_\lambda(x) = \phi_\lambda(\sqrt{2\mu}x)e^{-\mu x^2/2}$. Then, $\phi_\lambda(y)$ satisfies the differential equation

$$\phi_\lambda''(y) + \left(\frac{\lambda}{\mu} + \frac{1}{2} - \frac{y^2}{4}\right)\phi_\lambda(y) = 0, \tag{8.18}$$

with the boundary conditions $\phi_\lambda(y = \sqrt{2\mu}z) = 0$ and $\phi_\lambda(y) \sim e^{-y^2/4}$ as $y \to -\infty$. There are two linearly independent solutions to this equation, namely $D_{\lambda/\mu}(y)$ and $D_{\lambda/\mu}(-y)$, which are known as parabolic cylinder functions (Gradshteyn and Ryzhik, 2014). As $y \to -\infty$, the solution $D_{\lambda/\mu}(y) \sim e^{y^2/4}|y|^{-\lambda/\mu-1}$, while $D_{\lambda/\mu}(-y) \sim e^{-y^2/4}$. Hence, only the solution $D_{\lambda/\mu}(-y)$ satisfies the correct boundary condition at $y \to -\infty$. Therefore, the most general solution of the Fokker–Planck equation in Eq. (8.14) satisfying the boundary condition as $y \to -\infty$ can be expressed as

$$G_+(x, t \mid z) = \sum_\lambda a_\lambda e^{-\lambda t} D_{\lambda/\mu}(-\sqrt{2\mu}x)e^{-\mu x^2/2}. \tag{8.19}$$

We still need to satisfy the absorbing boundary condition $G_+(x = z, t \mid z) = 0$. This boundary condition imposes

$$D_{\lambda/\mu}(-\sqrt{2\mu}z) = 0, \tag{8.20}$$

which then fixes ("quantizes") the eigenvalues λ, which are necessarily positive.

For a fixed z, there are many positive roots for λ in Eq. (8.20), each of them providing a relaxation mode in Eq. (8.16). Of these multiple roots, the smallest positive root for λ, denoted by $\lambda_0(z)$, will provide the principal relaxation mode in

Eq. (8.16) at late times. For arbitrary z, it is difficult to solve $D_{\lambda/\mu}(-\sqrt{2\mu}z) = 0$ and determine the smallest eigenvalue $\lambda_0(z)$. However, for large z, one can make progress by perturbation theory as follows. We expect that, as $z \to \infty$, the lowest eigenvalue $\lambda_0(z) \to 0$, since this will then correspond to the stationary state. For finite but large z, we then expect $\lambda_0(z)$ to be small. To determine its z-dependence, we need to analyze the condition in Eq. (8.20) for large z and small $p = \lambda_0(z)/\mu$. The small-p behavior of $D_p(-y)$ can be derived from the differential equation that it satisfies, namely

$$D_p''(y) + \left(p + \frac{1}{2} - \frac{y^2}{4}\right) D_p(y) = 0, \tag{8.21}$$

with the boundary condition $D_p(-y) \sim e^{-y^2/4}$ as $y \to -\infty$. Note that for $p=0$, we have $D_0(y) = e^{-y^2/4}$. For small p we thus write a perturbation expansion of $D_p(-y)$ as

$$D_p(-y) = e^{-y^2/4}(1 + pG(y) + O(p^2)). \tag{8.22}$$

Substituting this into Eq. (8.21), and using $D_0''(-y) + (1/2 - y^2/4)D_0(-y) = 0$, we get, by matching the powers of p,

$$G''(y) - yG'(y) = -1, \tag{8.23}$$

with the boundary conditions $G(y) \to \ln|y|$ as $y \to -\infty$. This comes from the fact that when $y \to +\infty$, $D_p(-y) \approx |y|^p e^{-y^2/4}$. From Eq. (8.22) one has $1 + pG(y) = e^{y^2/4} D_p(-y) \approx |y|^p \approx 1 + p\ln|y|$. The solution of Eq. (8.23) satisfying this boundary condition is thus

$$G(y) = -\sqrt{\frac{\pi}{2}} \int_0^y e^{y'^2/2} \operatorname{erfc}\left(-\frac{y'}{\sqrt{2}}\right) dy' - \frac{1}{2}(\gamma_E + \ln 2). \tag{8.24}$$

As $y \to \infty$, one has, to leading order,

$$G(y) \approx \frac{\sqrt{2\pi}}{y} e^{y^2/2}. \tag{8.25}$$

Substituting this behavior into Eq. (8.22) we get, for large y (and to leading order in p for small p),

$$D_p(-y) \approx e^{-y^2/4}\left(1 - p\frac{\sqrt{2\pi}}{y} e^{y^2/2}\right). \tag{8.26}$$

Setting $D_p(-y) = 0$ as in Eq. (8.20) then gives

$$p = \frac{\lambda_0(z)}{\mu} \approx \frac{1}{\sqrt{2\pi}} y e^{-y^2/2}. \tag{8.27}$$

Using $y = \sqrt{2\mu}z$, we then get

$$\lambda_0(z) \approx \frac{1}{\sqrt{\pi}} \mu^{3/2} z e^{-\mu z^2}, \quad z \to \infty. \tag{8.28}$$

Using this result in Eqs. (8.19) and (8.15), we obtain, to leading order for large T and large z,

$$Q_{max}(z,T) \sim e^{-\lambda_0(z)T} \sim \exp\left\{-\exp\left[-\mu z^2 + \ln\left(\frac{T\mu^{3/2}z}{\sqrt{\pi}}\right)\right]\right\}$$

$$\rightarrow nF_I\left[\sqrt{4\mu\ln T}\left(z - \frac{1}{\sqrt{\mu}}\sqrt{\ln T}\right)\right], \qquad (8.29)$$

where $F_I(z) = \exp[-\exp[-z]]$ is the Gumbel distribution. As a result, for $\mu > 0$, the average value of the maximum grows very slowly for large t, as $\langle X_{max}\rangle \sim (1/\sqrt{\mu})\sqrt{\ln T}$, while its width around the mean decreases as $\sim 1/\sqrt{\ln T}$. In fact, for $\mu > 0$, a full analysis of the mean value of the maximum $\langle X_{max}\rangle$ for all t shows that initially it grows as $\sqrt{T}$ (for $T \ll 1/\mu$) where it does not feel the confining potential and hence behaves as a Brownian motion. But for $T \gg 1/\mu$, the particle feels the potential and the mean maximum crosses over to a slower growth as $\sqrt{\ln T}$. Hence, one has

$$\langle X_{max}\rangle \sim \begin{cases} \sqrt{T} & \text{for } T \ll 1/\mu, \\ \sqrt{\ln T} & \text{for } T \gg 1/\mu, \end{cases} \qquad (8.30)$$

which indicates a crossover from a strongly correlated regime at short time $T \ll 1/\mu$ to an IID regime at longer time $T \gg 1/\mu$.

This computation for the OU process thus presents one solvable example of a weakly correlated stationary time series where the distribution of the maximum can be computed exactly and one sees how the limiting Gumbel distribution emerges at long times—as argued by the heuristic RG argument presented in Chapter 4.

8.2 Extreme statistics in the freely expanding Jepsen gas

So far in this book we have discussed in detail the limiting distributions of the maximum X_{max} of a set of N IID random variables in the limit of large N. Here, we will study an example of a dynamical system where these distributions emerge as the stationary state of the dynamics in the long-time limit. We consider N particles on a line and assign their initial positions and velocities $\{X_i, V_i\}$. The positions are chosen independently from a uniform distribution over the interval $[-L, 0]$, with a density $n_0 = N/L$. The initial velocities V_i are each chosen independently from a distribution $\phi(V)$. For simplicity, we restrict ourselves to the case of positive velocities, i.e., such that $\phi(V) = 0$ for $V < 0$. Starting from its initial position, each particle starts moving ballistically with its initial velocity V. When two particles collide, they just exchange their velocities. This is the celebrated *Jepsen gas* (Jepsen, 1965). Note that this is a freely expanding gas on the infinite line, with no restricting boundaries, except at $t = 0$ where we assume that the gas is restricted in $[-L, 0]$ with a uniform density. As the system evolves in time, the particles collide elastically and exchange their velocities. For a given initial condition, the system up to time t is fully described by the set of trajectories $\{X_i + V_i t, i = 0, 1, \ldots, N\}$. Each of the

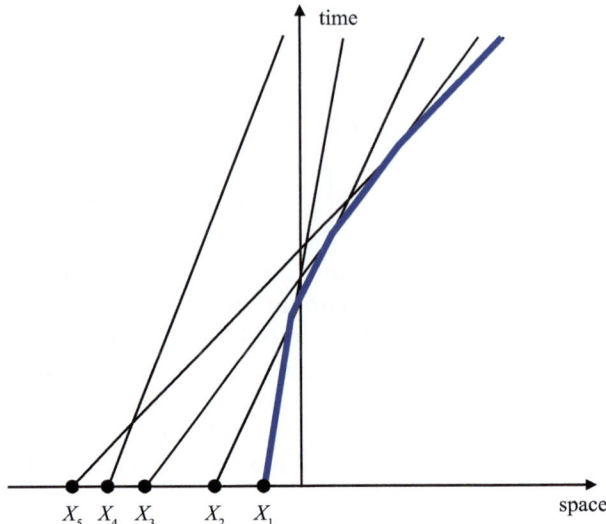

Fig. 8.1 Typical trajectories of $N = 5$ particles in the Jepsen gas where the particles are initially located at random positions on the negative x-axis and are assigned random positive velocities drawn independently from $\phi(V)$. When two particles cross, their velocities are exchanged. The solid blue curve shows the trajectory of the leader, which has changed lead 3 times in the current figure.

particles travels along such a trajectory until it collides with another particle, and each collision changes its trajectory (see Fig. 8.1).

Consider the leader, i.e., the rightmost particle in space at time t. Its initial velocity is distributed according to $\phi(V)$. The velocity of the leader increases whenever it collides with a particle with higher velocity coming from its left, see Fig. 8.1. For a fixed particle number N, it is obvious that if one waits for a long enough time, then the leader will acquire the largest velocity of the initial set $\{V_i,\ i = 0, 1, \ldots, N\}$. Depending on the tail of $\phi(V)$, we already know that the largest of the initial velocities will be distributed via either a Gumbel, Fréchet or Weibull distribution. Thus, as time evolves, the velocity distribution of the leader, starting from the initial velocity $\phi(V)$, crosses over to one of the three limiting distributions of the IID class. Thus, in this example, the IID limiting distributions emerge as the stationary velocity distribution of the leader (i.e., the current rightmost particle) in the Jepsen gas. Thus, as time evolves, the dynamics of the Jepsen gas orders the velocities of the initial independent set. Therefore, the maximum initial velocity emerges at long times as the velocity of the rightmost particle, i.e., the leader (Bena and Majumdar, 2007).

It turns out that there is a crossover time $t^*(N)$ which grows with N such that for $t \gg t^*(N)$ (the stationary regime) the leader velocity distribution is one of the three universality classes of IID extreme values depending on the tail of $\phi(V)$. In contrast, for $t \ll t^*(N)$ (growing regime), the distribution of the leader's velocity is time dependent. The crossover time $t^*(N)$ depends on the tail of $\phi(V)$ (Bena and Majumdar, 2007). As in Chapter 4, we consider three different types of distributions that lead to the three classes of extreme value distributions,

Case I   (Gumbel class) $\phi(V) \approx A_{\mathrm{I}} \mathrm{e}^{-V^{\delta}}$, $V \to \infty,\ \delta > 0$,   (8.31)

Case II   (Frechet class) $\phi(V) \approx A_{\mathrm{II}} V^{-1-\alpha}$, $V \to \infty,\ \alpha > 0$,   (8.32)

Case III   (Weibull class) $\phi(V) \approx A_{\mathrm{III}} (V_{\mathrm{c}} - V)^{\gamma-1}$, $V \to V_{\mathrm{c}},\ \gamma > 0$,   (8.33)

where in case III the support of $\phi(V)$ has an upper bound at $V = V_{\mathrm{c}}$. In these three cases, the crossover time $t^*(N)$, beyond which the leader velocity distribution becomes stationary, scales for large N as (Bena and Majumdar, 2007)

$$t^*(N) \sim \begin{cases} N & \text{Case I,} \\ N^{(\alpha-1)/\alpha} & \text{Case II,} \\ N^{(\gamma+1)/\gamma} & \text{Case III.} \end{cases} \qquad (8.34)$$

Note that in case II, for $0 < \alpha < 1$, $t^*(N) \to 0$ as $N \to \infty$, which indicates that the particle with the highest initial velocity emerges as the leader almost instantaneously.

Now that we know the crossover time $t^*(N)$ and the stationary leader velocity distributions for $t \gg t^*(N)$, it is natural to ask whether the time-dependent leader velocity distribution $P_{\mathrm{L}}(V, t)$ for $t \ll t^*(N)$ also has universal scaling forms. Indeed, it turns out that, for $1 \ll t \ll t^*(N)$, the distribution $P_{\mathrm{L}}(V, t)$ exhibits universal scaling behavior (N-independent as $N \to \infty$) of the type

$$P_{\mathrm{L}}(V, t) \approx \frac{1}{b_i(t)} g_i\left(\frac{V - a_i(t)}{b_i(t)}\right), \quad i = \mathrm{I, II, III}, \qquad (8.35)$$

where $a_i(t)$ and $b_i(t)$ are time-dependent non-universal scale factors (Bena and Majumdar, 2007). However, the scaling functions $g_i(z)$ are universal and read

$$g_{\mathrm{I}}(z) = -\mathrm{e}^{-z} \mathrm{Ei}(-\mathrm{e}^{-z}), \qquad (8.36)$$

$$g_{\mathrm{II}}(z) = \alpha(\alpha - 1) z^{-1-\alpha} \int_0^z \mathrm{e}^{-u^{1-\alpha}} \, du, \quad \alpha > 1, \qquad (8.37)$$

$$g_{\mathrm{III}}(z) = \gamma(\gamma + 1) |z|^{\gamma-1} \int_{|z|}^{\infty} \mathrm{e}^{-u^{\gamma+1}} \, du, \quad \gamma > 0, \qquad (8.38)$$

where $\mathrm{Ei}(-z) = -\int_z^{\infty} \mathrm{e}^{-t}/t \, dt$, and the exponents α and γ are defined in Eqs. (8.32) and (8.33) respectively. For a plot of these functions, see (Bena and Majumdar, 2007, Fig. 2). Thus, in the growing regime the leader velocity distributions, centered and scaled, are universal and are one of only three types, depending on the tail of $\phi(V)$. However, from Eqs. (8.36)–(8.38), it is clear that these limiting distributions in the growing regime are quite different from the three canonical extreme value distributions (Gumbel, Fréchet and Weibull) in the stationary regime (as detailed in Table 4.1). This is because, in the growing regime, the velocities and positions of the particles are correlated and the velocity of the particle that has the rightmost position is *not* the maximum of N IID random variables.

There are other interesting observables in this model that one can compute exactly. For instance, the average number of collisions that the leader undergoes before its velocity distribution becomes stationary can be computed exactly (Bena and Majumdar, 2007). Furthermore, we note that eventually, when all the initial velocities get completely ordered, the system reaches a "fan state" and there are no more collisions between the particles once this state is reached. One can then ask: How many collisions N_c take place in the full system until it reaches the fan state (Sabhapandit *et al.*, 2008)? It turns out that the statistics of N_c is completely universal, i.e., independent of the initial positions and the initial velocity distribution $\phi(V)$. For example, the mean and the variance, for any N, are given by (Sabhapandit *et al.*, 2008)

$$\langle N_c \rangle = \frac{N(N-1)}{4}, \qquad \langle N_c^2 \rangle - \langle N_c \rangle^2 = \frac{N(N-1)(2N+5)}{72}, \qquad (8.39)$$

and the full distribution of N_c converges, for large N, to a Gaussian distribution with the above mean and the variance given in Eq. (8.39). Finally, the ordering time, i.e., the time T_c needed to reach the fan state, also has a universal distribution, in the limit of large N, given by (Sabhapandit *et al.*, 2008)

$$P(T_c, N) \approx \frac{1}{bN^2} f_c\left(\frac{T_c}{bN^2}\right), \qquad (8.40)$$

where b is a non-universal scale factor, but the scaling function $f_c(z)$ is completely universal (i.e., independent of $\phi(V)$) and is given by

$$f_c(z) = \frac{1}{z^2} e^{-1/z}, \quad z > 0. \qquad (8.41)$$

8.3 Extreme statistics in the presence of a global constraint

Correlations between different degrees of freedom in a many-particle system without interaction can arise due to some global conservation laws satisfied by the dynamics of the system. Naturally, one would wonder how these correlations coming from these conservation laws would affect the extreme value statistics in such systems. In many such systems one encounters a condensation phase transition as one changes a control parameter. This transition is from a fluid phase to a condensed phase that is dominated by one single degree of freedom, which is the condensate. The EVS in such systems is important in understanding the behavior of the condensate. The global nature of these constraints and the absence of direct interaction often make these systems solvable for the extreme value statistics, even though the variables are correlated. In this section, we briefly discuss some specific examples.

The first example concerns a well-known interacting particle system known as the zero-range process (ZRP) (Evans and Hanney, 2005; Majumdar, 2010*a*) where one considers, for simplicity, a one-dimensional lattice of N sites with periodic boundary conditions. At each site, there is a certain number of particles $\ell_i \geq 0$. From the ith site, a single particle hops to a neighboring site with a rate $U(\ell_i)$

that depends only on the occupation number of the departure site, provided $\ell_i > 0$. Clearly, the dynamics conserves the total number of particles, $\sum_{i=1}^{L} \ell_i = L$. At long times, the system reaches a stationary state where the joint distribution of the occupation numbers at different sites is given by

$$P_{\text{ZRP}}(\ell_1, \ldots, \ell_N \mid L) = \frac{1}{Z_N(L)} \prod_{i=1}^{N} f(\ell_i) \delta_{\sum_{i=1}^{N} \ell_i, L}, \tag{8.42}$$

where $f(\ell) = 1/\prod_{k=1}^{\ell} U(k)$ and $Z_N(L)$ is the normalization constant. The δ function in Eq. (8.42) enforces the global constraint that the number of particles L is fixed. Without this δ function, the joint distribution would factorize and we would be back to the IID case. However, the presence of this δ function makes these variables correlated, albeit in a global manner. We are interested in the extreme statistics, i.e., the statistics of $\ell_{\max} = \max_{1 \leq i \leq N} \ell_i$. This corresponds to the largest number of particles carried by a single site, out of N sites. Of particular physical interest is the case when the transfer rate function $U(k)$ is such that the steady-state weight function has a tail such that $e^{-c\ell} \ll f(\ell) \ll 1/\ell^2$ for large ℓ, where c is any arbitrary constant (Majumdar, 2010a). A well-studied example is the case when $f(\ell)$ has a power-law tail, $f(\ell) \sim A/\ell^\gamma$ for large ℓ, with $\gamma > 2$ and $A > 0$. In this case, it is known that a condensation transition happens when the density $\rho = L/N$ exceeds a certain critical value ρ_c. For $\rho < \rho_c$ the particles are "democratically" distributed among all sites: this is the "fluid" phase. For $\rho > \rho_c$ a single condensate site emerges that carries a finite fraction of the total number of particles: this is the "condensed" phase. Clearly, in this condensed phase, $\ell_{\max}$ is identified with the number of particles carried by this single condensate.

This ZRP is one example where, despite the presence of correlations via a global constraint, the statistics of $\ell_{\max}$ can be computed explicitly, in the limit of large N and large L, but with the density $\rho = L/N$ fixed (Evans and Majumdar, 2008). Let us summarize the main results. For any fixed $\gamma > 2$, one finds that the centered and scaled distribution of $\ell_{\max}$ in the fluid phase $\rho < \rho_c$ converges to the Gumbel distribution discussed in Chapter 4. In this case, the global constraint enforced by the δ function can be relaxed (similar to changing from a canonical to a grand-canonical ensemble) and one can replace it in Eq. (8.42) by $e^{-\mu \sum_{i=1}^{N} \ell_i}$, where $\mu > 0$ is the chemical potential (or Lagrange multiplier). This factorizes the joint distribution in Eq. (8.42) into a product of effective IID random variables with an effective distribution $f_{\text{IID}}(\ell) = f(\ell) e^{-\mu \ell}$. The chemical potential is then fixed from the global conservation condition

$$\frac{\sum_{\ell \geq 0} \ell f(\ell) e^{-\mu \ell}}{\sum_{\ell \geq 0} f(\ell) e^{-\mu \ell}} = \rho. \tag{8.43}$$

Thus, for $\mu > 0$, the distribution of these IID variables has an effective exponential tail and, hence, the distribution of their maximum converges to the Gumbel distribution, as discussed in Chapter 4. Note that this exponential tail is induced here by the global constraint. When ρ approaches ρ_c from below, $\mu \to 0$ and we again obtain a factorization of the joint distribution in Eq. (8.42) but now the effective distribution has a power-law tail $f(\ell) \sim A/\ell^\gamma$ for large ℓ, with $A > 0$. Consequently, the distribution of the maximum, appropriately centered and scaled, approaches the Fréchet distribution, as discussed in Chapter 4, thus it is as if the global constraint does not play any role at all at the critical point $\rho = \rho_c$. However, for $\rho > \rho_c$, it turns out that the global constraint has a much stronger effect and the scaled distribution of $\ell_{\max}$ has a non-trivial distribution given by (Evans and Majumdar, 2008)

$$\text{Prob}(\ell_{\max} = x, N) \sim \frac{1}{N^\delta} V_\gamma \left(\frac{x - (\rho - \rho_c)N}{N^\delta} \right),\tag{8.44}$$

where the exponent δ is given by

$$\delta = \begin{cases} \dfrac{1}{\gamma - 1}, & 2 < \gamma < 3, \\ \dfrac{1}{2}, & \gamma > 3. \end{cases}\tag{8.45}$$

The scaling function $V_\gamma(z)$ also depends on γ. While for $\gamma > 3$ it is a simple Gaussian, for $2 < \gamma < 3$ it has a non-trivial form given by the integral representation

$$V_\gamma(z) = \frac{1}{\pi} \int_0^\infty dy\, e^{-c_3 y^{\gamma-1}} \cos\left[b \cos(\pi\gamma/2)\, y^{\gamma-1} + yz\right],\tag{8.46}$$

where $b = A\Gamma(1-\gamma)$, with A being the amplitude of the power-law tail of $f(\ell) \sim A/\ell^\gamma$ for large ℓ and $c_3 = -b\sin(\pi\gamma/2)$. The asymptotic behaviors of the scaling function $V_\gamma(z)$ are given by (Evans *et al.*, 2006)

$$V_\gamma(z) \simeq \begin{cases} A|z|^{-\gamma}, & z \to -\infty, \\ c_0, & z \to 0, \\ c_1 z^{(3-\gamma)/(2(\gamma-2))} e^{-c_2 z^{(\gamma-1)/(\gamma-2)}}, & z \to \infty, \end{cases}\tag{8.47}$$

where c_0, c_1 and c_2 are positive constants that can be explicitly computed (Evans *et al.*, 2006). This therefore represents an example where correlations affect the extreme value distribution non-trivially and yet it remains computable.

Another example with a global constraint concerns the largest time interval between returns to the origin of a one-dimensional lattice random walk (Frachebourg *et al.*, 1995), and more generally a renewal process (Godrèche *et al.*, 2009, 2015). In this case, let ℓ_i denote the time interval between the ith and the $(i-1)$th returns to the origin of a random walk of L steps. After every return to the origin, the process gets renewed due to the Markov nature of the walk. Consequently, the joint PDF of the intervals between successive returns and the total number of intervals N in L steps is given by

$$P_{\text{REN}}(\ell_1, \ldots, \ell_N, N \mid L) \propto \left[\prod_{n=1}^{N-1} f(\ell_n) \right] q(\ell_N) \delta_{\sum_{n=1}^{N} \ell_n, L}, \tag{8.48}$$

where L represents the total time interval. Here, the number of intervals N is also a random variable, unlike in the ZRP case of Eq. (8.42) where N was fixed. In addition, at variance with Eq. (8.42), while the first $N-1$ intervals in Eq. (8.48) have the same weight $f(\ell_n)$, the last one has a different weight, $q(\ell_N) = \sum_{\ell=\ell_N+1}^{\infty} f(\ell)$ (this is because the last interval is yet to be renewed). In the renewal processes as in Eq. (8.48), the weight $f(\ell)$ is taken as an input in the model. For example, for a one-dimensional random walk, $f(\ell) \sim 1/\ell^{3/2}$ for large ℓ. The statistics of $\ell_{\max} = \max_{1 \leq i \leq N} \ell_i$ has been studied in detail in Godrèche *et al.* (2009, 2015). The main results can be summarized as follows. If $f(\ell)$ decays faster than $1/\ell^2$ for large ℓ, one essentially recovers the IID results for the distribution of $\ell_{\max}$, indicating that the global constraint does not play any significant role. However, for $f(\ell) \sim 1/\ell^{\gamma}$ with $1 < \gamma < 2$, the constraint significantly modifies the distribution of $\ell_{\max}$, leading to a non-trivial distribution (Godrèche *et al.*, 2015). In addition, in Godrèche *et al.* (2009, 2015) another extreme observable was studied, namely the probability that the last interval is the longest.

We finish this section with yet another example of extreme statistics in the presence of a global constraint. It appeared in a recently introduced truncated long-range Ising model in a one-dimensional lattice of size L, where the couplings between two spins at sites i and j decay as an inverse square law, provided both spins belong to the same domain. The spins across domains do not have any long-range interaction (Bar and Mukamel, 2014*a,b*). In addition to this long-range interaction, there is also a short-range interaction between neighboring spins. This model was shown to exhibit a mixed-order transition (MOT) at a finite critical temperature where the correlation length diverges, as in a second-order phase transition, but the order parameter undergoes a jump as in a first-order phase transition. The configurations of the system can be labelled by the lengths of the spin domains $\{\ell_i\}_{1 \leq i \leq N}$, where N represents the number of domains. At finite temperature the Boltzmann weight associated with such a configuration can be written as

$$P_{\text{MOT}}(\ell_1, \ell_2, \ldots, \ell_N, N \mid L) = \frac{1}{Z(L)} y^N \prod_{n=1}^{N} \frac{1}{\ell_n^c} \delta_{\sum_{n=1}^{N} \ell_n, L}, \tag{8.49}$$

where $\delta_{i,j}$ is the usual Kronecker delta and $c > 1$ is related to both the temperature and the long-range coupling constant. The constant y represents the fugacity, which controls the number of domains N, which actually fluctuates from one configuration to another. Once again, there is a global constraint enforcing the sum rule that the domain lengths add up to the system size L. Again, this model differs slightly from the two models discussed above, namely the ZRP in Eq. (8.42) and the renewal process in Eq. (8.48). In the (c, y)-plane, this model has a phase transition across the critical line $y_c = 1/\zeta(c)$ where $\zeta(x)$ is the Riemann zeta function (see Fig. 8.2). For $y > y_c$ the system is in a paramagnetic phase with a large number of domains. In contrast, for $y < y_c$, the system is ferromagnetic with one large domain of size

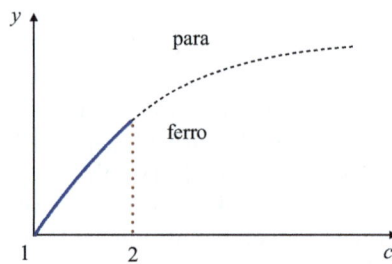

Fig. 8.2 Phase diagram in the (c, y)-plane, where $c > 1$. The two phases, paramagnetic and ferromagnetic, are separated by the critical line $y_c = 1/\zeta(c)$. The largest domain size scales with the system size as $\sim L$ for $1 < c < 2$ and its distribution has a non-trivial scaling form as in Eq. (8.50). For $c > 2$ and $y = y_c$, the appropriately scaled maximal domain size exhibits a Fréchet law.

$\propto L$. Very interesting properties emerge on the critical line $y = y_c$. In this case, for $1 < c < 2$, the largest domain size $\ell_{\max}$ scales extensively with the system size, $\ell_{\max} \propto L$ and the PDF of $\ell_{\max}$ exhibits the scaling form (Bar *et al.*, 2016)

$$\text{Prob}(\ell_{\max} = x, L) \approx \frac{1}{L} F_c\left(\frac{x}{L}\right), \tag{8.50}$$

where the scaling function $F_c(z)$ was computed exactly and was found to exhibit non-analytic behaviors at $z = 1/k$ where k is a positive integer. The scaling form in Eq. (8.50) also indicates that the fluctuations are anomalously large in this regime. This fact was shown to be related to the so-called fluctuation-dominated phase ordering (FDPO) regime where the spin–spin correlation function develops a cusp at short distance (Barma *et al.*, 2019). Thus, we see that, in the regime $1 < c < 2$ on the critical line $y = y_c$, the global constraint in Eq. (8.49) does affect the distribution of $\ell_{\max}$ in a non-trivial way. In contrast, for $y = y_c$ but $c > 2$, one recovers the Fréchet distribution of IID random variables, indicating that the global constraint is insignificant to a large extent—it only renormalizes the scale factors (Bar *et al.*, 2016).

These condensation transitions have also been observed in a series of works on run-and-tumble (RTP) models for active particles (Gradenigo and Majumdar, 2019; Mori *et al.*, 2021*b,a*). In the RTP model, a single particle starts from the origin with a random velocity $\mathbf{v}$ drawn from a distribution $W(\mathbf{v})$ and moves ballistically in that direction during an exponentially distributed run time. At the end of this run, the particle tumbles instantaneously and chooses a new velocity $\mathbf{v}$ (again from the same distribution $W(\mathbf{v})$) and a new run time, again from the same exponential distribution. Runs and tumbles thus alternate with time. If one measures the distribution of the position of the particle after a time T, one observes the signature of a condensation transition (for a class of $W(\mathbf{v})$) in the position distribution $P(\mathbf{r}, T)$. For an isotropic distribution $W(\mathbf{v})$, the distribution exhibits the large-deviation form $P(r, T) \sim e^{-T\Phi(r/T)}$, where the large-deviation function $\Phi(z)$ displays non-analytic behavior at $z = z_c$. For $z < z_c$, the run lengths are typically of order $O(1)$ and they are more or less equally distributed. However, for $z > z_c$, there exists a single run that subsumes a finite fraction of the total displacement: this corresponds to the condensed phase (Mori *et al.*, 2021*b*).

8.4 Hierarchically and logarithmically correlated random variables

So far, we have been discussing the extreme statistics of uncorrelated, weakly correlated and strongly correlated random variables. However, there exists a class of problems where the variables are *hierarchically* correlated. This is typically the case for disordered models defined on a tree. For example, one can consider the celebrated problem of a directed polymer on a tree (Derrida and Spohn, 1988). Consider a rooted Cayley tree where each node has two offspring. On each bond of the tree with n generations, starting from a single root at O, we assign a positive quenched random variable $\varepsilon_i \geq 0$ (see Fig. 8.3). A configuration of $\{\varepsilon_i\}$ specifies the disordered environment. We then consider a directed polymer of n steps that goes from the root to one of the leaf sites at the nth generation. The energy of such a directed path is given by the sum of the energies of all the bonds belonging to the path,

$$E_{\text{path}} = \sum_{i \in \text{ path}} \varepsilon_i. \qquad (8.51)$$

Thus, there are $N = 2^n$ possible paths, each with its own energy. This model was initially studied at finite temperature $T > 0$, where it was found that it exhibits a transition to a spin-glass-like ("frozen") phase at low temperature (reminiscent of a replica symmetry-breaking scenario found in mean-field spin-glasses; Derrida and Spohn, 1988).

This problem was then revisited at $T = 0$ (Majumdar and Krapivsky, 2000; Ben-Naim *et al.*, 2001; Dean and Majumdar, 2001) where one is interested in the path that has the *minimal* energy,

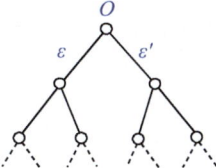

Fig. 8.3 A binary tree with a root at O. On each bond i, an energy ε_i is assigned, each drawn independently from a distribution $\rho(\varepsilon)$. For example, in the figure, we show the energies ε and ε' of the two daughter bonds of O. A directed polymer starts at O and goes down the tree to the leaves; the energy of the path is given by the sum of the disordered bond energies belonging to the path.

$$E_0 = \min[E_{\text{path}_1}, E_{\text{path}_2}, \dots, E_{\text{path}_N}]. \tag{8.52}$$

This is then a typical extreme value problem, and clearly, because of the partial overlaps between the paths, the energy variables $\{E_{\text{path}_i}\}$ are correlated. Indeed, the correlations between the energies of any two paths is proportional to the number of bonds they have in common: for these reasons, the $\{E_{\text{path}_i}\}$ are called *hierarchically correlated random variables*. It was shown in a series of works that the EVS of such random variables is often associated with a traveling-front structure (Majumdar and Krapivsky, 2000, 2002, 2003; Ben-Naim *et al.*, 2001; Dean and Majumdar, 2001; Majumdar, 2003). More precisely, in these cases the cumulative distribution of the minimum (or the maximum) satisfies a non-linear equation, which allows a traveling-front solution, as discussed below.

To proceed, consider the directed polymer on the Cayley tree with random energies ε_i's distributed according to a PDF $\rho(\varepsilon)$. One can then derive a recursion relation for the CDF of E_0 in Eq. (8.52), $P_n(x) = \text{Prob}(E_0 \geq x)$, which reads (Majumdar and Krapivsky, 2003)

$$P_{n+1}(x) = \left[\int \rho(\varepsilon) P_n(x-\varepsilon)\, d\varepsilon\right]\left[\int \rho(\varepsilon') P_n(x-\varepsilon')\, d\varepsilon'\right] = \left[\int \rho(\varepsilon) P_n(x-\varepsilon)\, d\varepsilon\right]^2, \tag{8.53}$$

together with the initial condition $P_0(x) = \Theta(-x)$. This relation follows from the following reasoning. Consider a particular realization of the disordered bond energies ε_i's on a tree, as in Fig. 8.3. The probability $P_n(x) = \text{Prob}(E_0 \geq x)$ is also the probability that all paths originating at O have energies bigger than x. Any path originating from the root O either passes through the left bond with energy ε or through the right bond with energy ε' as in Fig. 8.3. The left (respectively right) subtree starting with the left (respectively right) daughter of O must have energy bigger that $x - \varepsilon$ (respectively $x - \varepsilon'$). Hence, the probability of these joint events, using the fact that the subtrees are statistically independent, is given by the product $P_n(x - \varepsilon) P_n(x - \varepsilon')$; the subscript n indicates that the subtrees have n generations, as opposed to $n+1$ for the full tree. Averaging over the bond energies, each drawn independently from the distribution $\rho(\varepsilon)$, leads to the relation in Eq. (8.53).

To analyze the solution of the non-linear equation in Eq. (8.53), it is convenient to rewrite it in terms of $Q_n(x) = 1 - P_n(x) = \text{Prob}(E_0 \leq x)$. Substituting this form into Eq. (8.53), and using the fact that $\int \rho(\varepsilon)\, d\varepsilon = 1$, one obtains

$$Q_{n+1}(x) = 2\int \rho(\varepsilon) Q_n(x-\varepsilon)\, d\varepsilon - \left(\int \rho(\varepsilon) Q_n(x-\varepsilon)\, d\varepsilon\right)^2, \tag{8.54}$$

starting from the initial condition $Q_0(x) = \Theta(x)$. As n increases, the initial step function at $x = 0$ moves to the right linearly for large n with a constant speed, while the width of the step remains of order $O(1)$, thus retaining the shape of a front (see Fig. 8.4). To determine the speed, we focus on the left edge of the step where

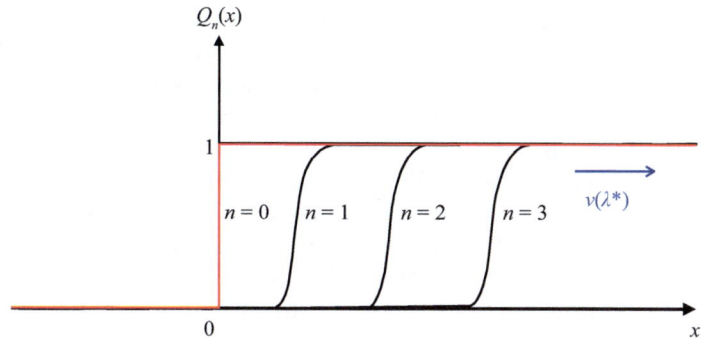

Fig. 8.4 The cumulative distribution of the ground-state energy $Q_n(x) = \text{Prob}(E_0 \leq x)$ plotted as a function of x for increasing values of n. For $n = 0$, it starts from a step function $Q_0(x) = \Theta(x)$. For a fixed $n > 0$, the function has a sigmoid shape and this front travels to the right with an asymptotic speed $v(\lambda^*)$ as $n \to \infty$.

$Q_n(x) \ll 1$. In this regime, one can neglect the non-linear term in Eq. (8.54) and keep only the linear term, which gives

$$Q_{n+1}(x) \approx 2 \int \rho(\varepsilon) Q_n(x - \varepsilon) \, d\varepsilon. \tag{8.55}$$

This linear equation admits a "traveling-front solution" of the form $Q_n(x) \approx e^{\lambda(x - v(\lambda)n)}$, where n plays the role of time and $v(\lambda)$ the speed of the front. Substituting this form into Eq. (8.55) provides a family of speeds $v(\lambda)$ parameterized by λ:

$$v(\lambda) = -\frac{1}{\lambda} \ln \left[2 \int \rho(\varepsilon) e^{-\lambda \varepsilon} \, d\varepsilon \right]. \tag{8.56}$$

For a generic distribution $\rho(\varepsilon)$, the function $v(\lambda)$ admits a unique maximum at $\lambda^* > 0$ with $v(\lambda^*) > 0$. For example, for $\rho(\varepsilon) = e^{-\varepsilon}$ with $\varepsilon > 0$, one gets $v(\lambda) = \ln((1 + \lambda)/2)/\lambda$. When plotted as a function of λ for $\lambda > 0$ it has a maximum at $\lambda^* = 3.311\,07\ldots$, where $v(\lambda^*) = 0.231\,961\ldots$ This "velocity selection principle," originally due to Kolmogorov in the case of the continuous Fisher–Kolmogorov–Petrovski–Piscounov (F-KPP) equation—for a comprehensive review, see Van Saarloos (2003); Brunet and Derrida (1997)—can be adapted here for this discrete non-linear equation (Majumdar and Krapivsky, 2003), and it states that this maximum velocity $v(\lambda^*)$ will be selected by the front for a sufficiently steep initial condition, as is the case here (Majumdar and Krapivsky, 2003). One can then compute the two leading terms of $\langle E_0 \rangle$ in the large-n limit, which generically read

$$\langle E_0 \rangle \approx v(\lambda^*)n + \frac{3}{2\lambda^*} \ln n. \tag{8.57}$$

This model of a directed polymer on a tree at zero temperature is also related to the so-called binary search tree (BST) problem in computer science. The study of BSTs is an important problem in computer science, with many applications related to sorting and search of data. For instance, when one stores data files in one's computer, usually one makes directories, subdirectories, sub-subdirectories, etc., which has a natural tree structure. The general idea is that if one searches for a particular item of data, it is much more efficient to search if the data is stored in a tree. Typically, if the number of data elements (e.g., the number of files) stored in the tree is of size N, then a binary search takes a time of order $O(\ln N)$, which is much smaller compared to $O(N)$ for a linear search. In fact, the search time is measured by the height H_N of the BST, which is the maximal depth of the nodes occupied on a tree with N occupied nodes (Majumdar and Krapivsky, 2002). It turns out that there is a direct mapping between the BST and the directed polymer on the Cayley tree (Majumdar and Krapivsky, 2002). The cumulative height distribution $P(H_N < n)$ for random data in the BST problem is precisely related to the distribution of the minimum (ground-state) energy of the directed problem discussed above. Th relation in Eq. (8.57), in the context of the BST, translates into a result for the average height (Majumdar and Krapivsky, 2002):

$$\langle H_N \rangle \approx \frac{1}{v(\lambda^*)} \ln N - \frac{3}{2\lambda^* v(\lambda^*)} \ln \ln N, \tag{8.58}$$

where $v(\lambda)$ is defined in Eq. (8.56) and λ^* is the unique maximum of $v(\lambda)$. Similar traveling-front solutions have also been found for other hierarchically correlated random variables, such as in a class of fragmentation models, the aggregation dynamics of growing random trees, or in simple models of dynamics of efficiency; for a review, see Majumdar and Krapivsky (2003). Interestingly, although the dispersion spectrum $v(\lambda)$, and consequently also λ^*, are non-universal (i.e., they will differ from one hierarchically correlated model to another), the coefficient $3/2$ in front of the logarithmic correction in Eq. (8.57) turns out to be universal.

For the directed polymer on the Cayley tree, it is also possible to study the full PDF of E_0 in the limit of large n (Dean and Majumdar, 2001). If the variables E_{path_i} were IID random variables [see, e.g., Eq. (2.22)], the PDF of E_0, properly centered and scaled, would be given (for an unbounded distribution $\rho(\varepsilon)$) by a Gumbel law $F_2'(-z) = \mathrm{e}^{z-\mathrm{e}^z}$ (note that we are considering here the minimum, at variance with the result in Eq. (4.26) that concerns the maximum, which explains the $-z$ in the argument). Instead of this, because the energy variables are actually hierarchically correlated, it was found that, for an unbounded distribution $\rho(\varepsilon)$, the limiting distribution is different from a Gumbel law. In particular, its right tail is non-universal and depends explicitly on $\rho(\varepsilon)$. For instance, it was found (Dean and Majumdar, 2001) that for $\rho(\varepsilon) = \mathrm{e}^{-|\varepsilon|}/2$, the limiting distribution has a simple exponential tail $\sim \mathrm{e}^{-z}$ for $z \to +\infty$, which is quite different from the super-exponential tail of the Gumbel law.

These behaviors of hierarchically correlated variables are reminiscent of the results obtained for *logarithmically* correlated random variables. For concreteness, let us indeed consider a set of N Gaussian random variables $V_1, V_2, \ldots, V_i, \ldots, V_N$, where the subscript i refers to a site index, which have logarithmic correlations, i.e.,

$$\langle (V_i - V_j)^2 \rangle \approx 4\sigma \ln |i - j|, \quad 1 \ll |i - j| \ll N, \tag{8.59}$$

where $\sigma > 0$, and write $V_{\min} = \min\{V_1, V_2, \ldots, V_N\}$. This model was first studied by Carpentier and Le Doussal (2001), who found that the average value $\langle V_{\min} \rangle$ indeed behaves as described in Eq. (8.57) for hierarchically correlated random variables (with the substitution $n \to \ln N$). In addition, they showed that the *left* tail of the limiting distribution of $V_{\min}$ (properly shifted and scaled) behaves as $|z|e^z$ as $z \to -\infty$, which is indeed different from the left tail of the Gumbel law $\sim e^z$. One can further consider the simple model of a single particle in the disordered potential V_i at site i, with $i = 0, 1, \ldots, N$, at thermal equilibrium at inverse temperature β. The probability p_i of finding the particle at site i is thus given by the Boltzmann weight,

$$p_i = \frac{1}{Z_N} e^{-\beta V_i}, \qquad Z_N = \sum_{i=1}^{N} e^{-\beta V_i}, \tag{8.60}$$

where the V_i are Gaussian random variables with logarithmic correlations like those in Eq. (8.59). Using RG techniques, it was shown that the model in Eq. (8.60) exhibits an interesting freezing phenomenon at a finite inverse temperature β_g, which is reminiscent of the transition found in Derrida's random energy model (REM; Derrida, 1981) and its generalizations (Derrida, 1985; Derrida and Gardner, 1986). Indeed, at high temperature, for $\beta < \beta_g$ the particle is delocalized over the whole sample, while for $\beta > \beta_g$ the Boltzmann weight is concentrated on the few local minima of the disordered potential (Carpentier and Le Doussal, 2001). Such a freezing transition was later confirmed by the exact solution of several models of logarithmically correlated random variables. These include, e.g., the equilibrium properties of a single particle in a random Gaussian high-dimensional landscape (Fyodorov and Bouchaud, 2007, 2008; Fyodorov and Sommers, 2007) or a circular variant of the REM with logarithmic correlations (Fyodorov and Bouchaud, 2008). For these different models, the full limiting distribution of the minimum is not universal, i.e., it depends on the details of the model, but the left tail turns out to be universal, and behaves as $|z|e^z$ as $z \to -\infty$ (Carpentier and Le Doussal, 2001). Since then, these models have generated a lot of interest in both physics and mathematics because of their relations with the extremes of the two-dimensional Gaussian free field (Fyodorov *et al.*, 2009; Bramson and Zeitouni, 2012; Bramson *et al.*, 2016). More recently, it was suggested that the same freezing transition also governs the extreme values taken by the characteristic polynomials of random matrices and the Riemann zeta function (Fyodorov *et al.*, 2012; Fyodorov and Keating, 2014). This, in turn, has generated a recent interest in mathematics (Arguin *et al.*, 2017; Paquette and Zeitouni, 2018), in connection with random matrices and branching processes.

These hierarchically correlated random variables also appear in another well-studied continuous-time process, namely branching Brownian motion in one dimension. Several extreme value observables have been studied extensively in this problem, and connections to traveling-front solutions have been established (McKean, 1975; Bramson, 1983; Brunet and Derrida, 2009, 2011). The extreme value statistics for branching Brownian motion in the presence of a finite death rate have also been studied extensively (Sawyer and Fleischman, 1979; Dumonteil *et al.*, 2013; Ramola *et al.*, 2014, 2015*a,b*). We have not discussed branching Brownian motion in detail here, and refer the interested reader to the original articles mentioned above.

8.5 Extreme statistics for fluctuating interfaces in one dimension: An example of a strongly correlated random variable

The extreme value statistics has also been studied for (1+1)-dimensional fluctuating interfaces. The relevant degrees of freedom here are the heights at different spatial points and, as we will see, they are strongly correlated. The most well-studied model of such a fluctuating interface is the so-called Kardar–Parisi–Zhang (KPZ) equation that describes the time evolution of the height $H(x,t)$ of an interface growing over a linear substrate of size L via the stochastic partial differential equation (Kardar *et al.*, 1986)

$$\frac{\partial H}{\partial t} = \frac{\partial^2 H}{\partial x^2} + \lambda \left(\frac{\partial H}{\partial x}\right)^2 + \eta(x,t), \tag{8.61}$$

where $\eta(x,t)$ is Gaussian white noise with zero mean and a correlator $\langle \eta(x,t)\eta(x',t')\rangle = 2\delta(x-x')\delta(t-t')$. For $\lambda = 0$, the equation becomes linear and is known as the Edwards–Wilkinson (EW) equation. Although Eq. (8.61) was originally introduced in the context of surface growth in $1+1$ dimensions, the same equation has emerged in a large variety of contexts, such as in fluid dynamics (noisy Burger's equation), directed polymers in a random medium and non-intersecting Brownian motions (vicious walkers), as well as Dyson's Brownian motion in random matrix theory, non-interacting fermions in a confining potential at finite temperature, random permutations, etc. (Halpin-Healy and Zhang, 1995; Krug, 1997; Halpin-Healy and Takeuchi, 2015; Spohn, 2016).

It is easy to see from Eq. (8.61) that the spatially averaged height $(1/L)\int_0^L H(x,t)\,dx$ keeps on growing with time, even in a system of finite size L. This is easily seen in the EW limit ($\lambda = 0$), where this average height performs a Brownian motion in time and thus typically grows as $\sqrt{t/L}$ at late times. In other words, there is a "zero mode" in the system. Consequently, the joint distribution of the heights at different spatial points never reaches a stationary state, even in a finite-sized system. Hence, one usually subtracts the zero mode by defining a relative height,

$$h(x,t) = H(x,t) - \frac{1}{L}\int_0^L H(y,t)\,dy. \tag{8.62}$$

Once the zero mode has been eliminated, it can be shown that the joint PDF of the relative height field $P(\{h\},t)$ reaches a stationary state as $t \to \infty$ in a finite system of size L. Indeed, the roughness of the interface is usually measured by the width, i.e., the standard deviation of the relative height $h(x,t)$:

$$W(t,L) = \sqrt{\langle h(x,t)^2 \rangle}. \tag{8.63}$$

If one starts from a flat interface at $t=0$, the width typically grows with time as a power law and finally saturates to an L-dependent constant when $t \gg L^z$, where z is known as the dynamical exponent, i.e.,

$$W(t,L) \sim \begin{cases} t^\beta, & t \ll L^z, \\ L^\chi, & t \gg L^z. \end{cases} \tag{8.64}$$

The former is called the "growing regime" while the latter is the "stationary regime." The exponents β and χ are called the growth and the roughness exponent respectively. These two regimes of $W(t,L)$ are connected via the Family–Vicseck scaling form $W(t,L) \sim L^\chi w(t/L^z)$, which provides a scaling relation between the three exponents:

$$\chi = \beta z. \tag{8.65}$$

For the EW case ($\lambda=0$), one can easily show that $\beta = 1/4$, $z=2$ and $\chi = 1/2$. In contrast, for the KPZ case ($\lambda>0$), it is known that $\beta = 1/3$, $z=3/2$ and $\chi = 1/2$. This crossover between the growing and the stationary regimes in Eq. (8.64) can be alternatively understood in terms of a time-dependent correlation length $\xi(t)$. At early times, the heights at different spatial points are essentially uncorrelated and $\xi(t) \ll a_0$, where a_0 is a microscopic length. As time grows, the heights at different spatial points get progressively more and more correlated and the correlation length grows as $\xi(t) \sim t^{1/z}$ for $t \ll L^z$. Finally, when time crosses L^z, the correlation length $\xi(t) \sim L$ becomes of the order of the system size and the heights at different spatial points become strongly correlated in the stationary state. Over the last two decades, there have been enormous activities in studying the statistics of the height $H(x,t)$ at a fixed position x, going beyond just the second moment. It is known (Halpin-Healy and Takeuchi, 2015; Spohn, 2016) that in the growing regime, when $a_0 \ll \xi(t) \ll L$, the full distribution of the height, at a fixed point in space, appropriately centered and scaled, is described by the Tracy–Widom distribution of random matrix theory, which will be discussed in the next section.

Here, our main interest is not on the relative height $h(x,t)$ itself but rather on the *maximal* relative height over all spatial points (Raychaudhuri *et al.*, 2001; Majumdar and Comtet, 2004, 2005),

$$h_{\mathrm{m}} = \max_x[h(x,t), 0 \leq x \leq L]. \tag{8.66}$$

This is our extreme observable of interest. Our goal is to compute the distribution of h_{m} at different times. From the general discussion above, we see that in the growing regime, when $t \ll L^z$, the correlation length $\xi(t) \ll L$. Hence, in this regime, the heights at different spatial points are weakly correlated. Following our general argument for weakly correlated variables (see Section 4.2), we would then expect that in the growing regime the maximal relative height, appropriately centered and scaled, should have the Gumbel distribution. In contrast, in the stationary regime, the height variables are strongly correlated and the maximal relative height h_m should have a different distribution. This distribution was first studied numerically in Raychaudhuri *et al.* (2001) and was then computed analytically in Majumdar and Comtet (2004, 2005). This, then, presents one of the rare solvable cases for the EVS of strongly correlated random variables. Below, we briefly outline the derivation of this distribution.

To compute the distribution of h_m in the stationary state, we need the explicit joint distribution of the relative heights at different spatial points. For the KPZ equation in Eq. (8.61) in 1+1 dimension, it is well known that, for periodic boundary conditions, where $h(0) = h(L)$, the joint distribution of the relative heights in the stationary state is independent of λ and thus is the same as in the EW case (Halpin-Healy and Zhang, 1995). In the latter case, it is known that, in the stationary state $t \gg L^z$, the actual height $H(x, t)$, as a function of x, locally performs a Brownian motion in space. However, this is not enough to write down the joint distribution of the relative heights $h(x, t)$ since it satisfies a global constraint from the definition in Eq. (8.62), namely

$$\int_0^L h(x, t) \, \mathrm{d}x = 0. \tag{8.67}$$

Combining the periodic boundary condition, the global constraint in Eq. (8.67) and the local Brownian nature of $h(x, t)$ as a function of x, the stationary joint distribution of the relative heights reads (Majumdar and Comtet, 2004, 2005)

$$P_{\mathrm{st}}[\{h\}] = C(L) \exp\left\{-\frac{1}{2}\int_0^L (\partial_x h)^2 \, \mathrm{d}x\right\} \delta[h(0) - h(L)]\delta\left[\int_0^L h(x, t) \, \mathrm{d}x\right], \tag{8.68}$$

where $C(L) = \sqrt{2\pi L^3}$ is the normalization constant and can be obtained by integrating over all the heights. The first term in Eq. (8.68) reflects the Brownian measure in space, the second term expresses the periodic boundary condition and the third term ensures the global zero-area constraint in Eq. (8.67). The stationary measure in Eq. (8.68) of the relative heights thus corresponds to a Brownian bridge, but with a global constraint that the area under the bridge is strictly zero (Majumdar and Comtet, 2004, 2005). This last fact plays a crucial role for the extreme statistics of relative heights and makes it non-trivial (Majumdar and Comtet, 2004, 2005).

To proceed further to compute the distribution of h_m, we first define the cumulative distribution of the maximum relative height $Q(z, L) = \mathrm{Prob}[h_m \leq z]$. The PDF of the maximum relative height is then $P(z, L) = Q'(z, L)$. Clearly, $Q(z, L)$ is also the probability that the heights at all points in $[0, L]$ are less than z and can be formally written in terms of the path integral (Majumdar and Comtet, 2004, 2005)

$$Q(z, L) = C(L) \int_{-\infty}^{z} du \int_{h(0)=u}^{h(L)=u} \mathcal{D}h(x) \exp\left\{-\frac{1}{2} \int_{0}^{L} (\partial_x h)^2 \, dx\right\}$$

$$\times \delta\left[\int_{0}^{L} h(x,t) \, dx\right] I(z, L), \tag{8.69}$$

where $I(z, L) = \prod_{x=0}^{L} \Theta[z - h(x)]$ is an indicator function which is 1 if all the heights are less than z and zero otherwise. Using path integral techniques—for the details, see Majumdar and Comtet (2004, 2005)—it was found that the PDF of h_m takes the scaling form, for all L,

$$P(h_m, L) = \frac{1}{\sqrt{L}} f\left(\frac{h_m}{\sqrt{L}}\right), \tag{8.70}$$

where the scaling function can be computed explicitly as (Majumdar and Comtet, 2004, 2005)

$$f(x) = \frac{2\sqrt{6}}{x^{10/3}} \sum_{k=1}^{\infty} e^{-b_k/x^2} b_k^{2/3} U\left(-\frac{5}{6}, \frac{4}{3}, \frac{b_k}{x^2}\right), \tag{8.71}$$

where $U(a, b, y)$ is the confluent hypergeometric function and $b_k = (2/27)\alpha_k^3$, where the α_k's are the absolute values of the zeros of the Airy function, $\mathrm{Ai}(-\alpha_k) = 0$. It is easy to obtain the small-x behavior of x since only the $k=1$ term dominates as $x \to 0$. Using $U(a, b, y) \sim y^{-a}$ for large y, we get, as $x \to 0$,

$$f(x) \approx \frac{8}{81} \alpha_1^{9/2} x^{-5} \exp\left[-\frac{2\alpha_1^3}{27x^2}\right], \qquad x \to 0. \tag{8.72}$$

The asymptotic behavior of $f(x)$ at large x can be obtained as (Majumdar and Comtet, 2004, 2005; Janson and Louchard, 2007; Rambeau *et al.*, 2011)

$$f(x) \approx 72\sqrt{\frac{6}{\pi}} x^2 e^{-6x^2}, \qquad x \to \infty. \tag{8.73}$$

It turns out, rather interestingly, that this same function has appeared before in several different problems in computer science and probability theory; it is known in the literature as the "Airy distribution function". In particular, it describes the distribution of the area under a Brownian excursion over the unit time interval; for a review of this function, its occurrence in different contexts and its generalizations, see Majumdar and Comtet (2005), Majumdar (2007*a*) and Janson (2007) and references therein.

The path integral technique mentioned above to compute the maximal relative height distribution of the EW/KPZ stationary interfaces with periodic boundary conditions have subsequently been generalized to different boundary conditions as well as to more complex interfaces (Györgyi *et al.*, 2007; Burkhardt *et al.*, 2007; Kearney *et al.*, 2007; Rambeau and Schehr, 2010; Rambeau *et al.*, 2011; Kearney and Majumdar, 2014).

8.6 Extreme statistics in random matrix theory

Another beautiful solvable example of the extremal statistics of strongly correlated variables can be found in the theory of random matrices (Tracy and Widom, 1994, 1996). Let us consider, for simplicity, an $N \times N$ Gaussian random matrix with real symmetric, complex Hermitian or quaternionic self-dual entries $X_{i,j}$ distributed via the joint Gaussian law (Mehta, 2004; Forrester, 2010)

$$P[\{X_{i,j}\}] \propto \exp\left[-\frac{\beta}{2} N \operatorname{Tr}(X^2)\right],$$ (8.74)

where β is the so-called Dyson index. The distribution is invariant respectively under orthogonal, unitary and symplectic transformations giving rise to the three classical ensembles: Gaussian orthogonal ensemble (GOE), Gaussian unitary ensemble (GUE) and Gaussian symplectic ensemble (GSE). The quantized values of β are respectively $\beta = 1$ (GOE), $\beta = 2$ (GUE) and $\beta = 4$ (GSE). The eigenvalues and eigenvectors turn out to be independent and their joint distribution thus factorizes. Integrating out the eigenvectors, we focus here only on the statistics of N eigenvalues $\lambda_1, \lambda_2, \ldots, \lambda_N$, which are all real. The joint PDF of these eigenvalues is given by the classical result (Mehta, 2004; Forrester, 2010)

$$P_{\text{joint}}(\lambda_1, \ldots, \lambda_N) = B_N(\beta) \exp\left[-\frac{\beta}{2} N \sum_{i=1}^{N} \lambda_i^2\right] \prod_{i<j} |\lambda_i - \lambda_j|^{\beta},$$ (8.75)

where $B_N(\beta)$ is the normalization constant. The presence of the Vandermonde term $\prod_{i<j} |\lambda_i - \lambda_j|^{\beta}$ in the joint distribution makes the eigenvalues $\{\lambda_i\}$ strongly correlated (it induces repulsion between every pair of eigenvalues). If this Vandermonde term were absent, the joint distribution would factorize as in Eq. (2.22) and the eigenvalues would be IID Gaussian variables. To understand the correlations between the eigenvalues a bit better, it is convenient to rewrite this joint distribution as

$$P_{\text{joint}}(\lambda_1, \ldots, \lambda_N) = B_N(\beta) \exp\left[-\beta\left(\frac{N}{2} \sum_{i=1}^{N} \lambda_i^2 - \frac{1}{2} \sum_{i \neq j} \ln |\lambda_i - \lambda_j|\right)\right].$$ (8.76)

Hence, this joint law can be interpreted as a Gibbs–Boltzmann weight (Dyson, 1962), $P_{\text{joint}}(\{\lambda_i\}) \propto \exp[-\beta E(\{\lambda_i\})]$, of an interacting gas of charged particles on a line where λ_i denotes the position of the ith charge and β plays the role of the inverse temperature. The energy $E(\{\lambda_i\})$ has two parts: each pair of charges repel each other via a two-dimensional Coulomb (logarithmic) repulsion (even though the charges are confined on the one-dimensional real line) and each charge is subject to an external confining parabolic potential. Note that while $\beta = 1$, 2 and 4 correspond to the three classical rotationally invariant Gaussian ensembles described by the measure in Eq. (8.74), it is possible to associate a matrix model with Eq. (8.76) for any value of $\beta > 0$, namely the tridiagonal random matrices introduced in Dumitriu and Edelman (2002). These ensembles defined as in Eq. (8.75) for generic β are sometimes called the "Gaussian β-ensembles". Here, we focus on the largest eigenvalue $\lambda_{\max} = \max_{1 \leq i \leq N} \lambda_i$: What can be said about its fluctuations, in particular when N is large? This is a non-trivial question as the interaction term, $\propto |\lambda_i - \lambda_j|^\beta$, renders the classical results of extreme value statistics for IID random variables discussed in Section 4.1 inapplicable.

The two terms in the energy of the Coulomb gas in Eq. (8.76), the pairwise Coulomb repulsion and the external harmonic potential, compete with each other. While the former tends to spread the charges apart, the latter tends to confine the charges near the origin. Note that we have scaled the amplitude of the first term in the energy in Eq. (8.76) by a factor of N such that both the first and the second terms in Eq. (8.76) are of the same order $O(N^2)$, and the typical value of λ_i is of order $O(1)$.

The average density of charges is given by

$$\rho_N(\lambda) = \frac{1}{N}\left\langle \sum_{i=1}^{N} \delta(\lambda - \lambda_i) \right\rangle, \tag{8.77}$$

where the angle brackets denote an average with respect to the joint PDF in Eq. (8.76). For Gaussian matrices as in Eq. (8.76), it is well known (Wigner, 1951; Mehta, 2004; Forrester, 2010) that as $N \to \infty$, the average density approaches an N-independent limiting form which has a semi-circular shape on the compact support $[-\sqrt{2}, +\sqrt{2}]$,

$$\lim_{N\to\infty} \rho_N(\lambda) = \tilde{\rho}_{\mathrm{sc}}(\lambda) = \frac{1}{\pi}\sqrt{2 - \lambda^2}, \tag{8.78}$$

where $\tilde{\rho}_{\mathrm{sc}}(\lambda)$ is called the Wigner semi-circular law. Hence, our first observation is that the maximum eigenvalue resides near the upper edge of the Wigner semi-circle:

$$\lim_{N\to\infty} \langle \lambda_{\max} \rangle = \sqrt{2}. \tag{8.79}$$

However, for finite but large N, the largest eigenvalue $\lambda_{\max}$ will fluctuate from sample to sample. One can estimate, for large N, the size s_N of these fluctuations, using a similar argument to the IID case, see Eq. (4.41). The argument goes as follows. We first recall that $\tilde{\rho}_{\mathrm{sc}}(\lambda)\,\mathrm{d}\lambda$ counts the fraction of eigenvalues in the interval $[\lambda, \lambda+\mathrm{d}\lambda]$. Therefore, if we integrate $\tilde{\rho}_{\mathrm{sc}}(\lambda)$ from $\sqrt{2} - s_N$ to $\sqrt{2}$, it should be $1/N$, since there is typically only one eigenvalue at the very upper edge close to $\sqrt{2}$:

$$\int_{\sqrt{2}-s_N}^{\sqrt{2}} \tilde{\rho}_{\mathrm{sc}}(\lambda)\,\mathrm{d}\lambda \approx \frac{1}{N}. \tag{8.80}$$

Using the expression of the Wigner semi-circular law from Eq. (8.78), and assuming $s_N \ll \sqrt{2}$, it is easy to see that, to leading order for large N,

$$s_N = O(N^{-2/3}). \tag{8.81}$$

This suggests that the typical fluctuations of the random variable $\lambda_{\max}$ can be expressed as

$$\lambda_{\max} \approx \sqrt{2} + \frac{1}{\sqrt{2}} N^{-2/3}\chi_\beta, \tag{8.82}$$

where χ_β is an N-independent random variable. The knowledge of the distribution of χ_β will tell us how the typical fluctuations of $\lambda_{\max}$, of order $\lambda_{\max} - \sqrt{2} = O(N^{-2/3})$, are distributed. Note that Eq. (8.82) describes the typical fluctuations of $\lambda_{\max}$ of order $O(N^{-2/3})$ around its mean, $\sqrt{2}$. However, in many applications one needs to know the probability of atypically large fluctuations of $\lambda_{\max}$ of, say, order $O(1)$ around the mean. These atypically large fluctuations are not included in the typical fluctuations in Eq. (8.82).

To compute the distribution of $\lambda_{\max}$ in both typical and atypical regimes, one first defines the cumulative distribution

$$Q_{\max}(w, N) = \text{Prob}[\lambda_{\max} < w], \tag{8.83}$$

which can be written as a ratio of two partition functions,

$$Q_{\max}(w, N) = \frac{Z_N(w)}{Z_N(w \to \infty)}, \tag{8.84}$$

$$Z_N(w) = \int_{-\infty}^{w} d\lambda_1 \cdots \int_{-\infty}^{w} d\lambda_N \, \exp\left[-\beta\left(\frac{N}{2}\sum_{i=1}^{N}\lambda_i^2 - \frac{1}{2}\sum_{i\neq j}\ln|\lambda_i - \lambda_j|\right)\right]. \tag{8.85}$$

The partition function $Z_N(w)$ in the numerator describes a two-dimensional Coulomb gas, confined on a one-dimensional line and subject to a harmonic potential, as in Eq. (8.76), but now in the presence of an impenetrable *hard wall* at w. This ratio of two partition functions in Eq. (8.84) is very hard to compute for finite N. However, for large N, it is possible to extract the behavior of $\lambda_{\max}$ in both the typical regime (where $|\lambda_{\max} - \sqrt{2}| = O(N^{-2/3})$) and the atypical regime (where $|\lambda_{\max} - \sqrt{2}| = O(1)$).

Let us summarize the main results, without providing details. In the regime of typical fluctuations, one can show, by analyzing the ratio of partition functions in Eq. (8.84), that Eq. (8.82) is indeed valid with the CDF of χ_β, i.e., $\mathcal{F}_\beta = \text{Prob}[\chi_\beta \leq x]$, known as the β-Tracy–Widom (TW) distribution. For $\beta = 1, 2, 4$, the CDF F_β can be written explicitly in terms of a special solution of a Painlevé II equation (Tracy and Widom, 1994, 1996), and for $\beta = 6$ in terms of an additional Painlevé transcendent (Rumanov, 2016; Grava *et al.*, 2016). For other values of β, it can be shown that $\mathcal{F}_\beta$ describes the fluctuations of the ground state of a one-dimensional random Schrödinger operator, called the "stochastic Airy operator" (Edelman and Sutton, 2007; Ramirez *et al.*, 2011). For arbitrary $\beta > 0$, it can be shown that the PDF $\mathcal{F}'_\beta(x)$ has asymmetric non-Gaussian tails (Tracy and Widom, 1994, 1996),

$$\mathcal{F}'_\beta(x) \approx \begin{cases} \exp\left[-\dfrac{\beta}{24}|x|^3\right], & x \to -\infty, \\[2ex] \exp\left[-\dfrac{2\beta}{3}x^{3/2}\right], & x \to +\infty. \end{cases} \tag{8.86}$$

Quite remarkably, the same TW distributions (in particular for $\beta = 1, 2$) have emerged in a number of a priori unrelated problems (Majumdar, 2007*b*) such as the longest increasing subsequence of random permutations (Baik *et al.*, 1999), directed polymers (Johansson, 2000; Baik and Rains, 2000) and growth models (Prähofer and Spohn, 2000; Gravner *et al.*, 2001; Majumdar and Nechaev, 2004; Imamura and Sasamoto, 2004) in the KPZ universality class in $(1 + 1)$ dimensions, as well as for the continuum $(1 + 1)$-dimensional KPZ equation (Sasamoto and Spohn, 2010; Calabrese *et al.*, 2010; Dotsenko, 2010; Amir *et al.*, 2011), sequence alignment problems (Majumdar and Nechaev, 2005), mesoscopic fluctuations in quantum dots (Vavilov *et al.*, 2001; Ostrovsky *et al.*, 2001; Silva and Ioffe, 2005; Lemarié *et al.*, 2013), height fluctuations of non-intersecting Brownian motions over a fixed time interval (Forrester *et al.*, 2011; Liechty, 2012), height fluctuations of non-intersecting interfaces in the presence of a long-range interaction induced by a substrate (Nadal and Majumdar, 2009) or more recently in the context of trapped fermions (Dean *et al.*, 2015, 2016), and in finance (Biroli *et al.*, 2007*b*). Remarkably, TW distributions have also been observed in experiments on nematic liquid crystals (Takeuchi and Sano, 2010, 2012; Takeuchi *et al.*, 2011), for $\beta = 1, 2$, in experiments involving coupled-fiber lasers (Fridman *et al.*, 2012), for $\beta = 1$, as well as in disordered superconductors (Lemarié *et al.*, 2013), for $\beta = 1$.

As mentioned before, while the β-TW distribution describes the typical fluctuations of $\lambda_{\max}$ around its mean $\langle \lambda_{\max} \rangle = \sqrt{2}$ on a small scale of $\mathcal{O}(N^{-2/3})$, it does not describe atypically large fluctuations, e.g., of order $\mathcal{O}(1)$, around $\langle \lambda_{\max} \rangle$. The probability of atypically large fluctuations, to leading order for large N, is described by two large-deviation (or rate) functions $\Phi_-(w)$ (for fluctuations to the left of the mean) and $\Phi_+(w)$ (for fluctuations to the right of the mean). These different regimes for the cumulative distribution $Q_{\max}(w, N)$ of $\lambda_{\max}$ in Eq. (8.83) can be summarized, at leading order for large N, as follows:

$$
Q_{\max}(w, N) \approx \begin{cases} \exp[-\beta N^2 \Phi_-(w)], & w < \sqrt{2} \text{ and } |w - \sqrt{2}| \sim \mathcal{O}(1), \\ \mathcal{F}_\beta[\sqrt{2} N^{2/3}(w - \sqrt{2})], & |w - \sqrt{2}| \sim \mathcal{O}(N^{-2/3}), \\ 1 - \exp[-\beta N \Phi_+(w)], & w > \sqrt{2} \text{ and } |w - \sqrt{2}| \sim \mathcal{O}(1). \end{cases} \tag{8.87}
$$

Equivalently, the PDF of $\lambda_{\max}$ obtained from $P_{\max}(w, N) = dQ_{\max}(w, N)/dw$ reads, at leading order for large N (see also Fig. 8.5),

$$
P_{\max}(w, \ N) \approx \begin{cases} \exp[-\beta N^2 \Phi_-(w)], & w < \sqrt{2} \text{ and } |w - \sqrt{2}| \sim \mathcal{O}(1), \\ \sqrt{2} N^{2/3} \mathcal{F}_\beta'[\sqrt{2} N^{2/3}(w - \sqrt{2})], & |w - \sqrt{2}| \sim \mathcal{O}(N^{-2/3}), \\ \exp[-\beta N \Phi_+(w)], & w > \sqrt{2} \text{ and } |w - \sqrt{2}| \sim \mathcal{O}(1). \end{cases}
$$
$$
\tag{8.88}
$$

Note that in Eqs. (8.87) and (8.88) the symbol $\approx$ means a logarithmic equivalent. We refer to Borot *et al.* (2011), Nadal and Majumdar (2011), Borot and Nadal (2012), and Forrester (2012) for the studies of the subleading corrections to the large-deviation forms in Eqs. (8.87) and (8.88) for both the left and right tails. For the left tail, the rate function $\Phi_-(w)$ was first explicitly computed in Dean

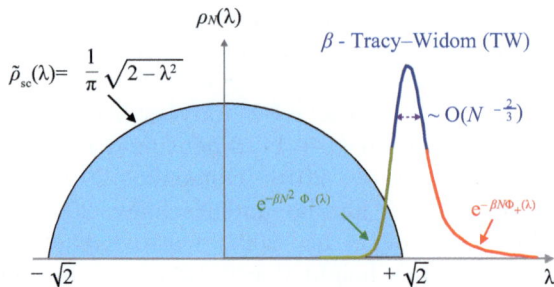

Fig. 8.5 Sketch of the PDF of $\lambda_{\max}$ with a peak around the right edge of the Wigner semi-circle given in Eq. (8.78) at $\langle \lambda_{\max} \rangle = \sqrt{2}$. The *typical* fluctuations around the mean are described by the β-Tracy–Widom distribution (blue), while the large deviations of order $\mathcal{O}(1)$ to the left and right of the mean are described by the left (green) and right (red) large-deviation tails [see Eq. (8.88)].

and Majumdar (2006, 2008) using Coulomb gas techniques. In particular, when w approaches the critical value $\sqrt{2}$ from below, it behaves as

$$\Phi_-(w) \sim \frac{1}{6\sqrt{2}}(\sqrt{2} - w)^3, \qquad w \underset{w<\sqrt{2}}{\longrightarrow} \sqrt{2}. \tag{8.89}$$

For the right tail, the large-deviation function $\Phi_+(w)$ was computed in Majumdar and Vergassola (2009). A rigorous derivation (but only valid for $\beta = 1$) of $\Phi_+(w)$ can be found in Ben Arous *et al.* (2001); see also Fyodorov (2004) for yet another derivation of $\Phi_+(w)$ for $\beta = 1$. In particular, near $w = \sqrt{2}$ it behaves as

$$\Phi_+(w) \sim \frac{2^{7/4}}{3}(w - \sqrt{2})^{3/2}, \qquad w \underset{w>\sqrt{2}}{\longrightarrow} \sqrt{2}. \tag{8.90}$$

By inserting the asymptotic behaviors of the rate functions given by Eqs. (8.89) and (8.90) into Eq. (8.88), one can check that there is a smooth matching between the left and right large-deviation tails and the tails of the central part described by the β-TW distribution in Eq. (8.86)—see Majumdar and Schehr (2014) for more details.

Interestingly, the study of the large deviations of $\lambda_{\max}$ reveals the existence of a phase transition, separating the left and right tails. Indeed, this corresponds to a thermodynamic phase transition for the free energy, $\propto \ln Q_{\max}(w, N)$, of a Coulomb gas in the presence of a wall, Eq. (8.84), as the position of the wall crosses the value $w_c = \sqrt{2}$. Indeed, from Eq. (8.87), one has

$$\lim_{N \to \infty} -\frac{1}{N^2} \ln Q_{\max}(w, N) = \begin{cases} \Phi_-(w), & w < \sqrt{2}, \\ 0, & w > \sqrt{2}. \end{cases} \tag{8.91}$$

Since $\Phi_-(w) \propto (\sqrt{2} - w)^3$ [see Eq. (8.89)], the third derivative of the free energy of this Coulomb gas is discontinuous at $w_c = \sqrt{2}$—this indicates a *third-order phase transition* (Nadal and Majumdar, 2011). It turns out that this transition is very similar to the so-called Gross–Witten–Wadia phase transition found in the 1980s in the context of two-dimensional lattice quantum chromodynamics (Gross and Witten, 1980; Wadia, 1980). As discussed in further detail in Majumdar and Schehr (2014), a similar third-order transition occurs in various physical systems, including non-intersecting Brownian motions (Forrester *et al.*, 2011; Schehr *et al.*, 2013), conductance fluctuations in mesoscopic physics (Vivo *et al.*, 2008, 2010; Damle *et al.*, 2011), random energy landscape (Fyodorov and Nadal, 2012) and entanglement in a bipartite system (Nadal *et al.*, 2010, 2011). In Majumdar and Schehr (2014), it was further argued that the universality of the TW distribution is inherited from the universality of phase transitions.

Here, we focused on the Gaussian β-ensembles in Eq. (8.75), but extreme value questions have also been studied for a wide variety of random matrix ensembles. This includes invariant ensembles, like the Laguerre–Wishart, Jacobi or Cauchy ensembles (for which all the eigenvalues are real), as well as non-invariant ensembles, and in particular the so-called Ginibre ensembles of RMT, for which the eigenvalues lie in the complex plane. In the latter case, the joint PDF of the complex eigenvalues can be identified with the Boltzmann weight of a two-dimensional Coulomb gas in the presence of an external quadratic confining potential, analogous to Eq. (8.76) but for complex eigenvalues $\lambda_i \to z_i$ where z_i are complex numbers. In these two-dimensional situations a natural extreme observable is the largest radius $r_{\max} = \max_{1 \leq i \leq N} |z_i|$ (Rider, 2003; Chafaï and Péché, 2014; Cunden *et al.*, 2016). Further extensions have also been considered, either by studying various matrix potentials (i.e., different from the quadratic one), which in some cases correspond to other interesting matrix models (Lacroix-A-Chez-Toine *et al.*, 2018), or by studying Coulomb gases in higher (> 2) dimensions (Cunden *et al.*, 2018). In the latter case many extreme value questions, like the distribution of typical fluctuations of the largest radius, remain open.

Another interesting extension where extreme observables have been studied corresponds to the so-called Riesz gas in one dimension. Here, one considers a system of N particles on a line in thermal equilibrium. Denoting by $\{x_1, x_2, \ldots, x_N\}$ the positions of the particles, the probability distribution in equilibrium of this gas is given by the Gibbs–Boltzmann distribution

$$P_{\text{joint}}(\{x_1, x_2, \ldots, x_N\}) = \frac{1}{Z_N} e^{-\beta E[\{x_i\}]}, \tag{8.92}$$

where $\beta = 1/(k_{\mathrm{B}}T)$ is the inverse temperature, Z_N is the normalizing partition function and the energy $E[\{x_i\}]$ is given by

$$E[\{x_i\}] = A \sum_{i=1}^{N} x_i^2 + J \operatorname{sgn}(k) \sum_{i \neq i} |x_i - x_j|^{-k}, \qquad k > -2. \tag{8.93}$$

The first term corresponds to an external confining ($A > 0$) harmonic potential, while the second term represents a repulsive interaction between any pair of particles. The exponent k characterizes the nature of the gas. Its value has to be $k > -2$, as otherwise the external harmonic potential is not able to confine the particles (the repulsion is strong enough to push the particles to $\pm\infty$ for $k < -2$). In the limit $k \to 0^+$, the interaction term can be expressed as a logarithmic repulsion and the Riesz gas reduces precisely to the log-gas discussed in Eq. (8.84). The other well-studied cases correspond to $k = -1$ (this is known as the "jellium model") and $k = +2$, which is known as the Calogero–Moser model. As in the log-gas case, one can investigate several EVS questions, in particular the distribution of the position $x_{\max}$ of the rightmost particle for arbitrary $k > -2$. We recall that in the limit $k \to 0$ this distribution, properly centered and scaled, is given by the Tracy–Widom distribution. Recently, the distribution of $x_{\max}$ was computed exactly for the jellium model, i.e., the $k = -1$ case (Dhar *et al.*, 2017, 2018). The limiting distribution is not known for other values of $k > -2$. However, the large-deviation functions associated with the distribution of $x_{\max}$ have been computed exactly for all $k > -2$ and shown to exhibit non-analytic behavior at some critical value (corresponding to the location of the right edge of the average density in the large-N limit), as in the log-gas case (Kethepalli *et al.*, 2022).

8.7 Extreme statistics in stochastic resetting systems

We have already discussed how extreme value statistics plays an important role in binary search tree problems where the random data is stored on a binary tree. EVS has also found applications in other stochastic search processes, such as the diffusive search for a target by an animal. In particular, a recently proposed search model, namely stochastic resetting (Evans and Majumdar, 2011*a*,*b*), has been quite successful in explaining animal foraging (Kuśmierz *et al.*, 2014; Falcón-Cortés *et al.*, 2017), biomolecular search (Roldán *et al.*, 2016) and chemical kinetics (Reuveni *et al.*, 2014); for a recent review, see Evans *et al.* (2020). Resetting is a stochastic process where an underlying process is intermittently reset to a preferred configuration with a given probability (Evans and Majumdar, 2011*a*,*b*). This simple yet pivotal setup has become an emerging and overarching topic in interdisciplinary science, since restarting a complex process again and again can expedite the completion of such a process and thus can be engineered as a useful tool. To understand this, let us consider a simple one-dimensional Brownian motion (starting from x_0) in the presence of a target at the origin. The motion is reset to its initial configuration with rate r, and one is interested in the mean first-passage time to the target. Although the mean first-passage time is divergent when $r = 0$ (Redner, 2001), a finite resetting rate renders it finite and thus facilitates a diffusion-mediated search which is otherwise detrimental (Evans and Majumdar, 2011*a*,*b*).

To see how EVS appears in this stochastic resetting problem, we consider one-dimensional diffusion with stochastic resetting at a constant rate r (Evans and Majumdar, 2011*a*,*b*). The searcher starts at position $x_0 > 0$ and we assume that there is an immobile target at the origin. The searcher performs normal diffusion with diffusion constant D that gets interrupted at rate r after which the searcher starts again from its initial position x_0. Let $Q_r(x_0, t)$ denote the survival probability of the target, i.e., the probability that the searcher has not found the target up to time t. This quantity has been exactly computed using both the backward Fokker–Planck approach and a renewal approach (Evans and Majumdar, 2011b), and a simple interpretation of this result in terms of EVS was provided (Evans and Majumdar, 2011b). The argument goes as follows. We consider this stochastic process up to a total observation time t. Since the mean time between reset events is typically $1/r$, the mean number of intervals between resets is of order $N = rt$. The target remains unhit by the searcher up to time t if the global minimum of this process, starting at x_0, remains positive (since the target is at the origin and $x_0 > 0$). To compute the global minimum, we can think of the total interval $[0, t]$ consisting of $N = rt$ uncorrelated blocks. For each block i there is a minimum m_i, and the m_i's are uncorrelated since the process is renewed after each resetting. The global minimum of the process is also the minimum of the m_i's, $i = 1, 2, \ldots, N = rt$. The cumulative PDF of m_i, i.e., $\mathrm{Prob}(m_i > m)$, is given by the Brownian motion result (since inside a block the particle performs normal diffusion), $\mathrm{Prob}(m_i > m) = \mathrm{erf}(|M - x_0|/\sqrt{4D\tau_i})$ [see, for instance, Eq. (1.11)], where τ_i is the duration of the ith block. Thus, the PDF of m_i is given by $P(m_i) = e^{-(m_i - x_0)^2/4D\tau_i}/\sqrt{\pi D\tau_i}$, with $m_i \leq x_0$. Since these time intervals between resets are taken from an exponential distribution with mean $1/r$, one has $P(\tau_i) = re^{-r\tau_i}$. Averaging over τ_i, one gets the effective distribution of the minimum of the ith block as

$$P_{\mathrm{eff}}(m_i) = r \int_0^t \mathrm{d}\tau_i e^{-r\tau_i} \frac{e^{-(m_i - x_0)^2/4D\tau_i}}{\sqrt{\pi D\tau_i}} \xrightarrow[t \to \infty]{} \alpha_0 e^{-\alpha_0(x_0 - m_i)}, \qquad m_i \leq x_0, \qquad (8.94)$$

where $\alpha_0 = \sqrt{r/D}$ is the typical inverse length traversed by the Brownian motion between two resetting events. Therefore, the probability that the minimum of all N intervals stays above 0 is simply given by $\left[\int_0^{x_0} P_{\mathrm{eff}}(m) \, \mathrm{d}m \right]^N$. Substituting the expression for $P_{\mathrm{eff}}(m)$ from Eq. (8.94), setting $N = rt$, one finds that, in terms of the rescaled length $z = \alpha_0 x_0$, the survival probability converges for large z and large t to the Gumbel form

$$Q_r(x_0, t) \approx \exp[-rt \exp(-z)]. \qquad (8.95)$$

Therefore, this constitues a nice application of the EVS for weakly correlated variables discussed in Section 4.2. The theory of EVS can also be useful for understanding the first-passage properties of other stochastic processes with resetting, going beyond Brownian motion. This includes, for example, Lévy flights (Kuśmierz *et al.*, 2014) and Lévy walks (Kuśmierz and Gudowska-Nowak, 2015) under resetting, branching processes under restart (Pal and Reuveni, 2017; Pal *et al., 2019*), active run-and-tumble particles with resetting (Evans and Majumdar, 2018), etc. Stochastic resetting is a rapidly developing field with many applications, including recent experiments in optical traps (Besga *et al.*, 2020; Tal-Friedman *et al.*, 2020; Faisant *et al.*, 2021).

Let us briefly mention an example of a stochastically resetting system where the EVS can be computed exactly, even though the underlying degrees of freedom are strongly correlated. Consider N independent Brownian motions on a line, all starting at the origin and *simultaneously resetting to the origin* with rate r. Even though

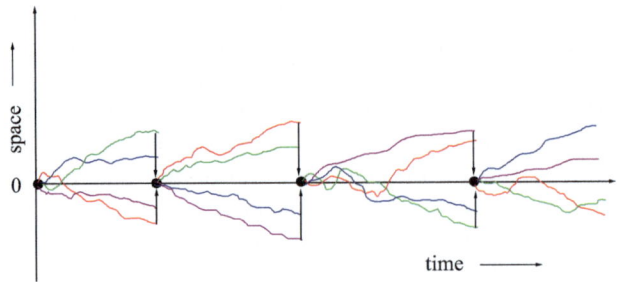

Fig. 8.6 Schematic space–time trajectories of N independent Brownian motions that start at the origin and reset simultaneously with rate r to the origin.

between two resetting events the particles evolve independently (see Fig. 8.6), the *simultaneous* resetting makes them strongly correlated. At long times, the system reaches a stationary state where the joint distribution reads (Biroli *et al.*, 2023)

$$P_{st}(x_1, x_2, \ldots, x_N) = r \int_0^\infty d\tau\, e^{-r\tau} \prod_{i=1}^N \frac{e^{-x_i^2/4Dt}}{\sqrt{4Dt}}. \tag{8.96}$$

Manifestly, the joint distribution does not factorize and one cannot use the results for the EVS of IID variables discussed in Chapter 4. Even though the particles are strongly correlated in the stationary state, the structure of the joint distribution in Eq. (8.96) allows one to compute several observables, including the extreme and order statistics. We refer the interested reader to the original article, but just mention one interesting result on EVS. For example, the maximum M_1, i.e., the position of the rightmost particle, takes the following scaling form in the large-N limit:

$$\text{Prob}(M_1 = w) \approx \frac{1}{L_N} f\left(\frac{w}{L_N}\right), \tag{8.97}$$

where $L_N = \sqrt{(4D \ln N)/r}$ and the scaling function $f(z)$, supported over $z \geq 0$, is given by (Biroli *et al.*, 2023)

$$f(z) = 2z e^{-z^2}, \qquad z \geq 0. \tag{8.98}$$

This is a new extreme value distribution for strongly correlated variables that has not appeared before. Going beyond this nice simple model, it has been shown that many such examples of solvable strongly correlated models can be constructed using the idea of stochastic resetting (Biroli *et al.*, 2024). This demonstrates that stochastic resetting will play an important role in constructing solvable models of EVS in the presence of strong correlations. Along with the recent reviews on stochastic resetting (Evans *et al.*, 2020; Pal *et al.*, 2022; Gupta and Jayannavar, 2022), we hope that the basic tools and techniques presented in this book on EVS will help the reader to tackle extreme value questions in future applications of stochastic resetting.

9
Conclusion and Perspectives

In this book, we have demonstrated the importance of correlations in a time series on the statistics of extremes. This includes not only the statistics of the values of the extremes, but also important extreme observables, such as the time at which the maximum of the time series occurs, the order statistics, record statistics, etc. To bring out the role of the correlations, we have mainly focused on two complementary models as illustrations: (i) the case of uncorrelated time series where the entries of the time series are independent and identically distributed random variables, and (ii) time series whose entries represent the positions of a random walk on a line, where the entries are strongly correlated.

The purpose of this book was to provide the reader with physical insights and basic technical know-how to deal with questions on extremes for a generic time series that frequently appear in many applications ranging from the physics of disordered systems all the way to finance and climate studies. Many results presented in this book were known in the literature in a rather scattered fashion, and one of the purposes of this book was to put them together in one place in a unified way.

The subject of extreme value statistics has become quite popular across disciplines due to its multitude of applications, and it is a rapidly evolving area of research. In Chapter 8 we went beyond the two basic models and discussed several recent developments and open problems. We hope that this will stimulate further new questions and lead to new models of extreme value statistics of correlated random variables.

Statistics of Extremes and Records in Random Sequences. Satya N. Majumdar and Grégory Schehr, Oxford University Press.
© Satya N. Majumdar and Grégory Schehr (2024). DOI: 10.1093/9780191838781.003.0009

References

Alessandro, B., Beatrice, C., Bertotti, G., and Montorsi, A. (1990). Domain-wall dynamics and Barkhausen effect in metallic ferromagnetic materials. I. Theory. *Journal of Applied Physics*, **68**(6), 2901–2907.

Amir, A. (2020a). An elementary renormalization-group approach to the generalized central limit theorem and extreme value distributions. *Journal of Statistical Mechanics: Theory and Experiment*, 2020(1), 013214.

Amir, A. (2020b). *Thinking Probabilistically: Stochastic Processes, Disordered Systems, and Their Applications*. Cambridge University Press.

Amir, G., Corwin, I., and Quastel, J. (2011). Probability distribution of the free energy of the continuum directed random polymer in 1+1 dimensions. *Communications on Pure and Applied Mathematics*, **64**(4), 466–537.

Anderson, A. and Kostinski, A. (2010). Reversible record breaking and variability: Temperature distributions across the globe. *Journal of Applied Meteorology and Climatology*, **49**(8), 1681–1691.

Anderson, P. W. (1958). Absence of diffusion in certain random lattices. *Physical Review*, **109**(5), 1492.

Arguin, L.-P., Belius, D., and Bourgade, P. (2017). Maximum of the characteristic polynomial of random unitary matrices. *Communications in Mathematical Physics*, **349**(2), 703–751.

Arnold, B., Balakrishnan, N., and Nagaraja, H. N. (2008). *A First Course in Order Statistics*. Society for Industrial and Applied Mathematics.

Arnold, B. C., Balakrishnan, N., and Nagaraja, H. N. (2011). *Records*. John Wiley & Sons.

Arratia, R., Barbour, A. D., and Tavaré, S. (2003). *Logarithmic Combinatorial Structures: A Probabilistic Approach*. Volume 1. European Mathematical Society.

Baik, J., Deift, P., and Johansson, K. (1999). On the distribution of the length of the longest increasing subsequence of random permutations. *Journal of the American Mathematical Society*, **12**(4), 1119–1178.

Baik, J. and Rains, E. M. (2000). Limiting distributions for a polynuclear growth model with external sources. *Journal of Statistical Physics*, **100**(3), 523–541.

Balakrishnan, N. (2013). *Handbook of the Logistic Distribution*. CRC Press.

Bar, A., Majumdar, S. N., Schehr, G., and Mukamel, D. (2016). Exact extreme-value statistics at mixed-order transitions. *Physical Review E*, **93**(5), 052130.

Bar, A. and Mukamel, D. (2014a). Mixed-order phase transition in a one-dimensional model. *Physical Review Letters*, **112**(1), 015701.

Bar, A. and Mukamel, D. (2014*b*). Mixed order transition and condensation in an exactly soluble one dimensional spin model. *Journal of Statistical Mechanics: Theory and Experiment*, 2014(11), P11001.

Barma, M., Majumdar, S. N., and Mukamel, D. (2019). Fluctuation-dominated phase ordering at a mixed order transition. *Journal of Physics A: Mathematical and Theoretical*, **52**(25), 254001.

Barndorff-Nielsen, O. E. and Shephard, N. (2001). Non-Gaussian Ornstein–Uhlenbeck-based models and some of their uses in financial economics. *Journal of the Royal Statistical Society: Series B (Statistical Methodology)*, **63**(2), 167–241.

Bassett Jr, G. W. (1992). Breaking recent global temperature records. *Climatic Change*, **21**(3), 303–315.

Båth, M. (1965). Lateral inhomogeneities of the upper mantle. *Tectonophysics*, **2**(6), 483–514.

Battilana, M., Majumdar, S. N., and Schehr, G. (2020). Universal gap statistics for random walks for a class of jump densities. *Markov Processes And Related Fields*, **26**(1), 57–94.

Bauer, M., Godrèche, C., and Luck, J.-M. (1999). Statistics of persistent events in the binomial random walk: Will the drunken sailor hit the sober man? *Journal of Statistical Physics*, **96**(5), 963–1019.

Ben Arous, G., Dembo, A., and Guionnet, A. (2001). Aging of spherical spin glasses. *Probability Theory and Related Fields*, **120**(1), 1–67.

Ben-Naim, E., Krapivsky, P. L., and Majumdar, S. N. (2001). Extremal properties of random trees. *Physical Review E*, **64**(3), 035101.

Ben-Naim, E., Redner, S., and Vazquez, F. (2007). Scaling in tournaments. *Europhysics Letters*, **77**(3), 30005.

Bena, I. and Majumdar, S. N. (2007). Universal extremal statistics in a freely expanding Jepsen gas. *Physical Review E*, **75**(5), 051103.

Benestad, R. E. (2003). How often can we expect a record event? *Climate Research*, **25**(1), 3–13.

Berman, S. M. (1964). Limit theorems for the maximum term in stationary sequences. *The Annals of Mathematical Statistics*, **35**(2), 502–516.

Bertin, E. and Clusel, M. (2006). Generalized extreme value statistics and sum of correlated variables. *Journal of Physics A: Mathematical and General*, **39**(24), 7607.

Bertin, E. and Györgyi, G. (2010). Renormalization flow in extreme value statistics. *Journal of Statistical Mechanics: Theory and Experiment*, 2010(8), P08022.

Besga, B., Bovon, A., Petrosyan, A., Majumdar, S. N., and Ciliberto, S. (2020). Optimal mean first-passage time for a Brownian searcher subjected to resetting: Experimental and theoretical results. *Physical Review Research*, **2**(3), 032029.

Biroli, G., Bouchaud, J.-P., and Potters, M. (2007*a*). Extreme value problems in random matrix theory and other disordered systems. *Journal of Statistical Mechanics: Theory and Experiment*, 2007(07), P07019.

Biroli, G., Bouchaud, J.-P., and Potters, M. (2007*b*). On the top eigenvalue of heavy-tailed random matrices. *Europhysics Letters*, **78**(1), 10001.

Biroli, M., Larralde, H., Majumdar, S. N., and Schehr, G. (2023). Extreme statistics and spacing distribution in a Brownian gas correlated by resetting. *Physical Review Letters*, **130**(20), 207101.

Biroli, M., Larralde, H., Majumdar, S. N., and Schehr, G. (2024). Exact extreme, order and sum statistics in a class of strongly correlated system. *Physical Review E* **109**, 014101.

Biroli, M., Mori, F., and Majumdar, S. N. (2022). Number of distinct sites visited by a resetting random walker. *Journal of Physics A: Mathematical and Theoretical*, **55**(24), 244001.

Bolech, C. J. and Rosso, A. (2004). Universal statistics of the critical depinning force of elastic systems in random media. *Physical Review Letters*, **93**(12), 125701.

Bonamy, D. (2017). Dynamics of cracks in disordered materials. *Comptes Rendus Physique*, **18**(5–6), 297–313.

Borot, G., Eynard, B., Majumdar, S. N., and Nadal, C. (2011). Large deviations of the maximal eigenvalue of random matrices. *Journal of Statistical Mechanics: Theory and Experiment*, 2011(11), P11024.

Borot, G. and Nadal, C. (2012). Right tail asymptotic expansion of Tracy–Widom-beta laws. *Random Matrices: Theory and Applications*, **1**(3), 1250006.

Bouchaud, J.-P. and Georges, A. (1990). Anomalous diffusion in disordered media: Statistical mechanisms, models and physical applications. *Physics Reports*, **195**(4–5), 127–293.

Bouchaud, J.-P. and Mézard, M. (1997). Universality classes for extreme-value statistics. *Journal of Physics A: Mathematical and General*, **30**(23), 7997.

Bramson, M. (1983). *Convergence of Solutions of the Kolmogorov Equation to Travelling Waves*. American Mathematical Society.

Bramson, M., Ding, J., and Zeitouni, O. (2016). Convergence in law of the maximum of the two-dimensional discrete Gaussian free field. *Communications on Pure and Applied Mathematics*, **69**(1), 62–123.

Bramson, M. and Zeitouni, O. (2012). Tightness of the recentered maximum of the two-dimensional discrete Gaussian free field. *Communications on Pure and Applied Mathematics*, **65**(1), 1–20.

Bray, A. J., Majumdar, S. N., and Schehr, G. (2013). Persistence and first-passage properties in nonequilibrium systems. *Advances in Physics*, **62**(3), 225–361.

Brunet, E. and Derrida, B. (1997). Shift in the velocity of a front due to a cutoff. *Physical Review E*, **56**(3), 2597.

Brunet, E. and Derrida, B. (2009). Statistics at the tip of a branching random walk and the delay of traveling waves. *Europhysics Letters*, **87**(6), 60010.

Brunet, E. and Derrida, B. (2011). A branching random walk seen from the tip. *Journal of Statistical Physics*, **143**(3), 420–446.

Buffet, E. (2003). On the time of the maximum of Brownian motion with drift. *Journal of Applied Mathematics and Stochastic Analysis*, **16**(3), 201–207.

Buldyrev, S. V., Gitterman, M., Havlin, S., Kazakov, A. Y., Luz, M. G. Da, Raposo, E. P., Stanley, H. E., and Viswanathan, G. M. (2001*a*). Properties of Lévy flights on an interval with absorbing boundaries. *Physica A: Statistical Mechanics and its Applications*, **302**(1–4), 148–161.

Buldyrev, S. V., Havlin, S., Kazakov, A. Y., Luz, M. G. Da, Raposo, E. P., Stanley, H. E., and Viswanathan, G. M. (2001*b*). Average time spent by Lévy flights and walks on an interval with absorbing boundaries. *Physical Review E*, **64**(4), 041108.

Bunge, J. and Goldie, C. M. (2001). Record sequences and their applications. *Handbook of Statistics*, **19**, 277–308.

Burkhardt, T. W., Györgyi, G., Moloney, N. R., and Rácz, Z. (2007). Extreme statistics for time series: Distribution of the maximum relative to the initial value. *Physical Review E*, **76**(4), 041119.

Calabrese, P., Le Doussal, P., and Rosso, A. (2010). Free-energy distribution of the directed polymer at high temperature. *Europhysics Letters*, **90**(2), 20002.

Carpentier, D. and Le Doussal, P. (2001). Glass transition of a particle in a random potential, front selection in nonlinear renormalization group, and entropic phenomena in Liouville and sinh-Gordon models. *Physical Review E*, **63**(2), 026110.

Chafaï, D. and Péché, S. (2014). A note on the second order universality at the edge of Coulomb gases on the plane. *Journal of Statistical Physics*, **156**, 368–383.

Chandrasekhar, S. (1943). Stochastic problems in physics and astronomy. *Reviews of Modern Physics*, **15**(1), 1.

Chaumont, L. (1999). A path transformation and its applications to fluctuation theory. *Journal of the London Mathematical Society*, **59**(2), 729–741.

Chicheportiche, R. and Bouchaud, J.-P. (2014). Some applications of first-passage ideas to finance. In *First-Passage Phenomena and Their Applications* (ed. R. Metzler, G. Oshanin, and S. Redner), pp. 447–476. World Scientific.

Coffman, E. G., Flajolet, Ph., Flatto, L., and Hofri, M. (1998). The maximum of a random walk and its application to rectangle packing. *Probability in the Engineering and Informational Sciences*, **12**(3), 373–386.

Comtet, A. and Majumdar, S. N. (2005). Precise asymptotics for a random walker's maximum. *Journal of Statistical Mechanics: Theory and Experiment*, 2005(06), P06013.

Cook, J. and Derrida, B. (1989). Polymers on disordered hierarchical lattices: A nonlinear combination of random variables. *Journal of Statistical Physics*, **57**, 89–139.

Cunden, F. D., Facchi, P., Ligabò, M., and Vivo, P. (2018). Universality of the weak pushed-to-pulled transition in systems with repulsive interactions. *Journal of Physics A: Mathematical and Theoretical*, **51**(35), 35LT01.

Cunden, F. D., Mezzadri, F., and Vivo, P. (2016). Large deviations of radial statistics in the two-dimensional one-component plasma. *Journal of Statistical Physics*, **164**, 1062–1081.

Damle, K., Majumdar, S. N., Tripathi, V., and Vivo, P. (2011). Phase transitions in the distribution of the Andreev conductance of superconductor–metal junctions with multiple transverse modes. *Physical Review Letters*, **107**(17), 177206.

Dassios, A. (1995). The distribution of the quantile of a Brownian motion with drift and the pricing of related path-dependent options. *The Annals of Applied Probability*, **5**(2), 389–398.

Dassios, A. (1996). Sample quantiles of stochastic processes with stationary and independent increments. *The Annals of Applied Probability*, **6**(3), 1041–1043.

David, H. A. and Nagaraja, H. N. (2004). *Order Statistics*. John Wiley & Sons.

De Bruyne, B., Majumdar, S. N., and Schehr, G. (2021). Expected maximum of bridge random walks & Lévy flights. *Journal of Statistical Mechanics: Theory and Experiment*, 2021(8), 083215.

De Bruyne, B., Majumdar, S. N., and Schehr, G. (2023). Universal order statistics for random walks & Lévy flights. *Journal of Statistical Physics*, **190**(1), 20.

Dean, D. S., Le Doussal, P., Majumdar, S. N., and Schehr, G. (2015). Finite-temperature free fermions and the Kardar–Parisi–Zhang equation at finite time. *Physical Review Letters*, **114**(11), 110402.

Dean, D. S., Le Doussal, P., Majumdar, S. N., and Schehr, G. (2016). Noninteracting fermions at finite temperature in a d-dimensional trap: Universal correlations. *Physical Review A*, **94**(6), 063622.

Dean, D. S. and Majumdar, S. N. (2001). Extreme-value statistics of hierarchically correlated variables deviation from Gumbel statistics and anomalous persistence. *Physical Review E*, **64**(4), 046121.

Dean, D. S. and Majumdar, S. N. (2006). Large deviations of extreme eigenvalues of random matrices. *Physical Review Letters*, **97**(16), 160201.

Dean, D. S. and Majumdar, S. N. (2008). Extreme value statistics of eigenvalues of Gaussian random matrices. *Physical Review E*, **77**(4), 041108.

Delorme, M., Rosso, A., and Wiese, K. J. (2017). Pickands' constant at first order in an expansion around Brownian motion. *Journal of Physics A: Mathematical and Theoretical*, **50**(16), 16LT04.

Delorme, M. and Wiese, K. J. (2016a). Extreme-value statistics of fractional Brownian motion bridges. *Physical Review E*, **94**(5), 052105.

Delorme, M. and Wiese, K. J. (2016b). Perturbative expansion for the maximum of fractional Brownian motion. *Physical Review E*, **94**(1), 012134.

Dembo, A., Ding, J., and Gao, F. (2013). Persistence of iterated partial sums. *Annales de l'Institut Henri Poincaré: Probabilités et statistiques*, **49**(3), 873–884.

Derrida, B. (1981). Random-energy model: An exactly solvable model of disordered systems. *Physical Review B*, **24**(5), 2613.

Derrida, B. (1985). A generalization of the random energy model which includes correlations between energies. *Journal de Physique Lettres*, **46**(9), 401–407.

Derrida, B. and Gardner, E. (1986). Solution of the generalised random energy model. *Journal of Physics C: Solid State Physics*, **19**(13), 2253.

Derrida, B. and Spohn, H. (1988). Polymers on disordered trees, spin glasses, and traveling waves. *Journal of Statistical Physics*, **51**(5), 817–840.

Dhar, A., Kundu, A., Majumdar, S. N., Sabhapandit, S., and Schehr, G. (2017). Exact extremal statistics in the classical 1D Coulomb gas. *Physical Review Letters*, **119**(6), 060601.

Dhar, A., Kundu, A., Majumdar, S. N., Sabhapandit, S., and Schehr, G. (2018). Extreme statistics and index distribution in the classical 1D Coulomb gas. *Journal of Physics A: Mathematical and Theoretical*, **51**(29), 295001.

Di Baldassarre, G., Elshamy, M., van Griensven, A., Soliman, E., Kigobe, M., Ndomba, P., Mutemi, J., Mutua, F., Moges, S., Xuan, Y., Solomatine, D., and Uhlenbrook, S. (2011). Future hydrology and climate in the River Nile basin: A review. *Journal des Sciences Hydrologiques*, **56**(2), 199–211.

Diaconis, P. and Pitman, J. (1986). Permutations, record values and random measures. Unpublished lecture notes, Statistics Department, University of California, Berkeley.

Dotsenko, V. (2010). Bethe ansatz derivation of the Tracy–Widom distribution for one-dimensional directed polymers. *Europhysics Letters*, **90**(2), 20003.

Dumitriu, I. and Edelman, A. (2002). Matrix models for beta ensembles. *Journal of Mathematical Physics*, **43**(11), 5830–5847.

Dumonteil, E., Majumdar, S. N., Rosso, A., and Zoia, A. (2013). Spatial extent of an outbreak in animal epidemics. *Proceedings of the National Academy of Sciences*, **110**(11), 4239–4244.

Dyson, F. J. (1962). Statistical theory of the energy levels of complex systems. I. *Journal of Mathematical Physics*, **3**(1), 140–156.

Edelman, A. and Sutton, B. D. (2007). From random matrices to stochastic operators. *Journal of Statistical Physics*, **127**(6), 1121–1165.

Embrechts, P., Rogers, L. C. G., and Yor, M. (1995). A proof of Dassios' representation of the *alpha*-quantile of Brownian motion with drift. *The Annals of Applied Probability*, **5**(3), 757–767.

Evans, M. R. and Hanney, T. (2005). Nonequilibrium statistical mechanics of the zero-range process and related models. *Journal of Physics A: Mathematical and General*, **38**(19), R195.

Evans, M. R. and Majumdar, S. N. (2008). Condensation and extreme value statistics. *Journal of Statistical Mechanics: Theory and Experiment*, 2008(05), P05004.

Evans, M. R. and Majumdar, S. N. (2011*a*). Diffusion with optimal resetting. *Journal of Physics A: Mathematical and Theoretical*, **44**(43), 435001.

Evans, M. R. and Majumdar, S. N. (2011*b*). Diffusion with stochastic resetting. *Physical Review Letters*, **106**(16), 160601.

Evans, M. R. and Majumdar, S. N. (2018). Run and tumble particle under resetting: A renewal approach. *Journal of Physics A: Mathematical and Theoretical*, **51**(47), 475003.

Evans, M. R., Majumdar, S. N., and Schehr, G. (2020). Stochastic resetting and applications. *Journal of Physics A: Mathematical and Theoretical*, **53**(19), 193001.

Evans, M. R., Majumdar, S. N., and Zia, R. K. P. (2006). Canonical analysis of condensation in factorised steady states. *Journal of Statistical Physics*, **123**(2), 357–390.

Faisant, F., Besga, B., Petrosyan, A., Ciliberto, S., and Majumdar, S. N. (2021). Optimal mean first-passage time of a Brownian searcher with resetting in one and two dimensions: Experiments, theory and numerical tests. *Journal of Statistical Mechanics: Theory and Experiment*, 2021(11), 113203.

Falcón-Cortés, A., Boyer, D., Giuggioli, L., and Majumdar, S. N. (2017). Localization transition induced by learning in random searches. *Physical Review Letters*, **119**(14), 140603.

Feller, W. (2008*a*). *An Introduction to Probability Theory and its Applications*. Volume 1. John Wiley & Sons.

Feller, W. (2008*b*). *An Introduction to Probability Theory and its Applications*. Volume 2. John Wiley & Sons.

Finch, S. R. (2003). *Mathematical Constants*. Cambridge University Press.

Fisher, D. S. (1998). Collective transport in random media: From superconductors to earthquakes. *Physics Reports*, **301**(1–3), 113–150.

Fisher, D. S. and Huse, D. A. (1991). Directed paths in a random potential. *Physical Review B*, **43**(13), 10728.

Fisher, R. A. and Tippett, L. H. C. (1928). Limiting forms of the frequency distribution of the largest or smallest member of a sample. *Mathematical Proceedings of the Cambridge Philosophical Society*, **24**(2), 180–190.

Forrester, P. J. (2010). *Log-Gases and Random Matrices*. Princeton University Press.

Forrester, P. J. (2012). Spectral density asymptotics for Gaussian and Laguerre β-ensembles in the exponentially small region. *Journal of Physics A: Mathematical and Theoretical*, **45**(7), 075206.

Forrester, P.-J., Majumdar, S. N., and Schehr, G. (2011). Non-intersecting Brownian walkers and Yang–Mills theory on the sphere. *Nuclear Physics B*, **844**(3), 500–526.

Forster, D., Nelson, D. R., and Stephen, M. J. (1977). Large-distance and long-time properties of a randomly stirred fluid. *Physical Review A*, **16**(2), 732.

Frachebourg, L., Ispolatov, I., and Krapivsky, P. L. (1995). Extremal properties of random systems. *Physical Review E*, **52**(6), R5727.

Franke, J., Klözer, A., de Visser, J. A. G. M., and Krug, J. (2011). Evolutionary accessibility of mutational pathways. *PLoS Computational Biology*, **7**(8), e1002134.

Franke, J. and Majumdar, S. N. (2012). Survival probability of an immobile target surrounded by mobile traps. *Journal of Statistical Mechanics: Theory and Experiment*, 2012(05), P05024.

Franke, J., Wergen, G., and Krug, J. (2012). Correlations of record events as a test for heavy-tailed distributions. *Physical Review Letters*, **108**(6), 064101.

Fréchet, M. (1927). Sur la loi de probabilité de l'écart maximum. *Annales de la Société Polonaise de Mathématique*, **6**, 93–116.

Fridman, M., Pugatch, R., Nixon, M., Friesem, A. A., and Davidson, N. (2012). Measuring maximal eigenvalue distribution of Wishart random matrices with coupled lasers. *Physical Review E*, **85**(2), 020101.

Frisch, H. (1988). Radiative transfer with frequency redistribution: Asymptotic methods, scaling laws and approximate solutions. In *Radiation in Moving Gaseous Media* (ed. R. P. Kudritzki, H. W. Yorke, and H. Frisch). Geneva Observatory.

Frisch, H. and Frisch, U. (1982). A method of Cauchy integral equation for non-coherent transfer in half-space. *Journal of Quantitative Spectroscopy and Radiative Transfer*, **28**(5), 361–375.

Frisch, U. and Frisch, H. (1995). Universality of escape from a half-space for symmetrical random walks. In *Lévy Flights and Related Topics in Physics* (ed. M. F. Shlesinger, G. M. Zaslavsky, and U. Frisch), pp. 262–268. Springer.

Fyodorov, Y. V. (2004). Complexity of random energy landscapes, glass transition, and absolute value of the spectral determinant of random matrices. *Physical Review Letters*, **92**(24), 240601.

Fyodorov, Y. V. and Bouchaud, J.-P. (2007). On the explicit construction of Parisi landscapes in finite dimensional Euclidean spaces. *JETP Letters*, **86**(7), 487–491.

Fyodorov, Y. V. and Bouchaud, J.-P. (2008). Statistical mechanics of a single particle in a multiscale random potential: Parisi landscapes in finite-dimensional Euclidean spaces. *Journal of Physics A: Mathematical and Theoretical*, **41**(32), 324009.

Fyodorov, Y. V., Hiary, G. A., and Keating, J. P. (2012). Freezing transition, characteristic polynomials of random matrices, and the Riemann zeta function. *Physical Review Letters*, **108**(17), 170601.

Fyodorov, Y. V. and Keating, J. P. (2014). Freezing transitions and extreme values: Random matrix theory, and disordered landscapes. *Philosophical Transactions of the Royal Society A: Mathematical, Physical and Engineering Sciences*, **372**(2007), 20120503.

Fyodorov, Y. V., Le Doussal, P., and Rosso, A. (2009). Statistical mechanics of logarithmic REM: Duality, freezing and extreme value statistics of $1/f$ noises generated by Gaussian free fields. *Journal of Statistical Mechanics: Theory and Experiment*, 2009(10), P10005.

Fyodorov, Y. V., Le Doussal, P., Rosso, A., and Texier, C. (2018). Exponential number of equilibria and depinning threshold for a directed polymer in a random potential. *Annals of Physics*, **397**, 1–64.

Fyodorov, Y. V. and Nadal, C. (2012). Critical behavior of the number of minima of a random landscape at the glass transition point and the Tracy–Widom distribution. *Physical Review Letters*, **109**(16), 167203.

Fyodorov, Y. V., Perret, A., and Schehr, G. (2015). Large time zero temperature dynamics of the spherical $p = 2$ spin glass model of finite size. *Journal of Statistical Mechanics: Theory and Experiment*, 2015(11), P11017.

Fyodorov, Y. V. and Sommers, H.-J. (2007). Classical particle in a box with random potential: Exploiting rotational symmetry of replicated Hamiltonian. *Nuclear Physics B*, **764**(3), 128–167.

Gembris, D., Taylor, J. G., and Suter, D. (2002). Trends and random fluctuations in athletics. *Nature*, **417**(6888), 506–506.

Gnedenko, B. (1943). Sur la distribution limite du terme maximum d'une série aléatoire. *Annals of Mathematics*, **44**(3), 423–453.

Godrèche, C. and Luck, J.-M. (2008). A record-driven growth process. *Journal of Statistical Mechanics: Theory and Experiment*, 2008(11), P11006.

Godrèche, C., Majumdar, S. N., and Schehr, G. (2009). Longest excursion of stochastic processes in nonequilibrium systems. *Physical Review Letters*, **102**(24), 240602.

Godrèche, C., Majumdar, S. N., and Schehr, G. (2014). Universal statistics of longest lasting records of random walks and Lévy flights. *Journal of Physics A: Mathematical and Theoretical*, **47**(25), 255001.

Godrèche, C., Majumdar, S. N., and Schehr, G. (2015). Statistics of the longest interval in renewal processes. *Journal of Statistical Mechanics: Theory and Experiment*, 2015(3), P03014.

Godrèche, C., Majumdar, S. N., and Schehr, G. (2016). Exact statistics of record increments of random walks and Lévy flights. *Physical Review Letters*, **117**(1), 010601.

Godrèche, C., Majumdar, S. N., and Schehr, G. (2017). Record statistics of a strongly correlated time series: Random walks and Lévy flights. *Journal of Physics A: Mathematical and Theoretical*, **50**(33), 333001.

Gourdon, X. (1997). *Combinatoire, algorithmique et géométrie des polynômes*. INRIA.

Gradenigo, G. and Majumdar, S. N. (2019). A first-order dynamical transition in the displacement distribution of a driven run-and-tumble particle. *Journal of Statistical Mechanics: Theory and Experiment*, 2019(5), 053206.

Gradshteyn, I. S. and Ryzhik, I. M. (2014). *Table of Integrals, Series, and Products*. Academic Press.

Grava, T., Its, A., Kapaev, A., and Mezzadri, F. (2016). On the Tracy–Widom β distribution for $\beta = 6$. *Symmetry, Integrability and Geometry: Methods and Applications*, **12**, 105.

Gravner, J., Tracy, C. A., and Widom, H. (2001). Limit theorems for height fluctuations in a class of discrete space and time growth models. *Journal of Statistical Physics*, **102**(5), 1085–1132.

Grebenkov, D. S., Lanoiselée, Y., and Majumdar, S. N. (2017). Mean perimeter and mean area of the convex hull over planar random walks. *Journal of Statistical Mechanics: Theory and Experiment*, 2017(10), 103203.

Gross, D. J. and Witten, E. (1980). Possible third-order phase transition in the large-N lattice gauge theory. *Physical Review D*, **21**(2), 446.

Gumbel, E. J. (2004). *Statistics of Extremes*. Courier Corporation.

Gupta, S. and Jayannavar, A. M. (2022). Stochastic resetting: A (very) brief review. *Frontiers in Physics*, **10**, 789097.

Györgyi, G., Moloney, N. R., Ozogány, K., and Rácz, Z. (2007). Maximal height statistics for $1/f$ α signals. *Physical Review E*, **75**(2), 021123.

Györgyi, G., Moloney, N. R., Ozogány, K., and Rácz, Z. (2008). Finite-size scaling in extreme statistics. *Physical Review Letters*, **100**(21), 210601.

Györgyi, G., Moloney, N. R., Ozogány, K., Rácz, Z., and Droz, M. (2010). Renormalization-group theory for finite-size scaling in extreme statistics. *Physical Review E*, **81**(4), 041135.

Halpin-Healy, T. and Takeuchi, K. A. (2015). A KPZ cocktail—shaken, not stirred.... *Journal of Statistical Physics*, **160**(4), 794–814.

Halpin-Healy, T. and Zhang, Y.-C. (1995). Kinetic roughening phenomena, stochastic growth, directed polymers and all that. Aspects of multidisciplinary statistical mechanics. *Physics Reports*, **254**(4–6), 215–414.

Hartmann, A. K., Majumdar, S. N., Schawe, H., and Schehr, G. (2020). The convex hull of the run-and-tumble particle in a plane. *Journal of Statistical Mechanics: Theory and Experiment*, 2020(5), 053401.

Hoyt, D. V. (1981). Weather records and climatic change. *Climatic Change*, **3**(3), 243–249.

Hughes, B. D. (1995). *Random Walks and Random Environments: Random Walks*. Volume 1. Oxford University Press.

Huse, D. A. and Henley, C. L. (1985). Pinning and roughening of domain walls in Ising systems due to random impurities. *Physical Review Letters*, **54**(25), 2708.

Imamura, T. and Sasamoto, T. (2004). Fluctuations of the one-dimensional polynuclear growth model with external sources. *Nuclear Physics B*, **699**(3), 503–544.

Imbrie, J. Z. and Spencer, T. (1988). Diffusion of directed polymers in a random environment. *Journal of Statistical Physics*, **52**, 609–626.

Ivanov, V. V. (1994). Resolvent method: Exact solutions of half-space transport problems by elementary means. *Astronomy and Astrophysics*, **286**, 328–337.

Janson, S. (2007). Brownian excursion area, Wright's constants in graph enumeration, and other Brownian areas. *Probability Surveys*, **4**, 80–145.

Janson, S. and Louchard, G. (2007). Tail estimates for the Brownian excursion area and other Brownian areas. *Electronic Journal of Probability*, **12**, 1600–1632.

Jensen, H. J. (2006). Evolution in complex systems: Record dynamics in models of spin glasses, superconductors and evolutionary ecology. In *Advances in Solid State Physics* (ed. B. Kramer), Volume 45, pp. 95–106. Springer.

Jepsen, D. W. (1965). Dynamics of a simple many-body system of hard rods. *Journal of Mathematical Physics*, **6**(3), 405–413.

Johansson, K. (2000). Shape fluctuations and random matrices. *Communications in Mathematical Physics*, **209**(2), 437–476.

Kardar, M. (1985). Roughening by impurities at finite temperatures. *Physical Review Letters*, **55**(26), 2923.

Kardar, M. (1987). Replica Bethe ansatz studies of two-dimensional interfaces with quenched random impurities. *Nuclear Physics B*, **290**, 582–602.

Kardar, M., Parisi, G., and Zhang, Y.-C. (1986). Dynamic scaling of growing interfaces. *Physical Review Letters*, **56**(9), 889.

Kardar, M. and Zhang, Y.-C. (1987). Scaling of directed polymers in random media. *Physical Review Letters*, **58**(20), 2087.

Katz, R. W., Parlange, M. B., and Naveau, P. (2002). Statistics of extremes in hydrology. *Advances in Water Resources*, **25**(8–12), 1287–1304.

Kearney, M. J. and Majumdar, S. N. (2014). Statistics of the first passage time of Brownian motion conditioned by maximum value or area. *Journal of Physics A: Mathematical and Theoretical*, **47**(46), 465001.

Kearney, M. J., Majumdar, S. N., and Martin, R. J. (2007). The first-passage area for drifted Brownian motion and the moments of the Airy distribution. *Journal of Physics A: Mathematical and Theoretical*, **40**(36), F863.

Kethepalli, J., Kulkarni, M., Kundu, A., Majumdar, S. N., Mukamel, D., and Schehr, G. (2022). Edge fluctuations and third-order phase transition in harmonically confined long-range systems. *Journal of Statistical Mechanics: Theory and Experiment*, 2022(3), 033203.

Klinger, J., Voituriez, R., and Bénichou, O. (2022). Splitting probabilities of symmetric jump processes. *Physical Review Letters*, **129**(14), 140603.

Koren, T., Lomholt, M. A., Chechkin, A., Aleksei, V., Klafter, J., and Metzler, R. (2007). Leapover lengths and first passage time statistics for Lévy flights. *Physical Review Letters*, **99**(16), 160602.

Krug, J. (1997). Origins of scale invariance in growth processes. *Advances in Physics*, **46**(2), 139–282.

Krug, J. and Halpin-Healy, T. (1998). Ground-state energy anisotropy for directed polymers in random media. *Journal of Physics A: Mathematical and General*, **31**(28), 5939.

Krug, J. and Jain, K. (2005). Breaking records in the evolutionary race. *Physica A: Statistical Mechanics and its Applications*, **358**(1), 1–9.

Kundu, A., Majumdar, S. N., and Schehr, G. (2013). Exact distributions of the number of distinct and common sites visited by N independent random walkers. *Physical Review Letters*, **110**(22), 220602.

Kuśmierz, L. and Gudowska-Nowak, A. (2015). Optimal first-arrival times in Lévy flights with resetting. *Physical Review E*, **92**(5), 052127.

Kuśmierz, L., Majumdar, S. N., Sabhapandit, S., and Schehr, G. (2014). First order transition for the optimal search time of Lévy flights with resetting. *Physical Review Letters*, **113**(22), 220602.

Lacroix-A-Chez-Toine, B., Grabsch, A., Majumdar, S. N., and Schehr, G. (2018). Extremes of 2D Coulomb gas: Universal intermediate deviation regime. *Journal of Statistical Mechanics: Theory and Experiment*, 2018(1), 013203.

Lacroix-A-Chez-Toine, B., Majumdar, S. N., and Schehr, G. (2019). Gap statistics close to the quantile of a random walk. *Journal of Physics A: Mathematical and Theoretical*, **52**(31), 315003.

Lamperti, J. (1961). A contribution to renewal theory. *Proceedings of the American Mathematical Society*, **12**(5), 724–731.

Le Doussal, P. and Monthus, C. (2003). Exact solutions for the statistics of extrema of some random 1D landscapes, application to the equilibrium and the dynamics of the toy model. *Physica A: Statistical Mechanics and its Applications*, **317**(1–2), 140–198.

Le Doussal, P. and Wiese, K. J. (2009). Driven particle in a random landscape: Disorder correlator, avalanche distribution, and extreme value statistics of records. *Physical Review E*, **79**(5), 051105.

Leadbetter, M. L., Lindgren, G., and Rootzén, H. (2012). *Extremes and Related Properties of Random Sequences and Processes*. Springer.

Lemarié, G., Kamlapure, A., Bucheli, D., Benfatto, L., Lorenzana, J., Seibold, G., Ganguli, S. C., Raychaudhuri, P., and Castellani, C. (2013). Universal scaling of the order-parameter distribution in strongly disordered superconductors. *Physical Review B*, **87**(18), 184509.

Lévy, P. (1940). Sur certains processus stochastiques homogènes. *Compositio mathematica*, **7**, 283–339.

Liechty, K. (2012). Nonintersecting Brownian motions on the half-line and discrete Gaussian orthogonal polynomials. *Journal of Statistical Physics*, **147**(3), 582–622.

Livan, G., Novaes, M., and Vivo, P. (2018). *Introduction to Random Matrices: Theory and Practice*. Springer.

Luković, M., Geisel, T., and Eule, S. (2013). Area and perimeter covered by anomalous diffusion processes. *New Journal of Physics*, **15**(6), 063034.

Majumdar, S. N. (2003). Traveling front solutions to directed diffusion-limited aggregation, digital search trees, and the Lempel–Ziv data compression algorithm. *Physical Review E*, **68**(2), 026103.

Majumdar, S. N. (2007*a*). Brownian functionals in physics and computer science. In *The Legacy Of Albert Einstein: A Collection of Essays in Celebration of the Year of Physics* (ed. N. Straumann), pp. 93–129. World Scientific.

Majumdar, S. N. (2007*b*). Course 4. Random matrices, the Ulam problem, directed polymers & growth models, and sequence matching. In *Complex Systems: Lecture Notes of the Les Houches Summer School 2006: Volume 85* (ed. J.-P. Bouchaud, M. Mézard, and J. Dalibard), pp. 179–216. Elsevier.

Majumdar, S. N. (2010*a*). Real-space condensation in stochastic mass transport models. In *Exact Methods in Low-dimensional Statistical Physics and Quantum Computing: Lecture Notes of the Les Houches Summer School: Volume 89, July 2008* (ed. J. Jacobsen, S. Ouvry, V. Pasquier, D. Serban, and L. Cugliandolo). Oxford University Press.

Majumdar, S. N. (2010*b*). Universal first-passage properties of discrete-time random walks and Lévy flights on a line: Statistics of the global maximum and records. *Physica A: Statistical Mechanics and its Applications*, **389**(20), 4299–4316.

Majumdar, S. N. and Bouchaud, J.-P. (2008). Optimal time to sell a stock in the black–scholes model: Comment on "Thou Shalt Buy and Hold," by A. Shiryaev, Z. Xu and X. Y. Zhou. *Quantitative Finance*, **8**(8), 753–760.

Majumdar, S. N. and Comtet, A. (2004). Exact maximal height distribution of fluctuating interfaces. *Physical Review Letters*, **92**(22), 225501.

Majumdar, S. N. and Comtet, A. (2005). Airy distribution function: From the area under a Brownian excursion to the maximal height of fluctuating interfaces. *Journal of Statistical Physics*, **119**(3), 777–826.

Majumdar, S. N., Comtet, A., and Randon-Furling, J. (2010*a*). Random convex hulls and extreme value statistics. *Journal of Statistical Physics*, **138**(6), 955–1009.

Majumdar, S. N., Comtet, A., and Ziff, R. M. (2006). Unified solution of the expected maximum of a discrete time random walk and the discrete flux to a spherical trap. *Journal of Statistical Physics*, **122**(5), 833–856.

Majumdar, S. N. and Kearney, M. J. (2007). Inelastic collapse of a ball bouncing on a randomly vibrating platform. *Physical Review E*, **76**(3), 031130.

Majumdar, S. N. and Krapivsky, P. L. (2000). Extremal paths on a random Cayley tree. *Physical Review E*, **62**(6), 7735.

Majumdar, S. N. and Krapivsky, P. L. (2002). Extreme value statistics and traveling fronts: Application to computer science. *Physical Review E*, **65**(3), 036127.

Majumdar, S. N. and Krapivsky, P. L. (2003). Extreme value statistics and traveling fronts: Various applications. *Physica A: Statistical Mechanics and its Applications*, **318**(1–2), 161–170.

Majumdar, S. N., Mallick, K., and Sabhapandit, S. (2009). Statistical properties of the final state in one-dimensional ballistic aggregation. *Physical Review E*, **79**(2), 021109.

Majumdar, S. N., Mori, F., Schawe, H., and Schehr, G. (2021*a*). Mean perimeter and area of the convex hull of a planar Brownian motion in the presence of resetting. *Physical Review E*, **103**(2), 022135.

Majumdar, S. N., Mounaix, Ph., Sabhapandit, S., and Schehr, G. (2021*b*). Record statistics for random walks and Lévy flights with resetting. *Journal of Physics A: Mathematical and Theoretical*, **55**(3), 034002.

Majumdar, S. N., Mounaix, Ph., and Schehr, G. (2014). On the gap and time interval between the first two maxima of long random walks. *Journal of Statistical Mechanics: Theory and Experiment*, 2014(9), P09013.

Majumdar, S. N., Mounaix, Ph., and Schehr, G. (2017). Survival probability of random walks and Lévy flights on a semi-infinite line. *Journal of Physics A: Mathematical and Theoretical*, **50**(46), 465002.

Majumdar, S. N., Mounaix, Ph., and Schehr, G. (2019*a*). Smoluchowski flux and lamb–lion problems for random walks and Lévy flights with a constant drift. *Journal of Statistical Mechanics: Theory and Experiment*, 2019(8), 083214.

Majumdar, S. N., Mounaix, Ph., and Schehr, G. (2021*c*). Universal record statistics for random walks and Lévy flights with a nonzero staying probability. *Journal of Physics A: Mathematical and Theoretical*, **54**(31), 315002.

Majumdar, S. N. and Nechaev, S. (2004). Anisotropic ballistic deposition model with links to the Ulam problem and the Tracy–Widom distribution. *Physical Review E*, **69**(1), 011103.

Majumdar, S. N. and Nechaev, S. (2005). Exact asymptotic results for the Bernoulli matching model of sequence alignment. *Physical Review E*, **72**(2), 020901.

Majumdar, S. N., Pal, A., and Schehr, G. (2020). Extreme value statistics of correlated random variables: A pedagogical review. *Physics Reports*, **840**, 1–32.

Majumdar, S. N., Randon-Furling, J., Kearney, M. J., and Yor, M. (2008). On the time to reach maximum for a variety of constrained Brownian motions. *Journal of Physics A: Mathematical and Theoretical*, **41**(36), 365005.

Majumdar, S. N., Rosso, A., and Zoia, A. (2010*b*). Time at which the maximum of a random acceleration process is reached. *Journal of Physics A: Mathematical and Theoretical*, **43**(11), 115001.

Majumdar, S. N. and Schehr, G. (2014). Top eigenvalue of a random matrix: Large deviations and third order phase transition. *Journal of Statistical Mechanics: Theory and Experiment*, 2014(1), P01012.

Majumdar, S. N., Schehr, G., and Wergen, G. (2012). Record statistics and persistence for a random walk with a drift. *Journal of Physics A: Mathematical and Theoretical*, **45**(35), 355002.

Majumdar, S. N. and Vergassola, M. (2009). Large deviations of the maximum eigenvalue for Wishart and Gaussian random matrices. *Physical Review Letters*, **102**(6), 060601.

Majumdar, S. N., von Bomhard, Ph., and Krug, J. (2019*b*). Exactly solvable record model for rainfall. *Physical Review Letters*, **122**(15), 158702.

Majumdar, S. N. and Ziff, R. M. (2008). Universal record statistics of random walks and Lévy flights. *Physical Review Letters*, **101**(5), 050601.

Matalas, N. C. (1997). Stochastic hydrology in the context of climate change. *Climatic Change*, **37**(1), 89–101.

McKean, H. P. (1975). Application of Brownian motion to the equation of Kolmogorov–Petrovskii–Piskunov. *Communications on Pure and Applied Mathematics*, **28**(3), 323–331.

Mehta, M. L. (2004). *Random Matrices*. Elsevier.

Metzler, R. and Klafter, J. (2000). The random walk's guide to anomalous diffusion: A fractional dynamics approach. *Physics Reports*, **339**(1), 1–77.

Metzler, R., Redner, S., and Oshanin, G. (2014). *First-passage phenomena and their applications*. World Scientific.

Mézard, M. (1990). On the glassy nature of random directed polymers in two dimensions. *Journal de Physique*, **51**(17), 1831–1846.

Mori, F. (2022, June). *Extreme Value Statistics of Stochastic Processes: From Brownian Motion to Active Particles*. Thesis, Université Paris-Saclay.

Mori, F., Gradenigo, G., and Majumdar, S. N. (2021*a*). First-order condensation transition in the position distribution of a run-and-tumble particle in one dimension. *Journal of Statistical Mechanics: Theory and Experiment*, 2021(10), 103208.

Mori, F., Le Doussal, P., Majumdar, S. N., and Schehr, G. (2021*b*). Condensation transition in the late-time position of a run-and-tumble particle. *Physical Review E*, **103**(6), 062134.

Mori, F., Majumdar, S. N., and Schehr, G. (2019). Time between the maximum and the minimum of a stochastic process. *Physical Review Letters*, **123**(20), 200201.

Mori, F., Majumdar, S. N., and Schehr, G. (2020). Distribution of the time between maximum and minimum of random walks. *Physical Review E*, **101**(5), 052111.

Mori, F., Majumdar, S. N., and Schehr, G. (2021*c*). Distribution of the time of the maximum for stationary processes. *Europhysics Letters*, **135**(3), 30003.

Mori, F., Majumdar, S. N., and Schehr, G. (2022). Time to reach the maximum for a stationary stochastic process. *Physical Review E*, **106**(5), 054110.

Morse, P. M. and Feshbach, H. (1953). *Methods of theoretical physics*. McGraw-Hill.

Mounaix, Ph., Majumdar, S. N., and Schehr, G. (2018). Asymptotics for the expected maximum of random walks and Lévy flights with a constant drift. *Journal of Statistical Mechanics: Theory and Experiment*, 2018(8), 083201.

Mounaix, Ph., Majumdar, S. N., and Schehr, G. (2020). Statistics of the number of records for random walks and Lévy flights on a 1D lattice. *Journal of Physics A: Mathematical and Theoretical*, **53**(41), 415003.

Nadal, C. and Majumdar, S. N. (2009). Nonintersecting Brownian interfaces and Wishart random matrices. *Physical Review E*, **79**(6), 061117.

Nadal, C. and Majumdar, S. N. (2011). A simple derivation of the Tracy–Widom distribution of the maximal eigenvalue of a Gaussian unitary random matrix. *Journal of Statistical Mechanics: Theory and Experiment*, 2011(04), P04001.

Nadal, C., Majumdar, S. N., and Vergassola, M. (2010). Phase transitions in the distribution of bipartite entanglement of a random pure state. *Physical Review Letters*, **104**(11), 110501.

Nadal, C., Majumdar, S. N., and Vergassola, M. (2011). Statistical distribution of quantum entanglement for a random bipartite state. *Journal of Statistical Physics*, **142**, 403–438.

Neuts, M. F. (1967). Waiting times between record observations. *Journal of Applied Probability*, **4**(1), 206–208.

Nevzorov, V. B. (2001). *Records: Mathematical Theory*. American Mathematical Society.

Novak, S. Y. (2011). *Extreme Value Methods with Applications to Finance*. CRC Press.

Oliveira, L. P., Jensen, H. J., Nicodemi, M., and Sibani, P. (2005). Record dynamics and the observed temperature plateau in the magnetic creep-rate of type-II superconductors. *Physical Review B*, **71**(10), 104526.

Ostrovsky, P. M., Skvortsov, M. A., and Feigel'Man, M. V. (2001). Density of states below the Thouless gap in a mesoscopic SNS junction. *Physical Review Letters*, **87**(2), 027002.

Pal, A., Eliazar, I., and Reuveni, S. (2019). First passage under restart with branching. *Physical Review Letters*, **122**(2), 020602.

Pal, A., Kostinski, S., and Reuveni, S. (2022). The inspection paradox in stochastic resetting. *Journal of Physics A: Mathematical and Theoretical*, **55**(2), 021001.

Pal, A. and Reuveni, S. (2017). First passage under restart. *Physical Review Letters*, **118**(3), 030603.

Pál, G., Raischel, F., Lennartz-Sassinek, S., Kun, F., and Main, I. G. (2016). Record-breaking events during the compressive failure of porous materials. *Physical Review E*, **93**(3), 033006.

Paquette, E. and Zeitouni, O. (2018). The maximum of the CUE field. *International Mathematics Research Notices*, **2018**(16), 5028–5119.

Park, S.-C., Szendro, I. G., Neidhart, J., and Krug, J. (2015). Phase transition in random adaptive walks on correlated fitness landscapes. *Physical Review E*, **91**(4), 042707.

Perret, A., Comtet, A., Majumdar, S. N., and Schehr, G. (2013). Near-extreme statistics of Brownian motion. *Physical Review Letters*, **111**(24), 240601.

Perret, A., Comtet, A., Majumdar, S. N., and Schehr, G. (2015). On certain functionals of the maximum of Brownian motion and their applications. *Journal of Statistical Physics*, **161**(5), 1112–1154.

Pitman, J. (2006). *Combinatorial Stochastic Processes: Ecole d'Eté de Probabilités de Saint-Flour XXXII – 2002.* Springer.

Pitman, J. and Tang, W. (2022). Hidden symmetries and limit laws in the extreme order statistics of the Laplace random walk. *The Annals of Probability*, **50**(4), 1647–1673.

Pitman, J. and Tang, W. (2023). Extreme order statistics of random walks. *Annales de l'Institut Henri Poincaré: Probabilités et statistiques*, **59**(1), 97–116.

Pitman, J. and Yor, M. (1997). The two-parameter Poisson–Dirichlet distribution derived from a stable subordinator. *The Annals of Probability*, **25**(2), 855–900.

Pollaczek, F. (1952). Fonctions caractéristiques de certaines répartitions définies au moyen de la notion d'ordre – application à la théorie des attentes. *Comptes Rendus Hebdomadaires des Séances de l'Académie des Sciences*, **234**(24), 2334–2336.

Port, S. C. (1963). An elementary probability approach to fluctuation theory. *Journal of Mathematical Analysis and Applications*, **6**(1), 109–151.

Prähofer, M. and Spohn, H. (2000). Universal distributions for growth processes in 1+1 dimensions and random matrices. *Physical Review Letters*, **84**(21), 4882.

Rajesh, R. and Majumdar, S. N. (2000a). Conserved mass models and particle systems in one dimension. *Journal of Statistical Physics*, **99**, 943–965.

Rajesh, R. and Majumdar, S. N. (2000b). Exact calculation of the spatiotemporal correlations in the Takayasu model and in the q model of force fluctuations in bead packs. *Physical Review E*, **62**(3), 3186.

Rajesh, R. and Majumdar, S. N. (2001). Exact phase diagram of a model with aggregation and chipping. *Physical Review E*, **63**(3), 036114.

Rambeau, J., Bustingorry, S., Kolton, A. B., and Schehr, G. (2011). Maximum relative height of elastic interfaces in random media. *Physical Review E*, **84**(4), 041131.

Rambeau, J. and Schehr, G. (2010). Extremal statistics of curved growing interfaces in 1+1 dimensions. *Europhysics Letters*, **91**(6), 60006.

Rambeau, J. and Schehr, G. (2011). Distribution of the time at which N vicious walkers reach their maximal height. *Physical Review E*, **83**(6), 061146.

Ramirez, J., Rider, B., and Virág, B. (2011). Beta ensembles, stochastic Airy spectrum, and a diffusion. *Journal of the American Mathematical Society*, **24**(4), 919–944.

Ramola, K., Majumdar, S. N., and Schehr, G. (2014). Universal order and gap statistics of critical branching Brownian motion. *Physical Review Letters*, **112**(21), 210602.

Ramola, K., Majumdar, S. N., and Schehr, G. (2015a). Branching Brownian motion conditioned on particle numbers. *Chaos, Solitons & Fractals*, **74**, 79–88.

Ramola, K., Majumdar, S. N., and Schehr, G. (2015b). Spatial extent of branching Brownian motion. *Physical Review E*, **91**(4), 042131.

Randon-Furling, J., Majumdar, S. N., and Comtet, A. (2009). Convex hull of N planar Brownian motions: exact results and an application to ecology. *Physical Review Letters*, **103**(14), 140602.

Raychaudhuri, S., Cranston, M., Przybyla, C., and Shapir, Y. (2001). Maximal height scaling of kinetically growing surfaces. *Physical Review Letters*, **87**(13), 136101.

Redner, S. (2001). *A Guide to First-Passage Processes*. Cambridge University Press.

Redner, S. and Petersen, M. R. (2006). Role of global warming on the statistics of record-breaking temperatures. *Physical Review E*, **74**(6), 061114.

Rényi, A. (1962). Théorie des éléments saillants d'une suite d'observations. *Annales scientifiques de l'Université de Clermont. Mathématiques*, **8**(2), 7–13.

Resnick, S. I. (2008). *Extreme Values, Regular Variation, and Point Processes*. Springer.

Reuveni, S., Urbakh, M., and Klafter, J. (2014). Role of substrate unbinding in Michaelis–Menten enzymatic reactions. *Proceedings of the National Academy of Sciences*, **111**(12), 4391–4396.

Reymbaut, A., Majumdar, S. N., and Rosso, A. (2011). The convex hull for a random acceleration process in two dimensions. *Journal of Physics A: Mathematical and Theoretical*, **44**(41), 415001.

Rider, B. (2003). A limit theorem at the edge of a non-Hermitian random matrix ensemble. *Journal of Physics A: Mathematical and General*, **36**(12), 3401.

Riordan, J. (2012). *Introduction to Combinatorial Analysis*. Courier Corporation.

Robe, D. M., Boettcher, S., Sibani, P., and Yunker, P. (2016). Record dynamics: Direct experimental evidence from jammed colloids. *Europhysics Letters*, **116**(3), 38003.

Roldán, E., Lisica, A., Sánchez-Taltavull, D., and Grill, S. W. (2016). Stochastic resetting in backtrack recovery by RNA polymerases. *Physical Review E*, **93**(6), 062411.

Rumanov, I. (2016). Painlevé representation of Tracy–Widom β distribution for $\beta = 6$. *Communications in Mathematical Physics*, **342**(3), 843–868.

Sabhapandit, S. (2011). Record statistics of continuous time random walk. *Europhysics Letters*, **94**(2), 20003.

Sabhapandit, S., Bena, I., and Majumdar, S. N. (2008). Statistics of the total number of collisions and the ordering time in a freely expanding hard-point gas. *Journal of Statistical Mechanics: Theory and Experiment*, 2008(05), P05012.

Sabhapandit, S. and Majumdar, S. N. (2007). Density of near-extreme events. *Physical Review Letters*, **98**(14), 140201.

Sabir, B. and Santhanam, M. S. (2014). Record statistics of financial time series and geometric random walks. *Physical Review E*, **90**(3), 032126.

Sadhu, T., Delorme, M., and Wiese, K. J. (2018). Generalized arcsine laws for fractional Brownian motion. *Physical Review Letters*, **120**(4), 040603.

Sasamoto, T. and Spohn, H. (2010). One-dimensional Kardar–Parisi–Zhang equation: An exact solution and its universality. *Physical Review Letters*, **104**, 230602.

Sawyer, S. and Fleischman, J. (1979). Maximum geographic range of a mutant allele considered as a subtype of a Brownian branching random field. *Proceedings of the National Academy of Sciences*, **76**(2), 872–875.

Schawe, H., Hartmann, A. K., Majumdar, S. N., and Schehr, G. (2018). Ground-state energy of noninteracting fermions with a random energy spectrum. *Europhysics Letters*, **124**(4), 40005.

Schehr, G. (2012). Extremes of N vicious walkers for large N: Application to the directed polymer and KPZ interfaces. *Journal of Statistical Physics*, **149**(3), 385–410.

Schehr, G. and Le Doussal, P. (2010). Extreme value statistics from the real space renormalization group: Brownian motion, Bessel processes and continuous time random walks. *Journal of Statistical Mechanics: Theory and Experiment*, 2010(01), P01009.

Schehr, G. and Majumdar, S. N. (2012). Universal order statistics of random walks. *Physical Review Letters*, **108**(4), 040601.

Schehr, G. and Majumdar, S. N. (2014). Exact record and order statistics of random walks via first-passage ideas. In *First-Passage Phenomena and their Applications* (ed. R. Metzler, G. Oshanin, and S. Redner), pp. 226–251. World Scientific.

Schehr, G., Majumdar, S. N., Comtet, A., and Forrester, P. J. (2013). Reunion probability of N vicious walkers: Typical and large fluctuations for large N. *Journal of Statistical Physics*, **150**(3), 491–530.

Schimmenti, V. M., Majumdar, S. N., and Rosso, A. (2021). Statistical properties of avalanches via the c-record process. *Physical Review E*, **104**(6), 064129.

Schmittmann, B. and Zia, R. K. P. (1999). Weather records: Musings on cold days after a long hot Indian summer. *American Journal of Physics*, **67**(12), 1269–1276.

Shepp, L. A. (1979). The joint density of the maximum and its location for a Wiener process with drift. *Journal of Applied Probability*, **16**(2), 423–427.

Shepp, L. A. and Lloyd, S. P. (1966). Ordered cycle lengths in a random permutation. *Transactions of the American Mathematical Society*, **121**(2), 340–357.

Shiryaev, A., Xu, Z., and Zhou, X. Y. (2008). Thou shalt buy and hold. *Quantitative Finance*, **8**(8), 765–776.

Shorrock, R. W. (1972). On record values and record times. *Journal of Applied Probability*, **9**(2), 316–326.

Sibani, P. (2007). Linear response in aging glassy systems, intermittency and the Poisson statistics of record fluctuations. *The European Physical Journal B*, **58**, 483–491.

Sibani, P., Brandt, M., and Alstrøm, P. (1998). Evolution and extinction dynamics in rugged fitness landscapes. *International Journal of Modern Physics B*, **12**(04), 361–391.

Sibani, P. and Littlewood, P. B. (1993). Slow dynamics from noise adaptation. *Physical Review Letters*, **71**(10), 1482.

Sibani, P., Rodriguez, G. F., and Kenning, G. G. (2006). Intermittent quakes and record dynamics in the thermoremanent magnetization of a spin-glass. *Physical Review B*, **74**(22), 224407.

Silva, A. and Ioffe, L. B. (2005). Subgap states in dirty superconductors and their effect on dephasing in Josephson qubits. *Physical Review B*, **71**(10), 104502.

Singh, P., Kundu, A., Majumdar, S. N., and Schawe, H. (2022). Mean area of the convex hull of a run and tumble particle in two dimensions. *Journal of Physics A: Mathematical and Theoretical*, **55**(22), 225001.

Sparre Andersen, E. (1954). On the fluctuations of sums of random variables II. *Mathematica Scandinavica*, **2**, 195–223.

Spitzer, F. (1956). A combinatorial lemma and its application to probability theory. *Transactions of the American Mathematical Society*, **82**(2), 323–339.

Spitzer, F. (1957). The Wiener–Hopf equation whose kernel is a probability density. *Duke Mathematical Journal*, **24**(3), 327–343.

Spohn, H. (2016). The Kardar–Parisi–Zhang equation: A statistical physics perspective. In *Stochastic Processes and Random Matrices: Lecture Notes of the Les Houches Summer School: Volume 104, July 2015* (ed. G. Schehr, A. Altland, Y. V. Fyodorov, N. O'Connell, and L. F. Cugliandolo), pp. 177–227. Oxford University Press.

Srivastava, S. C. L., Lakshminarayan, A., and Jain, S. R. (2013). Record statistics in random vectors and quantum chaos. *Europhysics Letters*, **101**(1), 10003.

Takeuchi, K. A. and Sano, M. (2010). Universal fluctuations of growing interfaces: Evidence in turbulent liquid crystals. *Physical Review Letters*, **104**(23), 230601.

Takeuchi, K. A. and Sano, M. (2012). Evidence for geometry-dependent universal fluctuations of the Kardar–Parisi–Zhang interfaces in liquid-crystal turbulence. *Journal of Statistical Physics*, **147**(5), 853–890.

Takeuchi, K. A., Sano, M., Sasamoto, T., and Spohn, H. (2011). Growing interfaces uncover universal fluctuations behind scale invariance. *Scientific Reports*, **1**(1), 1–5.

Tal-Friedman, O., Pal, A., Sekhon, A., Reuveni, S., and Roichman, Y. (2020). Experimental realization of diffusion with stochastic resetting. *The Journal of Physical Chemistry Letters*, **11**(17), 7350–7355.

Tankov, P. (2003). *Financial Modelling with Jump Processes*. Chapman and Hall / CRC.

Tracy, C. A. and Widom, H. (1994). Level-spacing distributions and the Airy kernel. *Communications in Mathematical Physics*, **159**(1), 151–174.

Tracy, C. A. and Widom, H. (1996). On orthogonal and symplectic matrix ensembles. *Communications in Mathematical Physics*, **177**(3), 727–754.

Van Saarloos, W. (2003). Front propagation into unstable states. *Physics Reports*, **386**(2–6), 29–222.

Vavilov, M. G., Brouwer, P. W., Ambegaokar, V., and Beenakker, C. W. J. (2001). Universal gap fluctuations in the superconductor proximity effect. *Physical Review Letters*, **86**(5), 874.

Vere-Jones, D. (1969). A note on the statistical interpretation of Båth's law. *Bulletin of the Seismological Society of America*, **59**(4), 1535–1541.

Vivo, P. (2015). Large deviations of the maximum of independent and identically distributed random variables. *European Journal of Physics*, **36**(5), 055037.

Vivo, P., Majumdar, S. N., and Bohigas, O. (2008). Distributions of conductance and shot noise and associated phase transitions. *Physical Review Letters*, **101**(21), 216809.

Vivo, P., Majumdar, S. N., and Bohigas, O. (2010). Probability distributions of linear statistics in chaotic cavities and associated phase transitions. *Physical Review B*, **81**(10), 104202.

Vogel, R. M., Zafirakou-Koulouris, A., and Matalas, N. C. (2001). Frequency of record-breaking floods in the united states. *Water Resources Research*, **37**(6), 1723–1731.

Wadia, S. R. (1980). $N = \infty$ phase transition in a class of exactly soluble model lattice gauge theories. *Physics Letters B*, **93**(4), 403–410.

Weibull, W. (1951). A statistical distribution function of wide applicability. *Journal of Applied Mechanics*, **18**(3), 293–297.

Weiss, G. H. and Rubin, J. R. (1983). Random walks: Theory and selected applications. *Advances in Chemical Physics*, **52**, 363–505.

Wendel, J. G. (1960). Order statistics of partial sums. *The Annals of Mathematical Statistics*, **31**(4), 1034–1044.

Wergen, G. (2013). Records in stochastic processes—theory and applications. *Journal of Physics A: Mathematical and Theoretical*, **46**(22), 223001.

Wergen, G. (2014). Modeling record-breaking stock prices. *Physica A: Statistical Mechanics and its Applications*, **396**, 114–133.

Wergen, G., Bogner, M., and Krug, J. (2011). Record statistics for biased random walks, with an application to financial data. *Physical Review E*, **83**(5), 051109.

Wergen, G., Hense, A., and Krug, J. (2014). Record occurrence and record values in daily and monthly temperatures. *Climate Dynamics*, **42**, 1275–1289.

Wergen, G. and Krug, J. (2010). Record-breaking temperatures reveal a warming climate. *Europhysics Letters*, **92**(3), 30008.

Wergen, G., Majumdar, S. N., and Schehr, G. (2012). Record statistics for multiple random walks. *Physical Review E*, **86**(1), 011119.

Wigner, E. P. (1951). On the statistical distribution of the widths and spacings of nuclear resonance levels. *Mathematical Proceedings of the Cambridge Philosophical Society*, **47**(4), 790–798.

Yor, M. (1995). The distribution of Brownian quantiles. *Journal of Applied Probability*, **32**(2), 405–416.

Zaburdaev, V., Denisov, S., and Klafter, J. (2015). Lévy walks. *Reviews of Modern Physics*, **87**(2), 483.

Zarfaty, L., Barkai, E., and Kessler, D. A. (2022). Discrete sampling of extreme events modifies their statistics. *Physical Review Letters*, **129**(9), 094101.

Ziff, R. M. (1991). Flux to a trap. *Journal of Statistical Physics*, **65**(5), 1217–1233.

Ziff, R. M., Majumdar, S. N., and Comtet, A. (2007). General flux to a trap in one and three dimensions. *Journal of Physics: Condensed Matter*, **19**(6), 065102.

Ziff, R. M., Majumdar, S. N., and Comtet, A. (2009). Capture of particles undergoing discrete random walks. *The Journal of Chemical Physics*, **130**(20), 204104.

Zoia, A., Rosso, A., and Kardar, M. (2007). Fractional Laplacian in bounded domains. *Physical Review E*, **76**(2), 021116.

Zumofen, G. and Klafter, J. (1995). Absorbing boundary in one-dimensional anomalous transport. *Physical Review E*, **51**(4), 2805.

Index